Elementary Particle Physics in a Nutshell

Elementary Particle Physics in a Nutshell

Christopher G. Tully

PRINCETON UNIVERSITY PRESS • PRINCETON AND OXFORD

Published by Princeton University Press, 41 William Street,
Princeton, New Jersey 08540

In the United Kingdom: Princeton University Press,
6 Oxford Street, Woodstock, Oxfordshire OX20 1TW

press.princeton.edu

Library of Congress Cataloging-in-Publication Data

Tully, Christopher G., 1970–
Elementary particle physics in a nutshell / Christopher G. Tully.
p. cm.
Includes bibliographical references and index.
ISBN 978-0-691-13116-0 (hardcover : alk. paper)
1. Particles (Nuclear physics) I. Title.
QC793.2.T85 2011
539.7'2—dc22

2011002143

British Library Cataloging-in-Publication Data is available

This book has been composed in Scala LF and Scala Sans

Printed on acid-free paper. ∞

Printed in the United States of America

10 9 8 7 6 5 4 3 2 1

Contents

Preface

The 21st century is a time of great change in particle physics. A new energy frontier recently opened up at the Large Hadron Collider (LHC) at CERN. It's a time of great excitement with the anticipation of unexpected outcomes. At the same time, the most widely used university-level texts on high-energy physics date back to the time leading up to the W and Z boson discoveries. Since then, the Standard Model of particle physics has been thoroughly explored at the Large Electron Positron (LEP) collider at CERN, the Tevatron at Fermilab, HERA at DESY and at two B-factories, KEKB and PEP-II. A decade of neutrino physics has brought an exciting new view on these elementary and light, but massive, particles. This text is an attempt to capture the modern understanding of particle physics in a snapshot of time leading up to the start-up of the LHC. I believe that the pause in the development of texts has been due in part to the anticipated discovery of the Higgs boson and the implications that the observed Higgs field properties will have in defining the high-energy unification of the fundamental interactions. However, it is difficult for a new generation of high-energy physics to prepare for the challenge of the LHC without having the perspective needed to look beyond the limitations of the current Standard Model. In this text, I attempt to introduce a complete working knowledge of the $SU(3)_C \times SU(2)_L \times U(1)_Y$ Standard Model as early as possible and then focus on the many experimental confirmations in the context of the full theory. Ultimately, this will lead us to the next generation of high-energy experiments with a focus on what we hope to learn.

The final editing of this book was completed while on sabbatical at the Institute for Advanced Study with support from the IBM Einstein Fellowship Fund.

Chris Tully
Princeton, 2010

Elementary Particle Physics in a Nutshell

1 Particle Physics: A Brief Overview

Particle physics is as much a science about today's universe as it is of the early universe. By discovering the basic building blocks of matter and their interactions, we are able to construct a language in which to frame questions about the early universe. What were the first forms of matter created in the early universe? What interactions were present in the early universe and how are they related to what we measure now? While we cannot return space-time to the initial configuration of the early universe, we can effectively turn back the clock when it comes to elementary particles by probing the interactions of matter at high energy. What we learn from studying high-energy interactions is that the universe is much simpler than what is observed at "room temperature" and that the interactions are a reflection of fundamental symmetries in Nature. An overview of the modern understanding of particle physics is described below with a more quantitative approach given in subsequent chapters and finally with a review of measurements, discoveries, and anticipated discoveries, that provide or will provide the experimental facts to support these theories.

We begin with the notion of a fundamental form of matter, an elementary particle. An elementary particle is treated as a pointlike object whose propagation through space is governed by a relativistically invariant equation of motion. The equation of motion takes on a particular form according to the intrinsic spin of the particle and whether the particle has a nonzero rest mass. In this introduction, we begin by assuming that elementary particles are massless and investigate the possible quantum numbers and degrees of freedom of elementary particle states.

1.1 Handedness in the Equation of Motion

A massless particle with nonzero intrinsic spin travels at the speed of light and has a definite handedness as defined by the sign of the dot product of the momentum and spin. The handedness of a massless particle, of which there are two possible values, is invariant and effectively decouples the elementary particles into two types, left-handed and right-handed. However, the association of handedness to a degree of freedom has to be extended to all solutions of the relativistic equation of motion. The time evolution of a solution to a wave equation introduces a time-dependent complex phase, where for a plane-wave solution of ordinary matter, we have

$$\exp(-i\mathcal{H}t/\hbar)\,\Psi_{\text{matter}} = \exp(-iEt/\hbar)\,\Psi_{\text{matter}} = \exp(-i\omega t)\,\Psi_{\text{matter}}. \tag{1.1}$$

Relativistic invariance and, in particular, causality introduces solutions that propagate with both positive and negative frequency. Relative to the sign of the frequency for "matter" solutions, a new set of solutions, the "antimatter" solutions, have the opposite sign of frequency, $\exp(i\omega t)$, so as to completely cancel contributions of the relativistic wave function outside the light cone. Therefore, in the relativistic equation of motion, there is always an antiparticle solution that is inseparable from the particle solution. In terms of handedness, an antiparticle solution has the sign of the dot product of momentum and spin reversed relative to the corresponding particle solution. We define a new quantity, called the chirality, that changes sign for antiparticles relative to particles. Therefore, a particle solution with left-handed chirality is relativistically linked through the equation of motion to an antiparticle solution that also has left-handed chirality. We can now separate in a relativistically invariant way two types of massless particles, left-handed and right-handed, according to their chirality.

1.2 Chiral Interactions

The existence of an interaction is reflected in the quantum numbers of the elementary particles. We introduce here a particular type of chiral interaction, one in which left-handed particle states can be transformed into one another in a manner similar to a rotation. However, unlike a spatial rotation, the chiral interaction acts on an internal space termed isospin, in analogy to a rotation of intrinsic spin. The smallest nontrivial representation of the isospin interaction is a two-component isospin doublet with three generators of isospin rotations. Left-handed particles interact under the chiral interaction, and, therefore, the symmetry associated with this interaction imposes a doubling of the number of left-handed elementary particles. There is an "up" and "down" type in each left-handed isospin doublet of elementary particles. If we further tailor our chiral interaction, we can begin to construct the table of known elementary particles. Namely, we do not introduce a right-handed chiral interaction. Furthermore, elementary particles that have right-handed chirality are not charged under the left-handed chiral interaction and are therefore singlets of the left-handed chiral symmetry group.

The evidence for the left-handed chiral interaction was initially observed from parity violation in the radionuclear decay of unstable isotopes emitting a polarized electron and an undetected electron antineutrino in the final state. While we have not introduced mass or an interaction for electric charge as would be expected for the electron, we can ignore these properties for now and construct a left-handed doublet from the elementary particles consisting of the electron (down-type) and the electron neutrino (up-type). The electron and neutrino are part of a general group of elementary particles known as the leptons.

1.3 Fundamental Strong Interaction

We now consider the force that leads to the formation of protons and neutrons, and is ultimately responsible for nuclear forces. This force is the fundamental strong interaction and, similar to the chiral interaction, is an interaction that acts on an internal space. In this case, the internal space is larger and has a smallest nontrivial representation given by a triplet with a set of eight generators of rotation. The triplet is referred to as a triplet of color, with components denoted red, green, and blue. As with the electron and electron neutrino, a left-handed triplet of color is also a doublet of the chiral interaction. The lightest down-type color triplet is called the down-quark. Correspondingly, the lightest up-type color triplet is called the up-quark. In contrast to the quarks, leptons are charge neutral with respect to the strong interaction.

1.4 Table of Elementary Particles

The chiral and fundamental strong interactions are a sufficient starting point to introduce the table of elementary particles, shown in figure 1.1. The quarks and leptons, shown on the left, have an intrinsic spin of $\hbar/2$. In the leftmost column, the up- and down-quarks and the electron and electron neutrino form what is known as the first generation of matter. The second and third generations are carbon copies of the first, ordered from left to right by the measured masses of the fermions. On the right-hand side of figure 1.1 are the particles that carry the forces or, according to our symmetry-based description of interactions, they are the "generators of rotation" for the internal spaces of color and isospin, called the eight gluons and the three weak bosons ($W^{\pm}$ and Z), respectively.

The properties of the elementary particles not explained by the chirality and color interactions are the electric charges and masses, and so too the unexplained presence of the photon. In order to explain the properties of mass and charge, here we look to a predicted and yet still elusive element in the particle table, the particle shown in the center of figure 1.1, the Higgs boson.

1.5 Mass and Electric Charge

While mass and electric charge are second nature from classical physics, they are highly nonobvious quantities in the elementary particles. In other words, their origin is believed to be linked to the properties of the physical vacuum rather than an inherent quantity that one would assign based on first principles, as explained below.

The "poor assumption" in the above discussions on elementary particles is the requirement of indistinguishability of the components of the particle doublets that represent the internal space of the chiral interaction. For the strong interaction, the components of the triplet of color are indistinguishable and hence are not explicitly labeled in figure 1.1. However, the electron and electron neutrino are different in their mass and in their

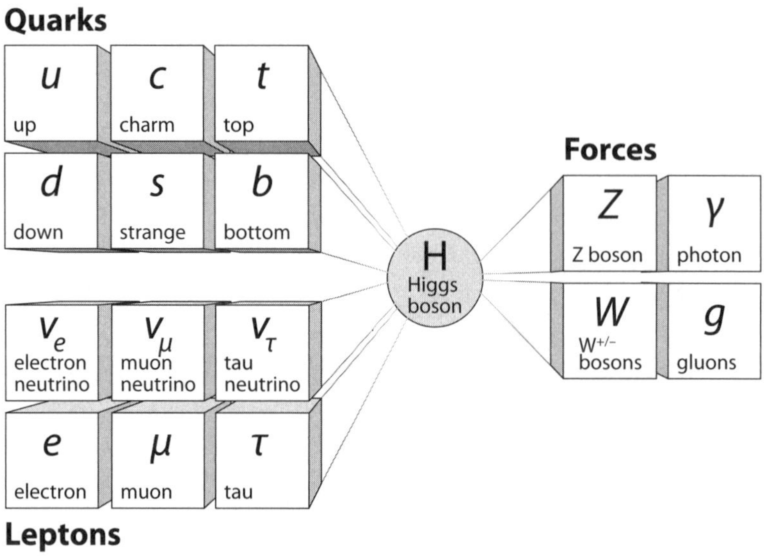

FIGURE 1.1. Elementary particles of the Standard Model. Leptons and quarks (spin-$\hbar/2$) are shown to the left. The gauge bosons (spin-$\hbar$) are on the right. The central particle, the Higgs boson (spin-0), has yet to be observed (Credit: Fermilab).

electric charge, similarly for the up- and down-quarks. The chiral interaction is therefore not an observed symmetry of Nature. Nevertheless, the chiral symmetry is or was clearly present in some form; otherwise, the particles would not interact with the $W^{\pm}$ and Z bosons in the way that they do. Moreover, for the chiral interaction to be based on an exact symmetry, the masses of elementary particles would have to be identically zero; otherwise, particles of left-handed chirality and right-handed chirality could be transformed into each other through a relativisitic transformation, changing the quantum numbers.

1.6 Hypercharge Interaction of the Standard Model

The Standard Model is a theory that solves the paradox of the "hidden symmetry" of the chiral interaction. To restore the indistinguishability of components of a chiral doublet, the Standard Model eliminates electromagnetism as an elementary interaction of massless fermions. The road to reintroducing electomagnetism as an interaction of massive fermions begins by postulating an alternative elementary interaction for massless fermions, called the hypercharge interaction. Hypercharge has every similarity to electric charge with the exception of the charge assignments to the elementary particles. A left-handed electron and electron neutrino are assigned the same hypercharge, and similarly for left-handed up- and down-quarks. Hypercharge assignments preserve the indistinguishability of the components of isospin doublets. The right-handed chiral particles are singlets in the chiral interaction and hence a right-handed up-quark can be assigned a different hypercharge than the left-handed up-quark or the right-handed down-quark without breaking the chiral symmetry. Thus, the critical step in constructing the Standard Model is to throw out mass and electric charge as a starting point for building a table of elementary particles. However, of what purpose is the hypercharge interaction in explaining the physically observed masses and charges of elementary particles and from where does the photon of electromagnetism originate? This brings us to the heart of the Standard Model theory, the electroweak symmetry breaking.

1.7 Higgs Mechanism

The central concept of the Standard Model is that the properties of the physical vacuum do not have the same symmetries as the fundamental interactions. This notion seems absurd at first, but physical examples of such systems, such as low-temperature superconductivity, clearly demonstrate such behavior in nonvacuous environments. Indeed, in a superconductor, the vacuum symmetry of zero electric charge is no longer present and photons do not propagate as massless particles in the space of a superconductor. The space of a low-temperature superconductor is filled with charge $2e$ electron pairs, known as Cooper pairs, that behave as a condensate of bosons. The photon is unable to propagate freely in a superconductor as it encounters nonzero electric charge at every point in space. The same type of mechanism can be postulated in the physical vacuum if there exists a condensate with nonzero hypercharge everywhere in space. Furthermore, the lack of an observed chiral symmetry in Nature would imply that a condensate of nonzero isospin is present in the vacuum ground state.

If there is a condensate in the ground state of the physical vacuum, what is it? One possible mechanism of electroweak symmetry breaking is known as the Higgs mechanism. The Higgs mechanism predicts the existence of a new type of elementary matter, called the Higgs bosons. The Higgs bosons are a set of spin-0 particle states with nonzero isospin and nonzero hypercharge quantum numbers. Similar to the Cooper pairs of superconductivity, the Higgs bosons interact with each other and do so in such a way so as to prefer a nonzero expectation value in the ground state. In other words, the Higgs condensate is present everywhere in space. Such an occurrence completely redefines our concept of the elementary particles and interactions. The table of elementary particles are those mass eigenstates that arise from particles interacting with the Higgs condensate. Similarly, the particles mediating the elementary interactions are hindered from propagating freely, resulting in a transformation of the elementary particle interactions. The photon of electromagnetism is the zero-mass eigenstate that propagates in the electrically neutral physical vacuum. It is in this way that the physical vacuum imposes the definition of electric charge and mass eigenstates in the elementary particles—these are not properties that come directly from the fundamental hypercharge and chiral interactions.

1.8 Program of Study

The Higgs mechanism of the Standard Model is a great leap beyond the notion of an empty vacuum and symmetry-preserving interactions. Indeed, the timeline of when the universe developed a vacuum filled with a symmetry-breaking condensate is not clear. Perhaps the most trying part of studying elementary particle physics is the lack of direct evidence to prove the existence of the Higgs mechanism or alternative electroweak symmetry-breaking mechanisms. Nevertheless, the experimental verification of the Standard Model is extensive with no apparent deviations with respect to all known predictions. Many internal consistencies of the Standard Model overwhelmingly support the concept of a symmetry-breaking physical vacuum, whether the fundamental source is the Higgs mechanism or something else. It is this predicament that has brought elementary particle physics into its most challenging and potentially the most revolutionary stage in its development. The energy scale that will confirm or refute the Higgs mechanism will be fully explored by the Large Hadron Collider (LHC) at the CERN laboratory in Geneva, Switzerland. The LHC experiments will be able to detect evidence for the Higgs bosons and probe possible extensions to the known physical symmetries in the elementary particles that would explain what stabilizes the electroweak scale.

The purpose of this text is to bring students a full understanding of the Standard Model, from relativistic kinematics and the Dirac equation through the concept of gauge interactions, and then to review in the context of the full theory the many areas of experimental investigation that have tested and subsequently confirmed the validity of the Standard Model predictions. Each chapter is concluded with a section that lists references that provide detailed background on the topics discussed. These references contain a wealth of interesting perspectives and lessons that were invaluable in the development of the Standard Model. In contrast, this book teaches the Standard Model as an established theory and applies the predictions to directly explain the mountain of experimental evidence that in retrospect was intended to challenge its validity. By presenting a fresh look at the Standard Model outside of the historically important questions that led to its creation, the intention is to prepare the ground for the next generation of exploration of elementary particle physics at the LHC.

1.9 Exercises

1. **Fundamental interactions.**

   (a) What are the interactions described by the Standard Model in a symmetry-preserving vacuum? Ignore the Higgs interactions.

   (b) What Standard Model interactions in part (a) are unaffected by the reduced symmetries of the physical vacuum?

2. **Elementary fermions.**

   (a) The right-handed up-type leptons are unusual fermion components in the Standard Model. These particles, known as right-handed neutrinos, are color singlets and isospin singlets and have zero hypercharge, and, hence, zero electric charge. In a symmetry-preserving vacuum, do right-handed neutrinos interact? What property of the left-handed neutrinos would imply that the right-handed neutrinos interact with the Higgs condensate?

   (b) If one counts the distinguishable and indistinguishable components of the strong and chiral multiplets, how many fermions make up the table of elementary particles, including the particles described in part (a)?

1.10 References and Further Reading

A selection of general introductions to the early universe, particle physics, experimental particle physics, and historical accounts can be found in the following reference texts: [1, 2, 3, 4, 5, 6, 7, 8, 9, 10, 11].

A selection of the original articles on the theory of the Standard Model of particle physics and the Higgs mechanism can be found here: [12, 13].

[1] Robert Cahn and Gerson Goldhaber. *The Experimental Foundations of Particle Physics.* Cambridge University Press, 2009. ISBN 978-0-521-52147-5.

[2] Stephen Hawking and Leonard Mlodinow. *A Briefer History of Time.* Bantam Dell Publishing, 2005. ISBN 978-0-553-38546-5.

[3] Don Lincoln. *Understanding the Universe.* World Scientific Publishing, 2004. ISBN 981-238-705-6.

[4] Martinus Veltman. *Facts and Mysteries in Elementary Particle Physics.* World Scientific Publishing, 2003. ISBN 981-238-149-X.

[5] Gerard 't Hooft. *In Search of the Ultimate Building Blocks.* Cambridge University Press, 1997. ISBN 0-521-57883-3.

[6] Harvey B. Newman and Thomas Ypsilantis , editors. *History of Original Ideas and Basic Discoveries in Particle Physics.* Plenum Press, 1996. ISBN 0-306-45217-0.

[7] Val L. Fitch and Jonathan L. Rosner. *Elementary Particle Physics in the Second Half of the Twentieth Century.* IOP Publishing, 1995. ISBN 1-56396-048-6.

[8] Gordon Kane. *The Particle Garden.* Perseus Publishing, 1995. ISBN 0-201-40826-0.

[9] Leon Lederman. *The God Particle.* Dell Publishing, 1993.ISBN 0-385-31211-3.

[10] Steven Weinberg. *The First Three Minutes.* BasicBooks, Harper Collins Publishers, 1993. ISBN 0-465-02437-8.

[11] Sheldon L. Glashow. *The Charm of Physics.* Simon and Schuster, 1991. ISBN 0-671-74013-X.

[12] S. L. Glashow, *Nucl. Phys.* **22** (1961) 579; S. Weinberg, *Phys. Rev. Lett.* **19** (1967) 1264; A. Salam, "Weak and Electromagnetic Interactions," in *Elementary Particle Theory,* Ed. N. Svartholm. Almquist and Wiksell, 1968, p. 367, LCCN 68055064; H. D. Politzer, *Phys. Rev. Lett.* **30** (1973) 1346; D. J. Gross and F. Wilczek, *Phys. Rev. Lett.* **30** (1973) 1343.

[13] P. W. Higgs, *Phys. Lett.* **12** (1964) 132, *Phys. Rev. Lett.* **13** (1964) 508, and *Phys. Rev.* **145** (1966) 1156; F. Englert and R. Brout, *Phys. Rev. Lett.* **13** (1964) 321.

2 Dirac Equation and Quantum Electrodynamics

Consider a spin-$\frac{\hbar}{2}$ particle, for example, an electron at rest. The rotations of the electron spin are described by the $SU(2)$ group and the states of the system are labeled by the values of the total spin, or specifically $\mathbf{S}^2$, and the projection of the spin onto an arbitrary axis of quantization, for example, S_z, the component along the z-direction. Now take the electron and Lorentz boost into a frame moving to the right (along the x-axis). How do we construct a wave function of a relativistic electron?

The answer to this question will lead us to one of the most profound equations in elementary particle physics, the Dirac equation. The Dirac equation introduces the first of several symmetry extensions to the structure of matter, namely, the antiparticles.

2.1 Natural Units and Conversions

By choosing *natural* units, where

$$\hbar = c = 1, \qquad [\text{Mass}] = [\text{Length}]^{-1} = [\text{Time}]^{-1}, \tag{2.1}$$

the quantities of mass, inverse length, and inverse time can be described by a single dimensional unit. The choice in this text is to measure all quantities in units of GeV. The conversion to units of meters and seconds is handled, for the most part, by inserting values of c and $\hbar c$ where needed:

$$\begin{aligned} c &\simeq 3 \times 10^8 \text{m/s}, \\ \hbar c &\simeq 0.2\,\text{GeV fm}. \end{aligned} \tag{2.2}$$

The charge of the electron is denoted e with $e < 0$. In general, the conventions of the Peskin and Schroeder [1] text are followed for all calculations.

2.2 Relativistic Invariance

A wave function in nonrelativistic quantum mechanics describes the probability of finding a particle at time t in a volume element d^3x centered at position $\boldsymbol{x}$ by the quantity

$|\psi(x,t)|^2 d^3x$. The normalization of the wave function is defined to be unity when integrated over all space, or a periodic box of volume V for plane-wave solutions,

$$\int_V \psi^* \psi d^3x = 1, \tag{2.3}$$

indicating that in a fixed snapshot of time the particle must be somewhere. The *generators of infinitesimal translations* in time and space are the familiar energy and momentum operators

$$\hat{E} = i\frac{\partial}{\partial t} \quad \text{and} \quad \hat{p} = -i\nabla, \tag{2.4}$$

respectively, giving rise to the commutation relation and hence the *Heisenberg uncertainty principle*,

$$[\hat{p}_i, \hat{x}_j] = -i\delta_{ij} \quad \text{and} \quad (\Delta p)(\Delta x) \gtrsim \hbar, \tag{2.5}$$

respectively.

Dirac wanted to describe the dynamics, infinitesimal time translations, of a particle with an equation that is first-order in time and Lorentz-invariant. To do this, he started with the general form:

$$\begin{aligned} i\frac{\partial \Psi(x,t)}{\partial t} &= (\boldsymbol{\alpha}\cdot\hat{p} + \beta m)\Psi(x,t) \\ &= (-i\boldsymbol{\alpha}\cdot\nabla + \hbar\beta m)\Psi(x,t) \\ &= \left[-i\left(\alpha^1\frac{\partial}{\partial x^1} + \alpha^2\frac{\partial}{\partial x^2} + \alpha^3\frac{\partial}{\partial x^3}\right) + \beta m\right]\Psi(x,t). \end{aligned} \tag{2.6}$$

The question then became, "What are the conditions on $\boldsymbol{\alpha}$ and β?" The relativistic energy-momentum relation for a free particle requires that

$$\begin{aligned} \hat{E}^2\Psi(x,t) &= \left(i\frac{\partial}{\partial t}\right)^2 \Psi(x,t) \\ &= (\boldsymbol{\alpha}\cdot\hat{p} + \beta m)(\boldsymbol{\alpha}\cdot\hat{p} + \beta m)\Psi(x,t) \\ &= \left[\sum_{i,j}\alpha^i\alpha^j p^i p^j + \sum_i(\alpha^i\beta + \beta\alpha^i)p^i m + \beta^2 m^2\right]\Psi(x,t) \\ &= (\hat{p}^2 + m^2)\Psi(x,t). \end{aligned} \tag{2.7}$$

This implies the anticommutation { } relations

$$\{\alpha^i, \beta\} = 0 \quad \text{and} \quad \{\alpha^i, \alpha^j\} = 2\delta_{ij} \tag{2.8}$$

and that the squares are unity

$$(\alpha^i)^2 = \beta^2 = 1 \qquad (i = 1,2,3). \tag{2.9}$$

As pure numbers $\boldsymbol{\alpha}$ and β would not satisfy the above conditions. Dirac proposed that $\boldsymbol{\alpha}$ and β are matrices and that the wave function Ψ a multicomponent column vector, known as a Dirac spinor.

2.3 Pauli-Dirac Representation and Connection with Nonrelativistic QM

We can find 4×4 matrix representations for $\boldsymbol{\alpha}$ and β that satisfy the Dirac conditions, equations (2.8) and (2.9). However, a four-component Dirac spinor, Ψ, has twice the number of degrees of freedom needed to describe a nonrelativistic spin-1/2 particle. We can therefore appeal to the low (kinetic) energy limit of the Dirac equation (2.6) to find the nonrelativistic correspondence of solutions to the Dirac equation and begin to associate physical meaning to the components of the spinor. The specific form for $\boldsymbol{\alpha}$ and β is not unique, and therefore there is the freedom to choose a particular representation. The *Pauli-Dirac* representation assigns

$$\alpha^i = \begin{pmatrix} 0 & \sigma^i \\ \sigma^i & 0 \end{pmatrix}, \quad \beta = \begin{pmatrix} I_2 & 0 \\ 0 & -I_2 \end{pmatrix} \tag{2.10}$$

where σ^i are the 2×2 Pauli spin matrices and I_2 the unit 2×2 matrix. Recall the Pauli representation for σ^i,

$$\sigma^1 = \begin{pmatrix} 0 & 1 \\ 1 & 0 \end{pmatrix}, \quad \sigma^2 = \begin{pmatrix} 0 & -i \\ i & 0 \end{pmatrix}, \quad \sigma^3 = \begin{pmatrix} 1 & 0 \\ 0 & -1 \end{pmatrix}. \tag{2.11}$$

We will see from equation (2.6) that the choice of β diagonal favors a nonrelativistic decoupling of the two 2-component solutions.

For an electron at rest, the Dirac equation (2.6) reduces to

$$i\frac{\partial \Psi}{\partial t} = \beta m \Psi \tag{2.12}$$

and has the following four solutions:

$$\begin{aligned}
&\Psi^1 = \exp(-imt)\begin{pmatrix} 1 \\ 0 \\ 0 \\ 0 \end{pmatrix} \quad \Psi^2 = \exp(-imt)\begin{pmatrix} 0 \\ 1 \\ 0 \\ 0 \end{pmatrix} \quad \text{(positive-frequency)} \\
&\Psi^3 = \exp(+imt)\begin{pmatrix} 0 \\ 0 \\ 1 \\ 0 \end{pmatrix} \quad \Psi^4 = \exp(+imt)\begin{pmatrix} 0 \\ 0 \\ 0 \\ 1 \end{pmatrix} \quad \text{(negative-frequency).}
\end{aligned} \tag{2.13}$$

Two of these solutions are positive-frequency solutions ($\Psi^{1,2}$) and two are negative ($\Psi^{3,4}$), corresponding to the sign on the right-hand side of equation (2.12). In the absence of an interaction, we do not know how to interpret the different free-particle solutions. We will therefore exercise the Dirac equation and introduce the electromagnetic interaction from an external four-potential

$$A^\mu : (\Phi, A) \tag{2.14}$$

via the "minimal coupling" substitution

$$p^\mu \to p^\mu - eA^\mu \tag{2.15}$$

where $e<0$ is the electron charge. The origin of the minimal coupling substitution is described later in the section on local gauge invariance.

The Dirac equation with the electromagnetic minimal coupling substitution becomes

$$i\frac{\partial \Psi}{\partial t} = \left(\boldsymbol{\alpha}\cdot(\hat{\boldsymbol{p}} - e\boldsymbol{A}) + \beta m + e\Phi\right)\Psi, \tag{2.16}$$

which we will use to study the interactions of a point charge with an applied electromagnetic field.

Initially, we have considered the matrices $\boldsymbol{\alpha}$ and β as introducing a static rearrangement of the components of Ψ. In the Heisenberg interpretation (H), these matrices become operators whose time evolution is governed by the Heisenberg equation of motion:

$$\frac{d\Omega^{(H)}}{dt} = i[\hat{E},\Omega^{(H)}] + \frac{\partial \Omega^{(H)}}{\partial t}. \tag{2.17}$$

The correspondence with the Schrödinger (S) time-independent operators is described as follows:

$$\begin{aligned}
\int_V \Psi'^{\dagger}(x,t)\,\Omega^{(S)}\,\Psi(x,t)d^3x &= \int_V \Psi'^{\dagger}(x,0)\,\Omega^{(H)}\,\Psi(x,0)d^3x,\\
\Psi(x,t) &= \exp(-i\hat{E}t)\Psi(x,0),\\
\Omega^{(H)} &= \exp(i\hat{E}t)\Omega^{(S)}\exp(-i\hat{E}t) \qquad \text{for } \Omega^{(H)}(0) = \Omega^{(S)}.
\end{aligned} \tag{2.18}$$

Therefore, applying these relations to the position operator $\hat{\boldsymbol{x}}$, we get

$$\frac{d\hat{\boldsymbol{x}}}{dt} = i[\hat{E},\hat{\boldsymbol{x}}] = \boldsymbol{\alpha}, \tag{2.19}$$

which by relativistic extension of the Ehrenfest classical correspondence principle indicates that $\boldsymbol{\alpha}$ is the velocity operator in Dirac theory. For the operator corresponding to the kinetic momentum $\boldsymbol{\pi} \equiv \hat{\boldsymbol{p}} - e\boldsymbol{A}$, we find

$$\begin{aligned}
\frac{d\boldsymbol{\pi}}{dt} &= i[\hat{E},\boldsymbol{\pi}] - e\frac{\partial}{\partial t}\boldsymbol{A}\\
&= i\sum_j \alpha^j[\pi^j,\boldsymbol{\pi}] + ie[\Phi,\boldsymbol{\pi}] - e\frac{\partial}{\partial t}\boldsymbol{A}\\
&= e[\boldsymbol{E} + \boldsymbol{\alpha}\times\boldsymbol{B}]
\end{aligned} \tag{2.20}$$

for the electric $\boldsymbol{E}$ and magnetic $\boldsymbol{B}$ fields

$$\boldsymbol{E} = -\frac{\partial \boldsymbol{A}}{\partial t} - \nabla\Phi \quad \text{and} \quad \boldsymbol{B} = \nabla\times\boldsymbol{A}, \tag{2.21}$$

reproducing the motion of a point charge e with $\boldsymbol{\alpha}$ as the velocity operator.

The nonrelativistic limit of equation (2.16) in the Pauli-Dirac representation can be conveniently solved by expressing the wave function in terms of two 2-component column matrices, $\tilde{\psi}_L$ and $\tilde{\psi}_S$, where the L and S refer to the large and small components, respectively. The relative magnitude of the two 2-component spinors results from specifying positive-frequency solutions, which we know from the free-particle rest-frame solutions, equation (2.13), come dominantly from the top two components of the four-component Dirac spinor. Therefore, writing

$$\Psi = \begin{pmatrix} \tilde{\psi}_L \\ \tilde{\psi}_S \end{pmatrix} \tag{2.22}$$

and substituting into equation (2.16) yields

$$i\frac{\partial}{\partial t}\begin{pmatrix} \tilde{\psi}_L \\ \tilde{\psi}_S \end{pmatrix} = \boldsymbol{\sigma}\cdot\boldsymbol{\pi}\begin{pmatrix} \tilde{\psi}_S \\ \tilde{\psi}_L \end{pmatrix} + e\Phi\begin{pmatrix} \tilde{\psi}_L \\ \tilde{\psi}_S \end{pmatrix} + m\begin{pmatrix} \tilde{\psi}_L \\ -\tilde{\psi}_S \end{pmatrix}. \tag{2.23}$$

If we now assume that the rest energy m is the largest energy in the system, the dominant time dependence of the positive-frequency solution can be factorized out,

$$\begin{pmatrix} \tilde{\psi}_L \\ \tilde{\psi}_S \end{pmatrix} \approx \exp(-imt)\begin{pmatrix} \psi_L \\ \psi_S \end{pmatrix}. \tag{2.24}$$

This approximation results in the following simplification:

$$i\frac{\partial}{\partial t}\begin{pmatrix} \psi_L \\ \psi_S \end{pmatrix} = \boldsymbol{\sigma}\cdot\boldsymbol{\pi}\begin{pmatrix} \psi_S \\ \psi_L \end{pmatrix} + e\Phi\begin{pmatrix} \psi_L \\ \psi_S \end{pmatrix} - 2m\begin{pmatrix} 0 \\ \psi_S \end{pmatrix}. \tag{2.25}$$

The time derivative of ψ will estimate the energy of the electron relative to its rest mass, as is the convention for energies in nonrelativistic mechanics. In the nonrelativistic limit where m is large compared to this correction and compared to $e\Phi$, the approximate solution for the lower two components of equation (2.25) is

$$\psi_S \approx \frac{\boldsymbol{\sigma}\cdot\boldsymbol{\pi}}{2m}\psi_L, \tag{2.26}$$

which when substituted back into the upper two components of equation (2.25) yields

$$i\frac{\partial \psi_L}{\partial t} = \left[\frac{(\boldsymbol{\sigma}\cdot\boldsymbol{\pi})(\boldsymbol{\sigma}\cdot\boldsymbol{\pi})}{2m} + e\Phi\right]\psi_L. \tag{2.27}$$

This is further reduced by the identity for Pauli spin matrices

$$(\boldsymbol{\sigma}\cdot\boldsymbol{a})(\boldsymbol{\sigma}\cdot\boldsymbol{b}) = \boldsymbol{a}\cdot\boldsymbol{b} + i\boldsymbol{\sigma}\cdot(\boldsymbol{a}\times\boldsymbol{b}), \tag{2.28}$$

which holds even if $\boldsymbol{a}$ and $\boldsymbol{b}$ are operators. Therefore, evaluating the operator cross-product term yields

$$\begin{aligned} \boldsymbol{\pi}\times\boldsymbol{\pi} &= (\hat{\boldsymbol{p}} - e\boldsymbol{A})\times(\hat{\boldsymbol{p}} - e\boldsymbol{A}) \\ &= ie(\nabla\times\boldsymbol{A}) = ie\boldsymbol{B}. \end{aligned} \tag{2.29}$$

Recalling that the operator $\hat{\boldsymbol{p}}$ will act on everything to the right,

$$\hat{\boldsymbol{p}}\times\boldsymbol{A} = -i(\nabla\times\boldsymbol{A}) - \boldsymbol{A}\times\hat{\boldsymbol{p}}, \tag{2.30}$$

while the operator ∇ is acting only on $\boldsymbol{A}$. With this simplification, equation (2.27) becomes

$$i\frac{\partial\psi_L}{\partial t} = \left[\frac{(\hat{p} - eA)^2}{2m} - \frac{e}{2m}\boldsymbol{\sigma}\cdot\boldsymbol{B} + e\Phi\right]\psi_L \tag{2.31}$$

where ψ_L is the Schrödinger-Pauli two-component wave function in nonrelativistic quantum mechanics. Replacing the spin $\hat{s} = \boldsymbol{\sigma}/2$ in the $-\boldsymbol{\mu}\cdot\boldsymbol{B}$ interaction term and writing the magnetic moment $\boldsymbol{\mu}$ as

$$\boldsymbol{\mu} = \frac{ge}{2m}\hat{s} \tag{2.32}$$

reveals the landmark result from the Dirac equation of $g = 2$ for the gyromagnetic ratio g of the electron.

2.3.1 *Constants of Motion*

The Heisenberg equation (2.17) of motion can be used to determine whether a given observable is a constant of the motion. If we apply this to the angular momentum operator

$$\hat{\boldsymbol{L}} = \hat{\boldsymbol{x}} \times \hat{\boldsymbol{p}}, \tag{2.33}$$

then we find for a free-particle Dirac Hamiltonian (2.6)

$$\frac{d\hat{\boldsymbol{L}}}{dt} = \boldsymbol{\alpha} \times \hat{\boldsymbol{p}}. \tag{2.34}$$

This result is contrary to that in nonrelativistic mechanics. For a free Dirac particle, $\boldsymbol{L}$ is not a constant of the motion.

How can angular momentum not be conserved in a free-particle Hamiltonian? We know that the Lorentz group incorporates rotational invariance for free-particle solutions of the Dirac equation, but recall that this is with respect to the total angular momentum $\boldsymbol{J}$. However, the free-particle Dirac Hamiltonian is inherently spin-1/2. Therefore, the intrinsic spin of the Dirac wave function is what gives rise to the nonconservation of $\boldsymbol{L}$.

Based on the expected low-energy behavior, the three-dimensional Dirac spin operator, denoted $\frac{1}{2}\boldsymbol{\Sigma}$, has the following form in the Pauli-Dirac representation:

$$\frac{1}{2}\boldsymbol{\Sigma} = \frac{1}{2}\begin{pmatrix} \boldsymbol{\sigma} & 0 \\ 0 & \boldsymbol{\sigma} \end{pmatrix}. \tag{2.35}$$

One can define a fifth matrix from the $\boldsymbol{\alpha}$ matrices, given by (in the Pauli-Dirac representation)

$$\gamma^5 \equiv -i\alpha^1\alpha^2\alpha^3 = \begin{pmatrix} 0 & I_2 \\ I_2 & 0 \end{pmatrix}, \tag{2.36}$$

that commutes with $\boldsymbol{\alpha}$, anticommutes with β, and satisfies the squared unity $(\gamma^5)^2 = 1$ condition. The Dirac spin operator can be conveniently written in terms of γ^5 and $\boldsymbol{\alpha}$ to simplify calculations:

$$\tfrac{1}{2}\boldsymbol{\Sigma} = \tfrac{1}{2}\gamma^5\boldsymbol{\alpha}. \tag{2.37}$$

More generally, the spin operator can be obtained from the representation of the Lorentz algebra for Dirac particles. Namely, a Dirac spinor is a four-component column vector Ψ that transforms under boosts and rotations according to the generators (in the Pauli-Dirac representation)

$$S^{0k} = \frac{i}{2}\begin{pmatrix} 0 & \sigma^k \\ \sigma^k & 0 \end{pmatrix} \quad \text{and} \quad S^{ij} = \frac{1}{2}\epsilon^{ijk}\begin{pmatrix} \sigma^k & 0 \\ 0 & \sigma^k \end{pmatrix}, \tag{2.38}$$

respectively, where the spatial part, S^{ij}, contains the three components of $\frac{1}{2}\boldsymbol{\Sigma}$. The method for deriving a representation for the Lorentz algebra is described later in the section on Lorentz transformations.

The Heisenberg equation of motion applied to $\frac{1}{2}\boldsymbol{\Sigma}$ precisely cancels the contribution from the angular momentum (2.34)

$$\frac{1}{2}\frac{d\boldsymbol{\Sigma}}{dt} = -\boldsymbol{\alpha} \times \hat{\boldsymbol{p}}, \tag{2.39}$$

showing explicitly that the total angular momentum $\boldsymbol{J} = \boldsymbol{L} + \frac{1}{2}\boldsymbol{\Sigma}$ is a constant of the motion for a free Dirac particle.

As suggested by the form of equation (2.39), we can test the quantity $\boldsymbol{\Sigma} \cdot \hat{\boldsymbol{p}}/|\boldsymbol{p}|$, known as the *helicity*, and for the free-particle Dirac Hamiltonian

$$\frac{d(\boldsymbol{\Sigma} \cdot \hat{\boldsymbol{p}})}{dt} = 0. \tag{2.40}$$

This shows that helicity is a constant of motion. We will return to the topic of helicity conservation when we discuss the Weyl representation of the Dirac matrices.

For an electron in the presence of an electromagnetic field, we can consider the quantity $\boldsymbol{\Sigma} \cdot \boldsymbol{\pi}$ where $\boldsymbol{\pi}$ is the kinetic momentum. For a set of static fields, A^μ time-independent, we find

$$\frac{d(\boldsymbol{\Sigma} \cdot \boldsymbol{\pi})}{dt} = i[\boldsymbol{\alpha} \cdot \boldsymbol{\pi} + \beta m + e\Phi, \boldsymbol{\Sigma} \cdot \boldsymbol{\pi}] = e\boldsymbol{\Sigma} \cdot \boldsymbol{E}, \tag{2.41}$$

using $\boldsymbol{\Sigma} = \gamma^5 \boldsymbol{\alpha}$ and similar results to the commutator given in equation (2.20). Therefore, for no electric field, the constancy of $\boldsymbol{\Sigma} \cdot \boldsymbol{\pi}$ means that a longitudinally polarized electron bent into a circular trajectory by a constant time-independent magnetic field will keep its spin steadfastly pointed in the direction of motion. In other words, the helicity of the electron will remain constant, as shown in figure 2.1. A left-handed helicity electron deflected by a static magnetic field has its spin rotated in such a way as to keep the helicity of the electron constant.

As a final remark, the preceding results assume that the gyromagnetic ratio of the electron is exactly $g = 2$, as predicted by the Dirac equation when treating the electromagnetic interaction as a classical field. An important discovery by Foley and Kusch in 1947 showed that magnetic moment of the electron deviates slightly from the $g = 2$ prediction. This deviation, parameterized as $g = 2(1 + a_e)$, was first explained by Schwinger as an anomalous magnetic moment arising from the virtual emission and absorption of photons [2]. Schwinger found

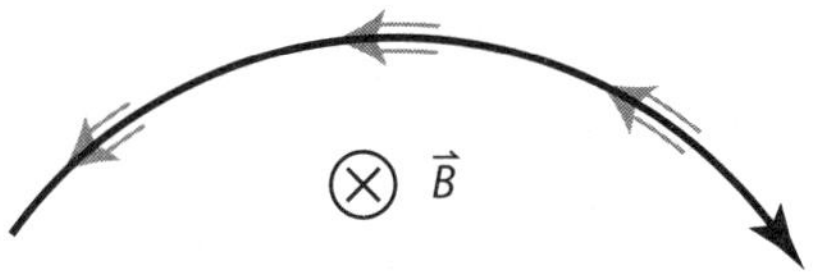

FIGURE 2.1. The left-handed electron orients its spin polarization (open arrow) to follow changes in the three-momentum direction (solid arrow) as the electron moves in a circular orbit in the magnetic field. The electron helicity is a constant of the motion.

$$a_e \approx \frac{\alpha_{\text{QED}}}{2\pi} \tag{2.42}$$

where $\alpha_{\text{QED}} = e^2/(4\pi)$ is called the fine structure constant. The fine structure constant is the only physical constant unique to the theory of the electromagnetic interaction of relativistic electrons and photons, also known as the theory of quantum electrodynamics (QED). QED has proven to have immensely precise predictive power. In fact, the magnetic moment of the electron is one of the most accurately predicted and measured quantities in physics, accurate to better than one part in a trillion. Measurements of g from the magnetic moment of the electron compared with QED predictions for the anomalous moment yield a value of the inverse of the fine structure constant of $\alpha_{\text{QED}}^{-1} = 137.035\ 999\ 679\ (94)$ where the uncertainty in the measurement and theory is in the last two digits, as given in parentheses [3].

2.3.2 *Velocity in Dirac Theory*

Equation (2.19) indicates that $\boldsymbol{\alpha}$ is to be interpreted as the velocity operator in Dirac theory. While we know the plane-wave solutions to the Dirac equation are eigenstates of momentum, it is natural to ask, "Is velocity a constant of the motion for a free Dirac particle?" The answer is clearly no. The time evolution of $\boldsymbol{\alpha}$ from the Heisenberg equation is given by

$$\frac{d\boldsymbol{\alpha}}{dt} = -2(\boldsymbol{\Sigma} \times \hat{\boldsymbol{p}}) - i2m\boldsymbol{\alpha}\beta. \tag{2.43}$$

The noncommutation of the velocity operator with the Hamiltonian is unexpected based on the classical relationship between velocity and momentum. It is, however, an intrinsic aspect of the relativistic quantum theory. The "quivering" motion predicted by the time dependence of $\boldsymbol{\alpha}$ is known as *Zitterbewegung*. The characteristic time scale for this motion corresponds to the Heisenberg uncertainty time $\Delta t \sim 1/2m$, the inverse of the electron-positron pair production threshold. The effect of the position averaging induced by Zitterbewegung is observable in atomic systems and is known as the *Darwin term*, a spatial averaging term for the potential.

An alternative way to view the origin of the quivering motion is predicted by equation (2.43). The time-dependent phase term in equation (2.43) comes from the β or mass term of the Dirac equation. In the high-energy or massless limit, the four-component Dirac

FIGURE 2.2. The mass term of the Dirac equation behaves as an interaction between a massless particle and an external scalar field that mixes left- and right-handed chiralities.

spinor decouples into two 2-component spinors that are termed the left- and right-handed *chiral* components, as will be discussed in the section on the Weyl representation. The β term can be viewed as a coupling term between two 2-component massless particles with definite chirality. Therefore, the mixing of left- and right-handed chirality by the mass term can be physically viewed as a particle interacting with an external scalar field reversing its chirality as it propagates through the vacuum, as sketched in figure 2.2. The Dirac velocity operator $\boldsymbol{\alpha}$ is telling us about the velocities of the chiral components of the Dirac spinor. This explains why the operator does not commute with the Dirac Hamiltonian. The $\boldsymbol{\alpha}$ operator has eigenvalues of ± 1 times the speed of light, representing the velocities of the chiral components in the massless limit.

2.4 Probability Current

The normalization condition from nonrelativistic mechanics (2.3) has the following positive-definite form when applied to the Dirac spinor:

$$\int_V \Psi^\dagger \Psi d^3x = \int_V \sum_{r=1}^{4} \psi_r^* \psi_r d^3x = 1. \tag{2.44}$$

We can construct a probability current from the Dirac equation (2.6) by multiplying on the left by $\Psi^\dagger$:

$$i\Psi^\dagger \frac{\partial \Psi}{\partial t} = -i\Psi^\dagger \boldsymbol{\alpha} \cdot \nabla \Psi + m\Psi^\dagger \beta \Psi. \tag{2.45}$$

Then, taking the Hermitian conjugate of (2.6) and multiplying on the right by Ψ gives

$$-i\frac{\partial \Psi^\dagger}{\partial t}\Psi = i\nabla \Psi^\dagger \cdot \boldsymbol{\alpha}\Psi + m\Psi^\dagger \beta \Psi. \tag{2.46}$$

Subtracting (2.46) from (2.45) yields the continuity equation for the Dirac probability

$$\frac{\partial}{\partial t}(\Psi^\dagger \Psi) + \nabla(\Psi^\dagger \boldsymbol{\alpha} \Psi) = 0 \tag{2.47}$$

with the three-dimensional probability current given by

$$j = \Psi^\dagger \boldsymbol{\alpha} \Psi. \tag{2.48}$$

Integrating equation (2.47) over all space and using Green's theorem gives

$$\frac{\partial}{\partial t}\int_V \Psi^\dagger \Psi d^3x = 0, \tag{2.49}$$

which suggests the interpretation of $\Psi^\dagger\Psi$ as a positive-definite probability density. It can be shown that

$$j^\mu = (\Psi^\dagger\Psi, \Psi^\dagger \boldsymbol{\alpha}\Psi) \tag{2.50}$$

forms a covariant Lorentz vector.

Implicit in the assumption that $\Psi^\dagger\Psi$ is to be interpreted as a probability density is that the solution Ψ to the Dirac equation is a single-particle wave function. We return to this point later in the section on antiparticles.

2.5 Free-Particle Solutions in the Pauli-Dirac Representation

The free-particle solutions to the Dirac equation (2.6) have the form

$$\Psi = \begin{pmatrix} \tilde{\psi}_A \\ \tilde{\psi}_B \end{pmatrix} = \begin{pmatrix} \psi_A(p) \\ \psi_B(p) \end{pmatrix} \exp(i\boldsymbol{p}\cdot\boldsymbol{x} - iEt) \tag{2.51}$$

where the sign of the exponent corresponds explicitly to a positive-frequency time evolution for positive-definite energy plane-wave solutions. With the spatial and time-dependent parts factored out, the two 2-component spinors are related by

$$E\begin{pmatrix} \psi_A(p) \\ \psi_B(p) \end{pmatrix} = \begin{pmatrix} 0 & \boldsymbol{\sigma} \\ \boldsymbol{\sigma} & 0 \end{pmatrix}\cdot\boldsymbol{p}\begin{pmatrix} \psi_A(p) \\ \psi_B(p) \end{pmatrix} + m\begin{pmatrix} I_2 & 0 \\ 0 & -I_2 \end{pmatrix}\begin{pmatrix} \psi_A(p) \\ \psi_B(p) \end{pmatrix} \tag{2.52}$$

giving

$$\psi_A(p) = \frac{1}{E-m}(\boldsymbol{\sigma}\cdot\boldsymbol{p})\psi_B(p) \quad \text{and} \quad \psi_B(p) = \frac{1}{E+m}(\boldsymbol{\sigma}\cdot\boldsymbol{p})\psi_A(p). \tag{2.53}$$

Recalling the form of the rest-frame solutions (2.13), we can find the two positive-frequency solutions by specifying the spin components of ψ_A up to a normalization factor as

$$\psi_A(p) \propto \begin{pmatrix} 1 \\ 0 \end{pmatrix} \quad \text{and} \quad \begin{pmatrix} 0 \\ 1 \end{pmatrix} \tag{2.54}$$

and then using the second equation of (2.53) to generate the lower two components of the four-component Dirac spinor, recalling

$$\boldsymbol{\sigma}\cdot\boldsymbol{p} = \begin{pmatrix} p_3 & p_1 - ip_2 \\ p_1 + ip_2 & -p_3 \end{pmatrix}. \tag{2.55}$$

This gives, including a normalization factor N,

$$u^{(1)}(p) = N\begin{pmatrix} 1 \\ 0 \\ p_3/(E+m) \\ (p_1+ip_2)/(E+m) \end{pmatrix} \text{ and}$$

$$u^{(2)}(p) = N\begin{pmatrix} 0 \\ 1 \\ (p_1-ip_2)/(E+m) \\ -p_3/(E+m) \end{pmatrix}. \tag{2.56}$$

The remaining two solutions are found by inserting the rest-frame behavior (2.13) for the lower two components

$$\psi_B(p) \propto \begin{pmatrix} 1 \\ 0 \end{pmatrix} \quad \text{and} \quad \begin{pmatrix} 0 \\ 1 \end{pmatrix}. \tag{2.57}$$

For nonzero $\boldsymbol{p}$, this gives

$$u^{(3)}(p) = N\begin{pmatrix} -p_3/(|E|+m) \\ -(p_1+ip_2)/(|E|+m) \\ 1 \\ 0 \end{pmatrix} \text{ and}$$

$$u^{(4)}(p) = N\begin{pmatrix} -(p_1-ip_2)/(|E|+m) \\ p_3/(|E|+m) \\ 0 \\ 1 \end{pmatrix}, \tag{2.58}$$

where we have reversed the sign of the energy $E \to -E$ and replaced $(E-m)$ by $-(|E|+m)$ using absolute values to indicate that a positive-definite energy convention is used. Here, the *Feynman-Stückelberg* prescription is applied to interpret these two solutions as having negative-frequency time evolution. We will return to the Feynman-Stückelberg formulation in the section on relativistic propagator theory. The prescription is to reverse the sign of all four components of the four-momentum, the energy, and three-momentum, and flip the spin direction. We define new solutions, $v^{(1)}(p)$ and $v^{(2)}(p)$, in terms of the $u^{(3)}(p)$ and $u^{(4)}(p)$ solutions, as follows:

$$\begin{aligned} v^{(1)}(p)\exp(-i\boldsymbol{p}\cdot\boldsymbol{x}+iEt) &\equiv u^{(4)}(-p)\exp(i(-\boldsymbol{p})\cdot\boldsymbol{x}-i(-E)t), \\ v^{(2)}(p)\exp(-i\boldsymbol{p}\cdot\boldsymbol{x}+iEt) &\equiv -u^{(3)}(-p)\exp(i(-\boldsymbol{p})\cdot\boldsymbol{x}-i(-E)t). \end{aligned} \tag{2.59}$$

The choice of relative phase in equation (2.59) is set by a discrete transformation defined in the section on charge-conjugation symmetry where the flipping of the spin components and sign reversal of the lower component can be shown to correspond to the $\bar{\mathbf{2}}$ representation of $SU(2)$. From equation (2.59), the explicit form of the $v^{(1)}(p)$ and $v^{(2)}(p)$ solutions are

$$
v^{(1)}(p) = N\begin{pmatrix} (p_1 - ip_2)/(E+m) \\ -p_3/(E+m) \\ 0 \\ 1 \end{pmatrix} \text{ and }
\\
v^{(2)}(p) = N\begin{pmatrix} -p_3/(E+m) \\ -(p_1 + ip_2)/(E+m) \\ -1 \\ 0 \end{pmatrix}, \tag{2.60}
$$

where E is positive-definite and, therefore, the absolute values on $|E|$ are no longer needed. The physical interpretation of the v solutions will become evident later in the section on charge-conjugation symmetry. The v solutions are written as positive-energy solutions, but with negative-frequency time evolution. As we will find, the interplay of the positive- and negative-frequency solutions acts much in the same way as the Fourier analysis of a function that must vanish over a finite region. In this case, causality restricts free-particle solutions to within the light cone. The cancellation outside of the light cone can only be achieved by summing over positive- and negative-frequency components.

The first two solutions are positive-frequency solutions and are indicated by $u^{(1)}(p)$ and $u^{(2)}(p)$. The positive-frequency solutions satisfy

$$(\boldsymbol{\alpha}\cdot\boldsymbol{p} + \beta m - E)\,u(p) = 0. \tag{2.61}$$

From this point on, we will restrict the u solutions to denote only the two positive-frequency solutions. The two negative-frequency solutions, written as positive-energy solutions, are denoted by v. Note that when applying the $(-E, -\boldsymbol{p})$ transform to the Dirac equation we get

$$(\boldsymbol{\alpha}\cdot(-\boldsymbol{p}) + \beta m - (-E))\,u(-p) = 0 \tag{2.62}$$

and, therefore, the v solutions satisfy

$$(\boldsymbol{\alpha}\cdot\boldsymbol{p} - \beta m - E)\,v(p) = 0. \tag{2.63}$$

The normalization of the spinor part of the Dirac wave function can be chosen to transform like the timelike component of a Lorentz four-vector

$$
u^{(r)\dagger}(p)u^{(s)}(p) = (2E)\delta_{rs},
\\
v^{(r)\dagger}(p)u^{(s)}(p) = (2E)\delta_{rs} \tag{2.64}
$$

with $r, s = 1, 2$, and therefore we find for the normalization factor N

$$
2E = [1 + |\boldsymbol{p}|^2/(E+m)^2]N^2,
\\
N = \sqrt{(E+m)} \tag{2.65}
$$

where the u and v solutions are orthogonal to each other. The choice of normalization in equation (2.64) allows $u^\dagger u$ and $v^\dagger v$ to have the transformation properties of $\Psi^\dagger\Psi$, as discussed in the section on probability current.

The normalized plane-wave solutions are therefore

$$\Psi^{(+)} = \frac{1}{\sqrt{2EV}} u^{(1 \text{ or } 2)}(p) \exp[i\mathbf{p}\cdot\mathbf{x} - iEt] \tag{2.66}$$

for the first two solutions, and

$$\Psi^{(-)} = \frac{1}{\sqrt{2EV}} v^{(1 \text{ or } 2)}(p) \exp[-i\mathbf{p}\cdot\mathbf{x} + iEt] \tag{2.67}$$

for the last two solutions. Note that the $\Psi^{(-)}$ solutions have negative-frequency dependence in the exponential. The square root factors in (2.66) and (2.67) compensate for the choice made in equation (2.64) so that the solutions, $\Psi^{(\pm)}$, satisfy the normalization condition (2.44) for the integral of the probability density $\Psi^\dagger\Psi$ over the volume V.

2.6 Antiparticles

The Dirac equation demands four solutions for the relativistic electron, although only two were expected. The nature of the additional solutions becomes evident when we try to localize a Dirac electron. To do this, we consider the scattering of a plane wave off of a steep electrostatic potential. Figure 2.3 shows a free electron of energy E in region I traveling to the right, where it encounters a step potential of height $V_0 > E + m$, with m equal to the mass of the electron. The positive-frequency solutions for the incident and reflected waves in region I are given by

$$\begin{aligned} \Psi_i^{(+)} &= a_i \exp(ik_z z - iEt)\begin{pmatrix} 1 \\ 0 \\ \frac{k_z}{E+m} \\ 0 \end{pmatrix}, \\ \Psi_r^{(+)} &= a_r \exp(-ik_z z - iEt)\begin{pmatrix} 1 \\ 0 \\ -\frac{k_z}{E+m} \\ 0 \end{pmatrix}, \end{aligned} \tag{2.68}$$

where only the spin-up reflected wave is written, as no spin flip will occur from a scalar electrostatic potential. The transmitted wave must satisfy the Dirac equation in the presence of a constant external potential $e\Phi = V_0$. The potential difference, therefore, changes the wave number in region II, giving

$$k_z'^2 = (E - V_0)^2 - m^2 = (E - V_0 - m)(E + m - V_0). \tag{2.69}$$

The positive-frequency spin-up transmitted wave is given by

$$\Psi_t^{(+)} = a_t \exp(ik_z' z - iEt)\begin{pmatrix} 1 \\ 0 \\ \frac{k_z'}{E - V_0 + m} \\ 0 \end{pmatrix} \tag{2.70}$$

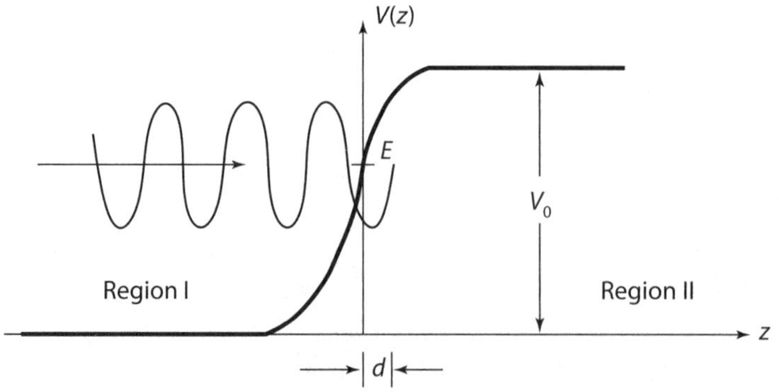

FIGURE 2.3. Electrostatic potential with a sharp boundary, with an incident free electron wave of energy E moving to the right in region I. For $V_0 > E + m$ the reflected current from the potential exceeds the incident one. This is an example of the *Klein paradox*.

where we substituted $E \rightarrow E - V_0$ and used the wave number k_z'.

To solve for this system, we make use of the probability current derived in the previous section. The amplitude of the transmitted wave a_t is fixed by continuity of the solution at $z = 0$ independent of time, which yields

$$\begin{aligned} &\text{first component:} \quad && a_i + a_r = a_t, \\ &\text{third component:} \quad && \frac{(a_i - a_r)\, k_z}{E + m} = \frac{a_t k_z'}{E - V_0 + m}. \end{aligned} \tag{2.71}$$

Solving for the reflected and transmitted amplitudes normalized to the incident, using (2.69) and defining r to be

$$r \equiv \left(\frac{k_z'}{k_z}\right) \frac{E + m}{E + m - V_0}, \tag{2.72}$$

gives

$$\frac{a_r}{a_i} = \frac{(1 - r)}{(1 + r)}, \qquad \frac{a_t}{a_i} = \frac{2}{(1 + r)}. \tag{2.73}$$

The incident, reflected, and transmitted probability currents can be directly computed using the z-component of equation (2.50)

$$j^3 = \Psi^\dagger \alpha^3 \Psi \tag{2.74}$$

giving

$$\begin{aligned} j_i &= a_i a_i^* \frac{2k_z}{E + m}, \\ j_r &= a_r a_r^* \frac{-2k_z}{E + m}, \\ j_t &= a_t a_t^* \frac{2k_z'}{E - V_0 + m}. \end{aligned} \tag{2.75}$$

The ratios of the reflected and transmitted currents relative to the incident current are given, respectively, by

$$\begin{aligned} \frac{j_r}{j_i} &= -\left(\frac{1-r}{1+r}\right)^2, \\ \frac{j_t}{j_i} &= \frac{4r}{(1+r)^2} = 1 - \frac{j_r}{j_i}. \end{aligned} \tag{2.76}$$

From the expression for k' in equation (2.69) we can see that k'_z is imaginary for $V_0 > 0$ and $|E - V_0| < m$, as expected from nonrelativistic quantum mechanics. The penetrating current decays away exponentially into the barrier. If we try to further localize the electron with a steeper potential such that $V_0 > E + m$, then the sign of k'^2_z becomes positive again and the transmitted wave oscillates. Evaluating r from equation (2.72) for $V_0 > E + m$ shows that $r < 0$ and therefore, from equation (2.76), the reflected current *exceeds* the incident current. This is known as the *Klein paradox*.

In probing the behavior of the positive-frequency solutions by localizing the penetration depth to a distance that is shorter than half the Compton wavelength $1/2m$, we have uncovered a new degree of freedom which is present only in the relativistic description of the electron. In the Klein paradox, what we were observing is pair production of electron-positron pairs out of the vacuum. The positron is the *antiparticle* of the electron. The existence of antiparticles is the explanation for the two extra degrees of freedom in the four-component Dirac spinor. The positivity of the positron charge will be examined in the next section on charge conjugation. Note that the failure of the probability current to correctly describe a normalized scattering probability is a limitation of the wave function approach to relativistic quantum mechanics that is ultimately addressed by a many-particle quantized Dirac field description of the theory found in texts on quantum field theory (QFT). A selection of QFT reference texts can be found in section 2.16.

A physical system that approaches the conditions of the Klein paradox is a tightly bound electron in a heavy, or high-Z, atom. The relativistic bound-state solutions to the hydrogen atom can be found in the standard texts [9], [10]. Quoting the result, the energy of the ground-state electron is given by

$$E_g = m\sqrt{1 - (Z\alpha_{\mathrm{QED}})^2}. \tag{2.77}$$

Note that the second term in the Taylor expansion of equation (2.77) faithfully reproduces the Rydberg energy for the hydrogen atom. For values of $Z\alpha_{\mathrm{QED}} > 1$, the negative-frequency solutions become oscillatory. For a uranium nucleus, the binding energy of the lowest energy electron is a substantial fraction of its rest-mass energy. Approximating a heavy nucleus by a pointlike Coulomb potential, a value of $(Z\alpha_{\mathrm{QED}})^2 \approx (92/137)^2 = 0.45$ corresponds to -26% of $m_e c^2$. Heavy nuclei provide an example of physical systems with strongly confined electron wave functions. There is, however, no evidence for bound nuclei with a charge larger than the inverse of the fine structure constant $\alpha^{-1}_{\mathrm{QED}} \approx 137$, and thereby directly exhibiting spontaneous electron-positron pair production.

The explanation of the Klein paradox directly reveals that antiparticles are an essential result of unifying quantum mechanics and relativity. A similar paradox arises from the

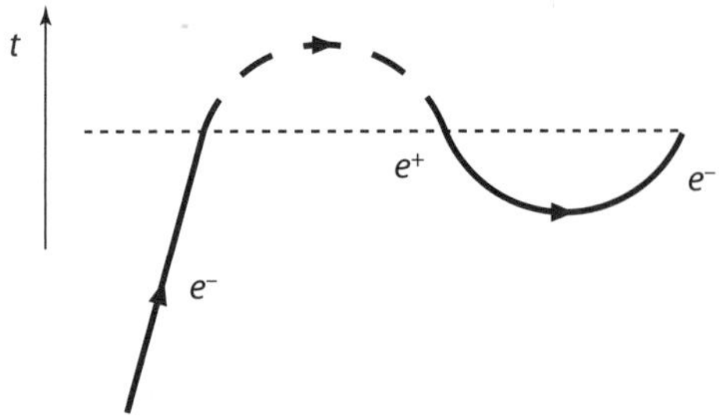

FIGURE 2.4. The Heisenberg uncertainty condition in the relativistic domain translates to the creation of multiple indistinguishable particles, an effect that can be diagrammatically viewed as a wiggle along the time axis.

constraint that Lorentz invariance imposes on the maximum velocity of a particle. The following commutation relations, which can be written (where $g^{\mu\nu}$ is conveniently inserted to account for the signs)

$$[\hat{p}^{\mu}, \hat{x}^{\nu}] = ig^{\mu\nu}, \qquad g^{\mu\nu} = \mathrm{diag}\{1, -1, -1, -1\}, \tag{2.78}$$

result in the Heisenberg uncertainty condition and prevent the Dirac wave function from being localized simultaneously in position and momentum. For an ultrarelativistic particle, any forward shift in position along the momentum direction would increase the velocity and conflict with the upper limit on velocity given by the speed of light. This paradox is resolved by the negative-frequency solutions. These additional solutions manifest as the creation of particle-antiparticle states, and therefore when one tries to localize a relativistic single-particle Dirac wave function, the result is a multiparticle state of indistiguishable particles. One can, therefore, diagrammatically view a wiggle along the time axis as the creation of multiple particle states, as drawn in figure 2.4.

2.6.1 *Charge-Conjugation Symmetry*

The Dirac equation inherently predicts the presence of antimatter. An intrinsic property of antimatter is that additive quantum numbers have the opposite sign, and, in particular, the electric charge is opposite to that of the matter particle. We can investigate this property by taking equation (2.16), reversing the sign of the charge, and denoting the solutions as Ψ^C:

$$i\frac{\partial \Psi^C}{\partial t} = \big(\boldsymbol{\alpha}\cdot(-i\nabla + eA) + \beta m - e\Phi\big)\Psi^C. \tag{2.79}$$

If the charge-conjugate solution Ψ^C is described by the same physical laws as the electron, then there must be a one-to-one correspondence between Ψ^C and Ψ. We begin by replacing Ψ^C with $C\Psi^*$ in equation (2.79), where C is a 4×4 matrix

$$i\frac{\partial (C\Psi^*)}{\partial t} = \big(\boldsymbol{\alpha}\cdot(-i\nabla + eA) + \beta m - e\Phi\big)(C\Psi^*). \tag{2.80}$$

Taking the complex conjugate of equation (2.80) gives

$$-i\frac{\partial(C\Psi^*)}{\partial t} = \left(\boldsymbol{\alpha}^* \cdot (i\nabla + eA) + \beta^* m - e\Phi\right)(C\Psi^*), \tag{2.81}$$

where β, Φ, and A are real, as are α^1 and α^3. Multiplying (2.81) on the left by $-(C^*)^{-1}$ and comparing with the original Dirac equation (2.16) indicates that C must satisfy

$$\begin{aligned}(C^*)^{-1}(\alpha^i)^* C^* &= \alpha^i,\\ (C^*)^{-1}\beta^* C^* &= -\beta\end{aligned} \tag{2.82}$$

for there to be equivalence under charge conjugation. If we rearrange (2.82) as

$$\begin{aligned}\alpha^i C - C(\alpha^i)^* &= 0,\\ \beta C + C\beta^* &= 0,\end{aligned} \tag{2.83}$$

then using the relations (2.8) and (2.9), we find that in the Pauli-Dirac representation the matrix C is given by $C = \eta_C \beta\alpha^2$ where η_C is an undetermined phase factor.

If we explicitly apply the charge-conjugation transform to a spin-up positive-frequency electron solution, then we find (setting $\eta_C = i$)

$$\begin{aligned}(i\beta\alpha^2)\left[u^{(1)}(p)\exp(i\boldsymbol{p}\cdot\boldsymbol{x} - iEt)\right]^* &= \sqrt{(E+m)}\begin{pmatrix}0 & 0 & 0 & 1\\ 0 & 0 & -1 & 0\\ 0 & -1 & 0 & 0\\ 1 & 0 & 0 & 0\end{pmatrix}\begin{pmatrix}1\\ 0\\ p_3/(E+m)\\ (p_1+ip_2)/(E+m)\end{pmatrix}^*\\ &\quad \exp(-i\boldsymbol{p}\cdot\boldsymbol{x} - iEt)\\ &= \sqrt{(E+m)}\begin{pmatrix}(p_1-ip_2)/(E+m)\\ -p_3/(E+m)\\ 0\\ 1\end{pmatrix}\exp(-i\boldsymbol{p}\cdot\boldsymbol{x} + iEt)\\ &= v^{(1)}(p)\exp(-i\boldsymbol{p}\cdot\boldsymbol{x} + iEt).\end{aligned} \tag{2.84}$$

Similarly, for the spin-down electron solution, we find

$$(i\beta\alpha^2)\left[u^{(2)}(p)\exp(i\boldsymbol{p}\cdot\boldsymbol{x} - iEt)\right]^* = v^{(2)}(p)\exp(-i\boldsymbol{p}\cdot\boldsymbol{x} + iEt). \tag{2.85}$$

From equation (2.79), we know that the charge-conjugate solution has the physical properties of a positively charged electron with the same four-momentum and same spin direction as its electron counterpart. This is the origin of the notation for the v solutions, which are the positron spinors.

The dependence of the form of C on the representation used for the Dirac matrices introduces the possibility of a representation such that $\Psi^C = \Psi^*$ with C the identity matrix. From equation (2.83), this is achieved by choosing β to be purely imaginary and the $\boldsymbol{\alpha}$ matrices to be real with β and $\boldsymbol{\alpha}$ satisfying the conditions (2.8) and (2.9). This form of the Dirac matrices, known as the Majorana representation, can be constructed as a permutation of the Pauli-Dirac representation, swapping $\beta \leftrightarrow \alpha^2$. In the Majorana representation, if Ψ is real, the particle and the antiparticle are identical, halving the number of independent solutions.

2.7 Lorentz Transformations

Lorentz transformations extend the three-dimensional rotation group to a four-dimensional Lorentz group. The generators for the rotation group are the total angular momentum operators $\boldsymbol{J} = \boldsymbol{L} + \boldsymbol{S}$. The angular momentum operators are given by

$$\hat{L} = \hat{x} \times \hat{p} = x \times (-i\nabla) = -i\epsilon^{ijk} x^i \frac{\partial}{\partial x^j} \tag{2.86}$$

with commutation relations

$$[L^i, L^j] = i\epsilon^{ijk} L^k. \tag{2.87}$$

Finite rotation operations are formed by exponentiating the total angular momentum operators

$$R = \exp[-i\theta_i J^i] \tag{2.88}$$

where θ_i is the rotation angle with respect to the i-axis. The angular momentum operators can also be written as asymmetric tensors,

$$L^{ij} = -i\left(x^i \frac{\partial}{\partial x^j} - x^j \frac{\partial}{\partial x^i}\right), \tag{2.89}$$

where the ij-indices indicate the axes of the rotation plane. The corresponding rotation angles for three dimensions are assembled in an antisymmetric tensor ω_{ij}

$$\omega_{ij} = \begin{pmatrix} 0 & \theta_3 & -\theta_2 \\ -\theta_3 & 0 & \theta_1 \\ \theta_2 & -\theta_1 & 0 \end{pmatrix} \tag{2.90}$$

such that the finite rotation operators have the form

$$R = \exp\left[-\frac{i}{2} \sum_{i,j=1}^{3} \omega_{ij} J^{ij}\right] \tag{2.91}$$

where for simplicity both upper and lower components of the asymmetric tensors are summed and a factor of $\frac{1}{2}$ is introduced. The asymmetric tensor form of the rotation operators is more convenient as it naturally generalizes to higher dimensions and to the Lorentz algebra.

The finite rotation angles and boost parameters of a Lorentz transformation are recorded in a 4×4 asymmetric tensor $\omega_{\mu\nu}$ by extending the three-dimensional expression to include planes defined by one time axis and one spatial axis:

$$\omega_{\mu\nu} = \begin{pmatrix} 0 & \eta_1 & \eta_2 & \eta_3 \\ -\eta_1 & 0 & \theta_3 & -\theta_2 \\ -\eta_2 & -\theta_3 & 0 & \theta_1 \\ -\eta_3 & \theta_2 & -\theta_1 & 0 \end{pmatrix}. \tag{2.92}$$

Here we use η, the *rapidity*, as the four-dimensional extension to the θ-spatial rotation angles. Also, the variable y is commonly used to represent rapidity and η to represent *pseudorapidity*, to be defined later. Rapidities have the advantage of being additive under successive Lorentz boosts in the same way that the θ-rotation angles are additive.

The generalization to four-dimensional Lorentz-invariant angular momentum operators is then

$$L^{\mu\nu} = i(x^\mu \partial^\nu - x^\nu \partial^\mu) \tag{2.93}$$

using $x^\mu = (t, \mathbf{x})$, $\partial^\mu = \left(\frac{\partial}{\partial t}, -\nabla\right)$, and $g^{\mu\nu} = \mathrm{diag}\{1, -1, -1, -1\}$ for the metric tensor, that is, $x^\mu = g^{\mu\nu} x_\nu$. Here we assume a good working knowledge of special relativity and four-vector Lorentz transformations. The commutation relations for the $L^{\mu\nu}$ are computed directly from the differential operators (2.93) and yield the expression

$$[L^{\mu\nu}, L^{\rho\sigma}] = i(g^{\nu\rho} L^{\mu\sigma} - g^{\mu\rho} L^{\nu\sigma} - g^{\nu\sigma} L^{\mu\rho} + g^{\mu\sigma} L^{\nu\rho}). \tag{2.94}$$

Equation (2.94) applies to all angular momentum operators. Therefore, the Lorentz transformations of Dirac spinors generated by operators denoted $S^{\mu\nu}$ satisfy the same commutation relations. Finding a matrix representation for $S^{\mu\nu}$ appears to be a daunting task; however, the lengthy expression (2.94) has a certain form that Dirac was able to assemble using 4×4 matrices satisfying the Dirac algebra,

$$\{\gamma^\mu, \gamma^\nu\} \equiv \gamma^\mu \gamma^\nu + \gamma^\nu \gamma^\mu = 2g^{\mu\nu} I_4 \qquad \text{(Dirac algebra)}. \tag{2.95}$$

Notice that the Dirac algebra is almost identical to the Dirac conditions (2.8) and (2.9) with the exception that $(\gamma^i)^2 = -1$.

An antisymmetric tensor $S^{\mu\nu}$ can be formed from the γ^μ matrices by taking the commutator

$$S^{\mu\nu} = \frac{i}{4}[\gamma^\mu, \gamma^\nu]. \tag{2.96}$$

Substituting $S^{\mu\nu}$ into equation (2.94) yields

$$\begin{aligned}
[S^{\mu\nu}, S^{\rho\sigma}] &= -\frac{1}{16}\left[[\gamma^\mu, \gamma^\nu], [\gamma^\rho, \gamma^\sigma]\right] \\
&= -\frac{1}{16}((\gamma^\mu\gamma^\nu - \gamma^\nu\gamma^\mu)(\gamma^\rho\gamma^\sigma - \gamma^\sigma\gamma^\rho) - (\gamma^\rho\gamma^\sigma - \gamma^\sigma\gamma^\rho)(\gamma^\mu\gamma^\nu - \gamma^\nu\gamma^\mu)) \\
&= -\frac{1}{16}((\{\mu\nu\rho\sigma\} - \{\sigma\rho\nu\mu\}) - (\{\nu\mu\rho\sigma\} - \{\sigma\rho\mu\nu\}) \\
&\quad -(\{\mu\nu\sigma\rho\} - \{\rho\sigma\nu\mu\}) + (\{\nu\mu\sigma\rho\} - \{\rho\sigma\mu\nu\}))
\end{aligned} \tag{2.97}$$

where the terms $\{\mu\nu\rho\sigma\} \equiv \gamma^\mu\gamma^\nu\gamma^\rho\gamma^\sigma$ have been ordered to correspond to the terms in equation (2.94). Applying the Dirac algebra to the central two indices yields

$$[S^{\mu\nu}, S^{\rho\sigma}] = \frac{i}{4}\left[2g^{\nu\rho}S^{\mu\sigma} - 2g^{\mu\rho}S^{\nu\sigma} - 2g^{\nu\sigma}S^{\mu\rho} + 2g^{\mu\sigma}S^{\nu\rho} + \cdots\right] \tag{2.98}$$

where the first terms have the same form as equation (2.94). The remaining terms in the expression will contribute the extra factor of 2 needed to satisfy the commutation relations. This establishes the $S^{\mu\nu}$ as the generators of the Lorentz transformations for Dirac spinors. Note that the Lorentz boost operators S^{0i} are not Hermitian ($S^{0i} \neq (S^{0i})^{\dagger}$); in fact, they are i times a Hermitian matrix, also known as *anti-Hermitian*.

The γ^{μ} matrices are related to the $\boldsymbol{\alpha}$ and β matrices of the Dirac equation in the following way:

$$\gamma^0 = \beta, \qquad \gamma^i = \beta\alpha^i, \qquad i = 1, 2, 3. \tag{2.99}$$

The conditions on the $\boldsymbol{\alpha}$ and β matrices from the construction of the Dirac equation precisely translate to the same Dirac algebra described above on the γ^{μ} matrices constructed from these four components, namely

$$\{\gamma^{\mu}, \gamma^{\nu}\} = 2g^{\mu\nu} I_4, \tag{2.100}$$

which also incorporates the squared values $(\gamma^{\mu})^2 = g^{\mu\mu}$ (no summation over μ).

If one takes the Dirac equation (2.6) in terms of $\boldsymbol{\alpha}$ and β matrices, repeated here,

$$i\frac{\partial \Psi}{\partial t} = (-i\boldsymbol{\alpha} \cdot \nabla + \beta m)\,\Psi,$$

and multiplies by $\gamma^0 (= \beta)$ on the left, then the Dirac equation can now be written in terms of the γ^{μ} matrices

$$\begin{aligned}\left[i\left(\gamma^0 \frac{\partial}{\partial x^0} + \gamma^i \frac{\partial}{\partial x^i}\right) - m\right]\Psi &= (i\gamma^{\mu}\partial_{\mu} - m)\Psi = (\gamma^{\mu} p_{\mu} - m)\Psi \\ &= (\not{p} - m)\Psi = 0.\end{aligned} \tag{2.101}$$

This is called the manifestly covariant form of the Dirac equation, as will be demonstrated below. The term $\not{p}$ is notation indicating the four-index contraction of the γ^{μ} matrices and the four-momentum operator. The motivation for the Dirac algebra can be seen by rewriting the square of the Dirac equation in the form

$$(\gamma^{\mu} p_{\mu})(\gamma^{\nu} p_{\nu})\Psi = \frac{1}{2} p_{\mu} p_{\nu} (\gamma^{\mu}\gamma^{\nu} + \gamma^{\nu}\gamma^{\mu})\Psi = m^2 \Psi, \tag{2.102}$$

which follows from the symmetric tensor $p_{\mu} p_{\nu}$.

To form Hermitian conjugate terms, we can use the convenient expression

$$(\gamma^{\mu})^{\dagger} = \gamma^0 \gamma^{\mu} \gamma^0, \tag{2.103}$$

which is a compact way of writing $(\gamma^i)^{\dagger} = -\gamma^i$ for $i = 1, 2, 3$ and $(\gamma^0)^{\dagger} = \gamma^0$. Then the fifth matrix, as given in equation (2.36) for the Pauli-Dirac representation, can be defined by a representation-independent formula to be

$$\gamma^5 \equiv \gamma_5 \equiv i\gamma^0\gamma^1\gamma^2\gamma^3 \tag{2.104}$$

with

$$\{\gamma^{\mu}, \gamma^5\} = 0, \qquad (\gamma^5)^2 = 1. \tag{2.105}$$

2.7.1 *Lorentz Invariance of the Dirac Equation*

A Lorentz transformation of the Dirac wave function $\Psi(x)$ affects the coordinates and the orientation of the spinor in the primed frame. The coordinate transform is a four-vector Lorentz transform $x' = \Lambda x$ described by a set of rotation and boost parameters $\omega_{\mu\nu}$ as given in (2.92). The finite Lorentz transformations of the Dirac spinor are described by an identical $\omega_{\mu\nu}$ and are given by

$$\Lambda_{\frac{1}{2}} = \exp\left(-\frac{i}{2}\omega_{\mu\nu} S^{\mu\nu}\right) \tag{2.106}$$

where $S^{\mu\nu}$ is given in equation (2.96). If the Dirac equation is Lorentz invariant, then a Dirac wave function $\Psi'(x')$ satisfying in the primed frame

$$\left[i\gamma^{\mu}\partial'_{\mu} - m\right]\Psi'(x') = 0 \tag{2.107}$$

can be written in terms of $\Psi(x)$ satisfying the Dirac equation in the unprimed frame. We can therefore write the relationship between the rest-frame wave function and the wave function in the primed frame as

$$\Psi'(x') = \Lambda_{\frac{1}{2}}\Psi(\Lambda^{-1}x') \tag{2.108}$$

and show that the Dirac equation retains its covariant form. In equation (2.108) the rest-frame wave function Ψ is evaluated at the point before boosting, namely $\Lambda^{-1}x'$, and the additional spinor Lorentz transform $\Lambda_{\frac{1}{2}}$ tells us how the orientation has changed, including the relative strength of the positive- and negative-frequency components. The substitution of equation (2.108) into the Dirac equation (2.107) yields

$$\left[i\gamma^{\mu}\partial'_{\mu} - m\right]\Psi'(x') \to \left[i\gamma^{\mu}\partial'_{\mu} - m\right]\Lambda_{\frac{1}{2}}\Psi(\Lambda^{-1}x'). \tag{2.109}$$

Recall that the spinor transformation $\Lambda_{\frac{1}{2}}$ acts in the space of the γ^{μ}, and so we transform the equation from the right, as follows:

$$\Lambda_{\frac{1}{2}}\Lambda_{\frac{1}{2}}^{-1}\left[i\gamma^{\mu}\partial'_{\mu} - m\right]\Lambda_{\frac{1}{2}}\Psi(\Lambda^{-1}x') = \Lambda_{\frac{1}{2}}\left[i\Lambda_{\frac{1}{2}}^{-1}\gamma^{\mu}\Lambda_{\frac{1}{2}}\partial'_{\mu} - m\right]\Psi(\Lambda^{-1}x'). \tag{2.110}$$

The action of the finite spinor transform on the γ^{μ} can be worked out explicitly starting with equations (2.96) and (2.106). The boost generators are given by $S^{0j} = \frac{i}{2}\gamma^{0}\gamma^{j}$ and thus the finite transformation is

$$\Lambda_{\frac{1}{2}}^{B,j} = \exp\left(-i\eta_j S^{0j}\right) = \exp\left(\gamma^{0}\gamma^{j}\eta_j/2\right) = I_4\cosh(\eta_j/2) + (\gamma^{0}\gamma^{j})\sinh(\eta_j/2) \tag{2.111}$$

with no summation over j. Using this expression to transform γ^{0} and γ^{j} yields

$$\begin{aligned}\left(\Lambda_{\frac{1}{2}}^{B,j}\right)^{-1}\gamma^{0}\Lambda_{\frac{1}{2}}^{B,j} &= \gamma^{0}\cosh\eta_j + \gamma^{j}\sinh\eta_j,\\ \left(\Lambda_{\frac{1}{2}}^{B,j}\right)^{-1}\gamma^{j}\Lambda_{\frac{1}{2}}^{B,j} &= \gamma^{0}\sinh\eta_j + \gamma^{j}\cosh\eta_j.\end{aligned} \tag{2.112}$$

A similar calculation for spatial rotations also shows that the operator transform of the γ^{μ} is equivalent to treating the γ^{μ} as a legitimate Lorentz four-vector

$$\Lambda_{\frac{1}{2}}^{-1}\gamma^{\mu}\Lambda_{\frac{1}{2}} = \Lambda^{\mu}_{\nu}\gamma^{\nu}. \tag{2.113}$$

A full derivation of equation (2.113) is a result known as *Pauli's fundamental theorem.* We can now apply this relation to equation (2.110) where $\partial_\mu = (\Lambda^{-1})^{\nu}_{\mu}\partial'_\nu$ gives

$$\begin{aligned}\left[i\gamma^{\mu}\partial'_{\mu} - m\right]\Psi'(x') &\to \Lambda_{\frac{1}{2}}\left[i\Lambda^{\mu}_{\nu}\gamma^{\nu}\partial_{\mu} - m\right]\Psi(\Lambda^{-1}x') \\ &= \Lambda_{\frac{1}{2}}\left[i\gamma^{\mu}(\Lambda^{-1})^{\nu}_{\mu}\partial_{\nu} - m\right]\Psi(\Lambda^{-1}x') \\ &= \Lambda_{\frac{1}{2}}\left[i\gamma^{\mu}\partial_{\mu} - m\right]\Psi(x) \\ &= 0,\end{aligned} \tag{2.114}$$

showing that the Dirac equation is invariant when applied to the Lorentz-transformed Dirac wave function, as given in equation (2.108). To some extent the γ^μ is reorienting to p_μ, as in the case of helicity as a constant of motion, so that the Lorentz-transformed spin components of Ψ satisfy the Dirac equation of motion. Note that the remaining $\Lambda_{\frac{1}{2}}$ on the left of equation (2.114) would be canceled if a corresponding object on the left transformed as $\Lambda_{\frac{1}{2}}^{-1}$. This leads us to consider Lorentz-invariant quantities formed from Dirac wave functions.

2.7.2 *Lorentz-Invariant Lagrangians and the Euler-Lagrange Equations*

As we will find, one of the most important quantities in elementary particle physics is Lorentz-invariant Lagrangians (or Lagrangian densities). Lagrangian mechanics applies equally well to relativistic quantum mechanics as it does to classical physics. A Lagrangian contains kinetic and potential energy terms. The forms of the kinetic energy terms are generic and depend only on the spin of the particle, such as the Dirac kinetic energy terms for spin-1/2 particles. The potential energy terms specify the interactions, and we will often refer to these terms as the interaction Lagrangian.

The action is defined to be a space-time integral of the Lagrangian density

$$S = \int^{\text{path}} \mathcal{L}(\Psi, \partial_\mu\Psi)d^4x \tag{2.115}$$

over a path connecting the endpoints of a particle's trajectory. In fact, one can consider the distance between endpoints as setting a mass scale. For a Lagrangian density that depends only on the Dirac wave function and its derivatives, $\mathcal{L}(\Psi, \partial_\mu\Psi)$, we can consider the small variation

$$\Psi(x) \to \Psi'(x) = \Psi(x) + \delta\Psi(x). \tag{2.116}$$

The *principle of least action* states that when a system evolves in time from one configuration to another, the path that the system follows is such that the action is an extremum. Therefore,

$$\begin{aligned}0 &= \delta S \\ &= \int d^4x \left\{\frac{\partial\mathcal{L}}{\partial\Psi}\delta\Psi + \frac{\partial\mathcal{L}}{\partial(\partial_\mu\Psi)}\delta(\partial_\mu\Psi)\right\}\end{aligned}$$

$$= \int d^4x \left\{ \frac{\partial \mathcal{L}}{\partial \Psi} \delta\Psi - \partial_\mu \left(\frac{\partial \mathcal{L}}{\partial(\partial_\mu \Psi)} \right) \delta\Psi + \partial_\mu \left(\frac{\partial \mathcal{L}}{\partial(\partial_\mu \Psi)} \delta\Psi \right) \right\}. \tag{2.117}$$

The third term is a surface term and can be taken sufficiently far away that the variations on Ψ are zero on the surface. However, the terms multiplying $\delta\Psi$ must vanish at all points, which when equated to zero result in the *Euler-Lagrange equations*

$$\partial_\mu \left(\frac{\partial \mathcal{L}}{\partial(\partial_\mu \Psi)} \right) - \frac{\partial \mathcal{L}}{\partial \Psi} = 0. \tag{2.118}$$

To form a Lorentz-invariant Lagrangian that results in the Dirac equation, we can start with the form of equation (2.101). If we consider first the mass term and the requirement that the Lagrangian be Hermitian, then we find that $\Psi^\dagger m \Psi$ does not work. First of all, under a Lorentz boost, this term becomes $\Psi^\dagger \Lambda_{\frac{1}{2}}^\dagger \Lambda_{\frac{1}{2}} \Psi$, but the boost matrix is not unitary, $\Lambda_{\frac{1}{2}}^\dagger \neq \Lambda_{\frac{1}{2}}^{-1}$. This is a general property of finite representations of the Lorentz group. We also know from our study of the probability current that $\Psi^\dagger \Psi$ should transform as the timelike component of a four-vector, i.e., not a Lorentz-invariant scalar. The solution is to go back to the original form of the Dirac equation mass term $\beta m = \gamma^0 m$ and to define

$$\bar{\Psi} \equiv \Psi^\dagger \gamma^0. \tag{2.119}$$

The Lorentz transformation of $\bar{\Psi}$ can be shown to be

$$\bar{\Psi} \to \bar{\Psi} \Lambda_{\frac{1}{2}}^{-1}. \tag{2.120}$$

Therefore, $\bar{\Psi} m \Psi$ is a Lorentz scalar. Similarly, $\bar{\Psi} \gamma^\mu \Psi$ can be shown to be a Lorentz vector, as follows from the covariant form of the Dirac probability current. The correct, Lorentz-invariant Dirac Lagrangian is therefore

$$\mathcal{L}_{\text{Dirac}} = \bar{\Psi} (i\gamma^\mu \partial_\mu - m) \Psi. \tag{2.121}$$

The Euler-Lagrange equation for $\bar{\Psi}$ (or $\Psi^\dagger$) applied to the Dirac Lagrangian (2.121) yields the Dirac equation (2.101). Similarly, the Euler-Lagrange equation applied to Ψ gives the same equation in Hermitian-conjugate form

$$-i\partial_\mu \bar{\Psi} \gamma^\mu - m\bar{\Psi} = 0. \tag{2.122}$$

Similarly, from the electromagnetic minimal coupling substitution introduced earlier and the Dirac probability current, we can identify the interaction Hamiltonian for electromagnetism to be

$$H_{\text{int}}^{\text{em}} = \int d^3x\, e j^\mu A_\mu. \tag{2.123}$$

2.8 Weyl Representation

An alternative representation for the α and β matrices assigns

$$\alpha^i = \begin{pmatrix} -\sigma^i & 0 \\ 0 & \sigma^i \end{pmatrix}, \qquad \beta = \begin{pmatrix} 0 & I_2 \\ I_2 & 0 \end{pmatrix}, \tag{2.124}$$

with

$$\gamma^5 = \begin{pmatrix} -I_2 & 0 \\ 0 & I_2 \end{pmatrix}. \tag{2.125}$$

This is known as the *Weyl representation* (following the sign convention of Peskin and Schroeder [1]). The most distinctive feature of the Weyl representation is that the Dirac spinor in the high-energy limit separates into two 2-component spinors according to states of definite chirality. The Dirac spinor Ψ can be decomposed into left- and right-handed chiral components using the *chiral projection operators* constructed with the γ_5 matrix, as follows:

$$\begin{aligned} \Psi &= \Psi_L + \Psi_R, \\ \Psi_L &= P_L\Psi = \tfrac{1}{2}(1-\gamma_5)\Psi = \begin{pmatrix} \xi^s \\ 0 \end{pmatrix}, \qquad (s = 1,2), \\ \Psi_R &= P_R\Psi = \tfrac{1}{2}(1+\gamma_5)\Psi = \begin{pmatrix} 0 \\ \eta^r \end{pmatrix}, \qquad (r = 1,2). \end{aligned} \tag{2.126}$$

The ξ^s is a left-handed chirality two-component spinor with one positive-frequency and one negative-frequency degree of freedom. The negative-frequency component of ξ^s has the opposite sign of helicity compared to the positive-frequency component. Similarly, the η^r describe the right-handed chirality solutions. The high-energy Weyl decomposition into chiral spinors is an alternative to the Pauli-Dirac representation, which, in contrast, separates in the low-energy limit into positive- and negative-frequency two-component spinors.

In the Weyl representation

$$\gamma^\mu = \begin{pmatrix} 0 & \sigma^\mu \\ \bar{\sigma}^\mu & 0 \end{pmatrix}, \qquad (\mu = 0,1,2,3), \tag{2.127}$$

where

$$\begin{aligned} \sigma^\mu &\equiv (I_2, \boldsymbol{\sigma}), \\ \bar{\sigma}^\mu &\equiv (I_2, -\boldsymbol{\sigma}) = \sigma_\mu. \end{aligned} \tag{2.128}$$

Note that the bar notation is doing the equivalent of flipping the covariant/contravariant sign convention for σ^μ without raising or lowering the index.

The Weyl representations of the boost and rotation generators are

$$S^{0k} = \frac{i}{4}[\gamma^0, \gamma^k] = -\frac{i}{2}\begin{pmatrix} \sigma^k & 0 \\ 0 & -\sigma^k \end{pmatrix} \tag{2.129}$$

and

$$S^{ij} = \frac{i}{4}[\gamma^i, \gamma^j] = \frac{1}{2}\epsilon^{ijk}\begin{pmatrix} \sigma^k & 0 \\ 0 & \sigma^k \end{pmatrix} \equiv \frac{1}{2}\epsilon^{ijk}\Sigma^k, \tag{2.130}$$

respectively. The three-dimensional spinor transformation matrix is precisely the Dirac spin operator and has the same form for Weyl and Pauli-Dirac representations.

2.8.1 *Weyl Spinor Two-Component Formalism*

Note that the γ^μ form of the Dirac equation (2.101) was developed before elementary particles were known to participate in chiral interactions. In hindsight, the mass term βm with the off-diagonal Weyl representation of β is somewhat more natural since it explicitly shows that the origin of mass can be equivalently formulated as the coupling of left-handed to right-handed chirality spinors, i.e., the ability to decelerate a left-handed particle down to rest and accelerate it back up to a right-handed particle state, something possible only if a particle has a finite mass.

Recall that the γ^μ form of the Dirac Lagrangian (2.121) introduces an additional γ^0 through the definition of $\bar{\Psi} \equiv \Psi^\dagger \gamma^0$, where multiplying to the right gives $(\gamma^0)^2 = I_4$, $\gamma^0\gamma^i = \alpha^i$, and $\gamma^0 m = \beta m$. An alternative to the γ^μ form is to define $\alpha^0 \equiv I_4$ and to use $j^\mu = \Psi^\dagger \alpha^\mu \Psi$ as a Lorentz vector representing the probability current and $\Psi^\dagger \beta m \Psi$ as a scalar interaction term. One can then construct a Lorentz-invariant Lagrangian using two-component chiral spinors in the Weyl representation

$$\bar{\Psi}(i\gamma^\mu \partial_\mu - m)\Psi = \Psi^\dagger i\alpha^\mu \partial_\mu \Psi - \Psi^\dagger \beta m \Psi = \xi^\dagger i\bar{\sigma}^\mu \partial_\mu \xi + \eta^\dagger i\sigma^\mu \partial_\mu \eta - m(\xi^\dagger \eta + \eta^\dagger \xi) \qquad (2.131)$$

where the bar notation for σ^μ and $\bar{\sigma}^\mu$ is given in equation (2.128). This approach is commonly used in theories beyond the Standard Model.

2.8.2 *Free-Particle Solutions via Lorentz Boost Transform*

If we first consider positive-frequency solutions, then the plane waves are of the form

$$\Psi^{(+)}(x) = Nu(p)\exp(-ip \cdot x) \quad \text{with} \quad p^2 = m^2 \qquad (2.132)$$

where $u(p)$ is a column vector that satisfies the Dirac equation

$$(\not{p} - m)u(p) = 0 \qquad (2.133)$$

and N is a normalization constant. The approach taken here is to determine the free-particle solutions in the rest frame of the particle, and then to use finite Lorentz transformations to construct a general free-particle solution.

In the rest frame of the particle, where $p = p_0 = (m, \mathbf{0})$, equation (2.133) becomes

$$(m\gamma^0 - mI_4)u(p_0) = m\begin{pmatrix} -I_2 & I_2 \\ I_2 & -I_2 \end{pmatrix} u(p_0) = 0 \qquad (2.134)$$

and the solutions can be written in terms of a single arbitrary two-component spinor ξ

$$u(p_0) = \sqrt{m}\begin{pmatrix} \xi \\ \xi \end{pmatrix}. \qquad (2.135)$$

The normalization of ξ is chosen as $\xi^\dagger \xi = 1$ and $\bar{u}(p_0)u(p_0) = 2m$. We will return to the choice of normalization below. Only two degrees of freedom specify the solution, and these are precisely the spin components of the positive-frequency spin-1/2 particle.

We can obtain $u(p)$ in any other frame by boosting. According to equations (2.106), (2.108), and (2.129), a boosted Dirac spinor is given by

$$u^s(p) = \exp\left(-\frac{1}{2}\omega_{0k}\begin{pmatrix}\sigma^k & 0\\ 0 & -\sigma^k\end{pmatrix}\right)\sqrt{m}\begin{pmatrix}\xi^s\\ \xi^s\end{pmatrix}, \qquad s = 1,2, \tag{2.136}$$

where the s index labels the two spin states in the rest frame and $\omega_{01}, \omega_{02}, \omega_{03}$ are the rapidity parameters for the Lorentz boost. The spinor normalization is given by

$$u^{r\dagger}(p)u^s(p) = 2E\delta^{rs} \quad \text{or} \quad \bar{u}^r(p)u^s(p) = 2m\delta^{rs} \tag{2.137}$$

where the $2E$ normalization is used in the case of massless particles. Similarly, the negative-frequency solutions can be obtained using the same method

$$\Psi^{(-)}(x) = Nv(p)\exp(ip\cdot x) \quad \text{with} \quad p^2 = m^2 \tag{2.138}$$

where $v(p)$ is a column vector that satisfies the Dirac equation, recalling that the positron solutions are obtained using

$$(\not{p} + m)v(p) = 0. \tag{2.139}$$

This gives

$$(m\gamma^0 + mI_4)v(p_0) = m\begin{pmatrix}I_2 & I_2\\ I_2 & I_2\end{pmatrix}v(p_0) = 0 \tag{2.140}$$

and the solutions are

$$v(p_0) = \sqrt{m}\begin{pmatrix}\eta\\ -\eta\end{pmatrix} \tag{2.141}$$

where η^s is another basis of two-component spinors. This gives

$$v^s(p) = \exp\left(-\frac{1}{2}\omega_{0k}\begin{pmatrix}\sigma^k & 0\\ 0 & -\sigma^k\end{pmatrix}\right)\sqrt{m}\begin{pmatrix}\eta^s\\ -\eta^s\end{pmatrix}, \qquad s = 1,2. \tag{2.142}$$

These solutions are normalized according to

$$v^{r\dagger}(p)v^s(p) = 2E\delta^{rs} \quad \text{or} \quad \bar{v}^r(p)v^s(p) = -2m\delta^{rs}. \tag{2.143}$$

The orthogonality of the u and v solutions is given by the Lorentz-invariant scalars

$$\bar{u}^r(p)v^s(p) = \bar{v}^r(p)u^s(p) = 0 \tag{2.144}$$

where $u^\dagger\gamma^0 = \bar{u}$ and $v^\dagger\gamma^0 = \bar{v}$, as used above in the normalization conditions.

2.9 Projection Operators and Completeness Relations

It is often the case that only a subset of Dirac solutions participate in a given scattering process. In the four-component Dirac formalism, projecting out components of the solutions to understand the overlap between initial and final states in an elementary particle interaction is particularly revealing. We begin by reviewing various types of projection operators that can be constructed in Dirac theory.

The general property of projection operators is that projecting twice has the same effect as projecting once, $P^2 = P$, and that the projection operators span the possible final states, $P_+ + P_- = 1$, while being orthogonal $P_\pm P_\mp = 0$. One example encountered previously are the *chiral projection operators*

$$\begin{aligned} P_L &= \tfrac{1}{2}(1-\gamma_5), \\ P_R &= \tfrac{1}{2}(1+\gamma_5), \end{aligned} \tag{2.145}$$

for left- and right-handed chiralities, respectively. The forms of the chiral projection operators are independent of the representation of the Dirac matrices. The relationship between chirality and helicity in the high-energy limit can be seen in the following way. The Dirac spin operator is given by $\boldsymbol{\Sigma}/2$ defined in (2.35), which in terms of Dirac matrices is

$$\frac{1}{2}\boldsymbol{\Sigma} = \frac{1}{2}\gamma^5\gamma^0\boldsymbol{\gamma}. \tag{2.146}$$

For a massless Dirac particle, the evaluation of $\boldsymbol{\Sigma}\cdot\boldsymbol{p}$ on a positive-frequency solution gives

$$(\boldsymbol{\Sigma}\cdot\boldsymbol{p})u = \gamma^5\gamma^0(\boldsymbol{\gamma}\cdot\boldsymbol{p})u = \gamma^5\gamma^0|\boldsymbol{p}|\gamma^0 u = |\boldsymbol{p}|\gamma^5 u \tag{2.147}$$

and for the negative-frequency solutions

$$(\boldsymbol{\Sigma}\cdot\boldsymbol{p})v = \gamma^5\gamma^0(\boldsymbol{\gamma}\cdot\boldsymbol{p})\,v = -\gamma^5\gamma^0|\boldsymbol{p}|\gamma^0 v = -|\boldsymbol{p}|\gamma^5 v, \tag{2.148}$$

showing that the γ^5 eigenvalues are the helicity eigenvalues in the massless limit with the sign reversed for antiparticle solutions. This is why a right-handed helicity antiparticle has left-handed chirality for massless particles.

The projection operators for spin can be generalized to an arbitrary spin quantization axis. In the rest frame of the particle, the spin direction is a three-vector. In order to treat spin in a covariant way, a four-vector for the spin polarization s^μ is defined. If $\hat{\mathbf{s}}$ is a unit vector along the spin direction, then in the rest frame $s^\mu = (0, \hat{\mathbf{s}})$, $s^2 = -1$, and $s^\mu p_\mu = 0$ since $p_\mu = (m, \mathbf{0})$. By Lorentz covariance, $s^2 = -1$ and $s^\mu p_\mu = 0$ are true in any frame and s^μ transforms to other frames according to the Lorentz transformation properties of four-vectors. The four-vector spin properties are equivalently described as being "spacelike" and "transverse to the four-momentum." We can construct the *spin projection operators* as

$$\begin{aligned} \Sigma(s) &= \frac{1+\gamma_5 \not{s}}{2}, \\ \Sigma(-s) &= \frac{1-\gamma_5 \not{s}}{2}, \end{aligned} \tag{2.149}$$

for the decomposition of Dirac spinors into spin-up and spin-down components along the $\hat{\mathbf{s}}$ direction. Notice that the $\gamma^5\not{s}$ term can be rewritten as $\gamma^5\not{s} = \gamma^0(\boldsymbol{\Sigma}\cdot\hat{\mathbf{s}})$ where $\boldsymbol{\Sigma}/2$ is the three-dimensional Dirac spin operator.

The properties of the spin projection operators are

$$\begin{aligned} \Sigma(s)u(p,s) &= u(p,s), \\ \Sigma(s)v(p,s) &= v(p,s), \end{aligned}$$

$$
\begin{aligned}
&\Sigma(-s)u(p,s) = \Sigma(-s)\,v(p,s) = 0,\\
&\Sigma(\pm s)\,\Sigma(\pm s) = \left(\frac{1 \pm \gamma_5 \not{s}}{2}\right)\left(\frac{1 \pm \gamma_5 \not{s}}{2}\right) = \frac{1 \pm 2\gamma_5 \not{s} - \gamma_5^2 s^2}{4} = \Sigma(\pm s),\\
&\Sigma(\pm s)\,\Sigma(\mp s) = \left(\frac{1 \pm \gamma_5 \not{s}}{2}\right)\left(\frac{1 \mp \gamma_5 \not{s}}{2}\right) = \frac{1 + \gamma_5^2 s^2}{4} = 0,\\
&\Sigma(\pm s) + \Sigma(\mp s) = 1.
\end{aligned}
\tag{2.150}
$$

The *frequency projection operators* are defined as

$$
\begin{aligned}
\Lambda_+ &= \frac{\not{p} + m}{2m},\\
\Lambda_- &= \frac{-\not{p} + m}{2m},
\end{aligned}
\tag{2.151}
$$

so that $\Lambda_+ + \Lambda_- = 1$, and

$$
\Lambda_+ u = u, \qquad \Lambda_- u = 0, \qquad \Lambda_+ v = 0, \qquad \Lambda_- v = v,
\tag{2.152}
$$

where the u and v solutions are projected out, respectively. Note that the $\Lambda_\pm$ are referred to as the *energy projection operators* when operating on the four original free-particle solutions of the Dirac equation before the definition of the v solutions. When operating on the u and v solutions, which have strictly positive energies, the $\pm$ subscript of the $\Lambda_\pm$ projection operators corresponds to the sign of the time evolution of the Dirac solutions according to $\exp(-i(\pm)\,p \cdot x)$.

If we evaluate the commutator of the frequency and spin projection operators, then we find that they commute

$$
\begin{aligned}
\Sigma(\pm' s), \Lambda_\pm(p)] &= \left(\frac{1 \pm' \gamma_5 \not{s}}{2}\right)\left(\frac{\pm \not{p} + m}{2m}\right) - \Lambda_\pm(p)\,\Sigma(\pm' s)\\
&= \frac{\pm \not{p} + m \pm' (\pm)\,\gamma_5 \gamma^\mu \gamma^\nu s_\mu p_\nu \pm' m \gamma_5 \not{s}}{4m} - \Lambda_\pm(p)\,\Sigma(\pm' s)\\
&= \frac{\pm \not{p} + m \pm' (\pm)\,\gamma_5 (2g^{\mu\nu} - \gamma^\nu \gamma^\mu)\, s_\mu p_\nu \pm' m \gamma_5 \not{s}}{4m} - \Lambda_\pm(p)\,\Sigma(\pm' s)\\
&= \frac{\pm \not{p} + m \pm' (\pm)\,(2g^{\mu\nu} \gamma_5 + \gamma^\nu \gamma_5 \gamma^\mu)\, s_\mu p_\nu \pm' m \gamma_5 \not{s}}{4m} - \Lambda_\pm(p)\,\Sigma(\pm' s)\\
&= \frac{\pm \not{p} + m \pm' (\pm \not{p} + m)\,\gamma_5 \not{s}}{4m} - \Lambda_\pm(p)\,\Sigma(\pm' s)\\
&= \left(\frac{\pm \not{p} + m}{2m}\right)\left(\frac{1 \pm' \gamma_5 \not{s}}{2}\right) - \Lambda_\pm(p)\,\Sigma(\pm' s) = 0.
\end{aligned}
\tag{2.153}
$$

Therefore, it is possible to project from a given Dirac wave function the four independent solutions corresponding to positive and negative frequency and to spin up and spin down along a given direction.

Now we come to an important practical result that applies to the evaluation of a scattering process. The projection operators can be thought of in the context of the *completeness relations.* Namely, we can write a solution to the Dirac equation in terms of a complete orthonormal set of eigenfunctions

$$\Psi(p) = \sum_{s=1,2} c^s u^s(p) + \sum_{r=1,2} d^r v^r(p) \tag{2.154}$$

with c^s and d^r being complex expansion coefficients and where the orthogonality and normalization conditions are

$$\begin{aligned} \bar{u}^r(p)v^s(p) &= \bar{v}^r(p)u^s(p) = 0, \\ \bar{u}^r(p)u^s(p) &= 2m\delta^{rs}, \\ \bar{v}^r(p)v^s(p) &= -2m\delta^{rs}. \end{aligned} \tag{2.155}$$

If these solutions span the Hilbert space, then by completeness

$$\sum_{s=1,2} \frac{u\bar{u}}{2m} + \sum_{r=1,2} \frac{v\bar{v}}{-2m} = I_4, \tag{2.156}$$

which, as a reminder, is not a number but rather the identity matrix for the dimensionality of the spinor representation, I_4. The completeness relations are simply another way of stating equation (2.154). Explicit evaluation shows that

$$\begin{aligned} \sum_{s=1,2} \frac{u\bar{u}}{2m} &= \frac{\not{p}+m}{2m} = \Lambda_+, \\ \sum_{r=1,2} \frac{v\bar{v}}{-2m} &= \frac{-\not{p}+m}{2m} = \Lambda_-, \end{aligned} \tag{2.157}$$

giving us precisely the same frequency projection operators defined above. The equations in (2.157) are commonly used in scattering calculations to account for all spin states for positive- and negative-frequency solutions. For example, unpolarized cross sections are formed by averaging over initial spins and total cross sections by summing over the final spins. Similarly, the specific contribution of a particular spin component in the calculation of a scattering process can be isolated by substituting in the appropriate term

$$\begin{aligned} (u\bar{u})_{\pm s} &= (\not{p}+m)(1 \pm \gamma_5 \not{s})/2, \\ (v\bar{v})_{\pm s} &= (\not{p}-m)(1 \pm \gamma_5 \not{s})/2. \end{aligned} \tag{2.158}$$

In the massless limit, the component breakdown in equation (2.158) is related to the decomposition by combined frequency and chirality, given by

$$\begin{aligned} (u\bar{u})_{R,L} &= (\not{p}+m)(1 \pm \gamma_5)/2, \\ (v\bar{v})_{R,L} &= (\not{p}-m)(1 \pm \gamma_5)/2. \end{aligned} \tag{2.159}$$

2.10 Discrete Lorentz Transformations

A Lorentz transformation is a real linear transformation of the coordinates conserving the norm of the intervals between any two points in space-time. The coordinates x^μ of an event observed in frame $\mathcal{O}$ correspond to the coordinates x'^μ in frame $\mathcal{O}'$ (with a common origin) via the transformation

$$(x')^\nu \equiv \Lambda^\nu_\mu x^\mu \tag{2.160}$$

with the inverse transformation given by

$$x^{\mu} = (x')^{\nu} \Lambda^{\mu}_{\nu}. \tag{2.161}$$

The distance between two space-time points is invariant, implying the following orthogonality relation for the Lorentz transformation:

$$\begin{aligned} ds^2 &= dx^{\mu} dx_{\mu} = (dx')^{\nu} (dx')_{\nu} = \Lambda^{\nu}_{\mu} \Lambda^{\sigma}_{\nu} dx^{\mu} dx_{\sigma} \\ &= \delta^{\sigma}_{\mu} dx^{\mu} dx_{\sigma} \quad \text{and, therefore,} \quad \Lambda^{\nu}_{\mu} \Lambda^{\sigma}_{\nu} = \delta^{\sigma}_{\mu} \end{aligned} \tag{2.162}$$

with

$$\delta^{\sigma}_{\mu} = g_{\mu\rho} g^{\rho\sigma} = \begin{cases} 1 & \text{if } \mu = \sigma, \\ 0 & \text{if } \mu \neq \sigma. \end{cases} \tag{2.163}$$

Taking the determinant gives

$$\begin{aligned} \det(\Lambda^{\nu}_{\mu} \Lambda^{\sigma}_{\nu}) &= \det(\Lambda^{\nu}_{\mu}) \det(\Lambda^{\sigma}_{\nu}) = \left[\det(\Lambda^{\nu}_{\mu})\right]^2 \\ &= \det(\delta^{\sigma}_{\mu}) = 1 \end{aligned} \tag{2.164}$$

with the result that $\det(\Lambda^{\nu}_{\mu}) = \pm 1$.

The proper, orthochronous Lorentz transformations correspond to $\det(\Lambda^{\nu}_{\mu}) = +1$ and $\operatorname{sign}(\Lambda^{0}_{0}) = +1$, and these are the continuous transformations of rotations and boosts. There are three additional connected components of the full Lorentz group depending on $\det(\Lambda^{\nu}_{\mu})$ and $\operatorname{sign}(\Lambda^{0}_{0})$, but these are reachable only via the discrete Lorentz transformations of *parity* (P), *time-reversal* (T), and the combined space-time inversion PT, as shown in figure 2.5. Relativistic Lagrangians are, in general, invariant under the restricted Lorentz group, $\boldsymbol{L}^{\uparrow}_{+}$. The other three components of the full Lorentz group do not contain the identity transform and hence do not form groups. However, the proper Lorentz group is the union of $\boldsymbol{L}^{\uparrow}_{+} \cup \boldsymbol{L}^{\downarrow}_{+}$ and contains the identity transform of the complex Lorentz group, i.e., in the complex Lorentz group the identity of $\boldsymbol{L}^{\uparrow}_{+}$ can be continuously transformed to

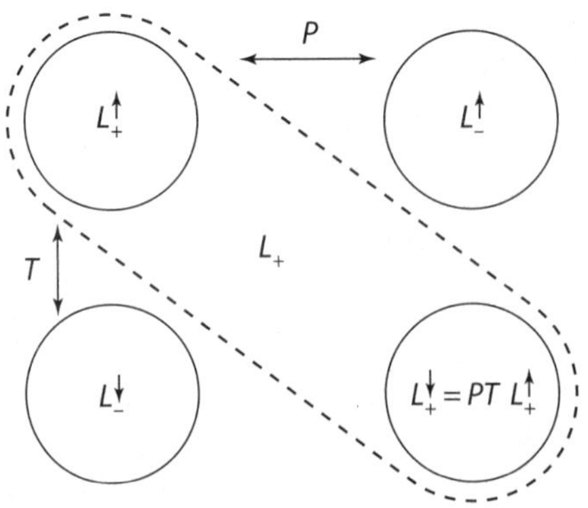

FIGURE 2.5. The four components of the full Lorentz group are connected via the discrete transforms of parity (P), time reversal (T), and the combined space-time inversion PT. Relativistic Lagrangians are, in general, invariant under the restricted Lorentz group $\boldsymbol{L}^{\uparrow}_{+}$. The CPT theorem is related to the identity transform of the proper Lorentz group, $\boldsymbol{L}_{+}$, in the complex Lorentz group.

the space-time inversion PT. The doubling of $\boldsymbol{L}_+^\uparrow \to \boldsymbol{L}_+$ is similar to the doubling of degrees of freedom in the Dirac equation to describe antiparticles, where the invariance under the discrete transform of charge-conjugation (C), the swapping of matter and antimatter, was investigated earlier in the text. In fact, particles with distinct antiparticles are generally represented by complex wave functions. The reason for mentioning the basic structure of the full and complex Lorentz groups is to provide a context for quoting one of the far-reaching results of invariance in elementary particle interactions, the CPT theorem.

The CPT theorem states that the combined properties of locality, Lorentz invariance of the Lagrangian under $\boldsymbol{L}_+^\uparrow$, and hermiticity of the Hamiltonian require that scattering processes are invariant under the discrete transform of CPT. The CPT transform interchanges initial and final states, exchanges particles and antiparticles, and reverses spin components and helicities while keeping energies and momenta the same

$$\mathcal{M}(\alpha_{R,L} \to \beta_{R,L}) = \mathcal{M}(\hat{\beta}_{L,R} \to \hat{\alpha}_{L,R}) \tag{2.165}$$

where $\mathcal{M}$ is a scattering matrix element with the initial(final) states labeled by $\alpha_{R,L}(\beta_{R,L})$ on the left and $\hat{\beta}_{L,R}(\hat{\alpha}_{L,R})$ on the right. The $\hat{x}_{L,R}$ notation indicates that particles and antiparticles are exchanged and the helicities are reversed. The CPT transformation is the only fundamental discrete symmetry that is respected by all known elementary particle interactions. A full derivation of the theorem can be found in [4].

As with charge conjugation, explicit representations exist for the discrete transforms of parity and time reversal and, hence, the combined operation of CPT. Parity transforms $(t, \boldsymbol{x}) \to (t, -\boldsymbol{x})$, changing a right-handed coordinate system to a left-handed one. In terms of particle properties, the parity operation will flip the three-momentum of a particle while the spin direction remains unchanged. Time reversal transforms $(t, \boldsymbol{x}) \to (-t, \boldsymbol{x})$, causing a flip of the three-momentum, but in this case spin will also change direction. The combined operation of PT will therefore keep the three-momentum of the particle intact while changing the direction of the spin and, therefore, the particle helicity. For nonrelativistic spin-1/2 particles, the spin must be rotated 4π to return to its initial orientation in $SU(2)$. A similar double rotation is required for Dirac spinors, and, hence, the time-reversal transform is connected with the spin statistics of Dirac particles, as will be shown below.

The application of the parity transform on the Dirac equation (2.101) yields

$$\begin{aligned}\left[i\gamma^\mu \partial_\mu - m\right]\Psi(x) &\to \left[i\gamma^\mu (\Lambda^{-1})^\nu_\mu \partial_\nu - m\right] U_P \Psi(\Lambda^{-1}x) \\ &= \left[i\left(\gamma^0 \frac{\partial}{\partial x^0} - \gamma^i \frac{\partial}{\partial x^i}\right) - m\right] U_P \Psi(t, -\boldsymbol{x}).\end{aligned} \tag{2.166}$$

To indicate the parity-reversed space-time coordinates, we introduce $\tilde{x}^\mu \equiv (t, -\boldsymbol{x})$. To show that the same Dirac equation governs $U_P \Psi(\tilde{x})$, we multiply on the left by $U_P U_P^{-1}$ and determine that for invariance of the free Dirac equation under parity, we must find a unitary transformation of the spinors that satifies

$$\begin{aligned} U_P^{-1} \gamma^0 U_P &= \gamma^0, \\ U_P^{-1} \boldsymbol{\gamma} U_P &= -\boldsymbol{\gamma}. \end{aligned} \tag{2.167}$$

This is not a surprising result as γ^μ has been shown to transform like a four-vector and hence the spatial part has been reversed. From the Dirac algebra (2.95), it is seen that these conditions are satisfied by the assignment

$$U_P = \eta_P \gamma^0 \tag{2.168}$$

where U_p is independent of the representation of the Dirac matrices and η_P is an arbitrary phase. Applying the parity transform to the Dirac spinors (2.136) and (2.142) using (2.127) yields

$$\begin{aligned} u(p) &\to U_P u(\tilde{p}) = \eta_P u(\tilde{p}), \\ v(p) &\to U_P v(\tilde{p}) = -\eta_P v(\tilde{p}), \end{aligned} \tag{2.169}$$

showing that the intrinsic parity of a particle is the opposite of the corresponding antiparticle.

Time reversal swaps initial and final states and therefore involves the complex conjugation of the operator observables in a given process. Such a transformation is called antiunitary. The time-reversal operation on the Dirac spinor flips the direction of the spin. In general, we can write the transformed Dirac solution as $U_T \Psi^*(-\tilde{x})$. One way to construct an explicit representation of U_T is to construct a spin rotation operator that is real, to be unaffected by the complex conjugation. This can be done by rotating in the 13-plane in the Weyl (and Pauli-Dirac) representation where γ^1 and γ^3 are real. The spinor rotation is constructed using (2.96) and (2.106) and gives

$$\Lambda_{\frac{1}{2}}^{R^{13}} = \exp\left(i\theta_2 S^{13}\right) = \exp\left(-\gamma^1\gamma^3\theta_2/2\right) = I_4 \cos(\theta_2/2) - (\gamma^1\gamma^3)\sin(\theta_2/2), \tag{2.170}$$

which for $\theta_2 = \pi$ will produce a spin flip

$$\Lambda_{\frac{1}{2}}^{R^{13}}\big|_{\theta_2=\pi} = -\gamma^1\gamma^3. \tag{2.171}$$

Therefore, assigning the time-reversal spinor transform with the arbitrary phase η_T

$$U_T = \eta_T \gamma^1 \gamma^3 \tag{2.172}$$

satisfies a set of transformation requirements similar to those in (2.167)

$$\begin{aligned} (U_T^{-1} i\gamma^0 U_T)^* &= -i\gamma^0, \\ (U_T^{-1} i\boldsymbol{\gamma} U_T)^* &= i\boldsymbol{\gamma}. \end{aligned} \tag{2.173}$$

The complex conjugation makes the time-reversal transform antiunitary as required to reverse initial and final states. Notice that a $\theta_2 = 2\pi$ rotation (2.170) gives

$$\Lambda_{\frac{1}{2}}^{R^{13}}\big|_{\theta_2=2\pi} = -I_4, \tag{2.174}$$

reproducing the result that for spin-1/2 particles, a 4π rotation is required to return the spin to its original direction. This means that two successive time reversals on a Dirac spinor gives $TT = -1$, a property that is central to the origin of *Fermi-Dirac spin statistics.*

The threefold operation of CPT will be equal up to a phase to the successive operation of each discrete transformation

$$\Psi(x) \xrightarrow{T} \eta_T \gamma^1 \gamma^3 \Psi^*(-\tilde{x}) \xrightarrow{P} \eta_{PT} \gamma^0 \gamma^1 \gamma^3 \Psi^*(-x) \xrightarrow{C} \eta_{CPT} \gamma^5 \Psi(-x) \tag{2.175}$$

where we have used the definition of γ^5 in (2.104) to simplify the *CPT* spinor transform.

Since time reversal inverts initial and final states, the operators acting on the states are subject to complex conjugation. Although charge conjugation involves complex conjugation, there is no inversion of initial and final states and therefore the operation is a unitary transform. However, *CPT* is necessarily antiunitary due to the time-reversal operation and through *CPT* invariance, *CP* invariance is an equivalent test of *T* invariance. Therefore, scattering processes that involve complex phases are capable of producing measureable *CP*-violating effects. This important effect will be discussed later in the text.

The effect of the discrete transformations on the four-vector potential of electromagnetism can be summarized according to the known properties of charge and current sources for electromagnetic fields:

$$\begin{aligned} C&: \quad (A_0, \boldsymbol{A}) \to (-A_0, -\boldsymbol{A})_{t,x}, \\ P&: \quad (A_0, \boldsymbol{A}) \to (A_0, -\boldsymbol{A})_{t,-x}, \\ T&: \quad (A_0, \boldsymbol{A}) \to (A_0, -\boldsymbol{A})_{-t,x}, \\ CPT&: \quad (A_0, \boldsymbol{A}) \to (-A_0, -\boldsymbol{A})_{-t,-x}. \end{aligned} \tag{2.176}$$

The corresponding transformation properties of the Dirac charged current j^μ are identical and thus the interaction Hamiltonian $H_{\text{int}}^{\text{em}} = \int d^3x e j^\mu A_\mu$ from equation (2.123) is invariant under any combination of *C*, *P*, and *T* discrete Lorentz transformations. We will find that the strong interaction is also invariant under any combination of *C*, *P*, and *T* while the electroweak interaction is invariant only under the threefold combination of *CPT*.

2.11 Covariant Form of the Electromagnetic Interaction

In earlier examples, the electromagnetic potential was treated classically. The relativistic treatment of the electromagnetic interaction necessitates a photon equation of motion. To write an equation of motion for the four-vector potential A^μ of the photon, we first need to construct a manifestly Lorentz-invariant form for the electromagnetic interaction. The two Maxwell equations involving charge and current sources written in rationalized Heaviside-Lorentz units are given by

$$\begin{aligned} \nabla \cdot \boldsymbol{E} &= \rho, \\ \nabla \times \boldsymbol{B} &= \boldsymbol{j} + \frac{\partial \boldsymbol{E}}{\partial t}, \end{aligned} \tag{2.177}$$

with the $\boldsymbol{E}$ and $\boldsymbol{B}$ fields given by

$$\begin{aligned} \boldsymbol{E} &= -\frac{\partial \boldsymbol{A}}{\partial t} - \nabla \Phi, \\ \boldsymbol{B} &= \nabla \times \boldsymbol{A}. \end{aligned} \tag{2.178}$$

The equations (2.178) incorporate the remaining two Maxwell equations. By successive substitution of equations (2.178) into (2.177), we find

$$
\begin{aligned}
-\frac{\partial}{\partial t}(\nabla\cdot A)-\nabla^2\Phi=\rho,\\
\frac{\partial^2 A}{\partial t^2}-\nabla^2 A+\nabla\left(\frac{\partial\Phi}{\partial t}+\nabla\cdot A\right)=j.
\end{aligned}
\tag{2.179}
$$

Rearranging the terms in equation (2.179) yields

$$
\begin{aligned}
\left(\frac{\partial^2\Phi}{\partial t^2}-\nabla^2\Phi\right)-\frac{\partial}{\partial t}\left(\frac{\partial\Phi}{\partial t}+\nabla\cdot A\right)=\rho,\\
\left(\frac{\partial^2 A}{\partial t^2}-\partial^2 A\right)+\nabla\left(\frac{\partial\Phi}{\partial t}+\nabla\cdot A\right)=j.
\end{aligned}
\tag{2.180}
$$

Recall that the conservation of charge is expressed by the Lorentz-invariant continuity equation $\partial_\mu j^\mu_{em}=0$ with $j^\mu_{em}=(\rho,\boldsymbol{j})$. In elementary particle interactions, we will treat the source of the electric current to be a Dirac particle with elementary charge e with $e<0$. We will therefore extract the charge e from the four-vector current density $j^\mu_{em}=ej^\mu$, such that j^μ is the probability current density for Dirac electrons. Equation (2.180) can be written in terms of the four-vector potential $A^\mu=(\Phi,\boldsymbol{A})$ and the electric current density ej^μ

$$
\partial_\mu\partial^\mu A^\nu-\partial^\nu(\partial_\mu A^\mu)=ej^\nu
\tag{2.181}
$$

or, equivalently,

$$
\partial_\mu F^{\mu\nu}=ej^\nu
\tag{2.182}
$$

where $F^{\mu\nu}$ is the *electromagnetic field tensor* defined as

$$
F^{\mu\nu}=\partial^\mu A^\nu-\partial^\nu A^\mu .
\tag{2.183}
$$

Writing $F^{\mu\nu}$ in terms of $\boldsymbol{E}$ and $\boldsymbol{B}$ fields gives

$$
F^{\mu\nu}=\begin{pmatrix} 0 & -E_x & -E_y & -E_z\\ E_x & 0 & -B_z & B_y\\ E_y & B_z & 0 & -B_x\\ E_z & -B_y & B_x & 0\end{pmatrix}.
\tag{2.184}
$$

One can transform from the field values in one inertial frame to those seen in another by Lorentz transforming the two four-vector indices $(F')^{\mu\nu}=\Lambda^\mu_\sigma\Lambda^\nu_\rho F^{\sigma\rho}$. The construction of $F^{\mu\nu}$ mirrors in some sense the construction of the Dirac wave function Ψ in that in order to write the Maxwell equations first order in the time derivative and in a Lorentz-invariant form, we ended up increasing the dimensionality of the object satisfying the equations. To explictly write the two Maxwell equations not involving j^μ in terms of $F^{\mu\nu}$ directly, it is useful to construct the dual of the Maxwell tensor using the *Levi-Civita tensor* $\epsilon^{\mu\nu\alpha\beta}$

$$
\mathcal{F}^{\mu\nu}=\frac{1}{2}\epsilon^{\mu\nu\alpha\beta}F_{\alpha\beta}=\begin{pmatrix} 0 & -B_x & -B_y & -B_z\\ B_x & 0 & E_z & -E_y\\ B_y & -E_z & 0 & E_x\\ B_z & E_y & -E_x & 0\end{pmatrix}
\tag{2.185}
$$

with the remaining Maxwell equations specified by

$$\partial_\mu \mathcal{F}^{\mu\nu} = 0. \tag{2.186}$$

The four-potential solutions (Φ, A) to Maxwell's equations (2.182) are not unique, and potentials A' and Φ' written as

$$\begin{aligned} A' &= A + \nabla\chi, \\ \Phi' &= \Phi - \partial\chi/\partial t, \end{aligned} \tag{2.187}$$

where χ is an arbitrary function of x and t, also satisfy equations (2.182). Such a transformation (2.187) is known as a *local gauge transformation*. The arbitrariness in the A^μ solutions seems innocuous at this point, but later we will find that local gauge transformations are a defining structure of all elementary particle interactions. The gauge transformation (2.187) can be written

$$(A')^\mu = A^\mu - \partial^\mu \chi, \tag{2.188}$$

where χ satisfies $\partial_\mu \partial^\mu \chi = \Box \chi = 0$. Here we have introduced the notation $\Box \equiv \partial_\mu \partial^\mu$ for the "box" operator, or *d'Alembertian*. When χ is chosen such that

$$\partial_\mu A^\mu = 0, \tag{2.189}$$

this condition is known as the *Lorentz gauge*. The Lorentz gauge, unlike the *Coulomb gauge* $(\nabla \cdot A = 0)$, is Lorentz-covariant. Starting from equation (2.181), the equation of motion for A^μ in the Lorentz gauge simplifies to

$$\partial_\mu \partial^\mu A^\nu = \Box A^\nu = e j^\nu. \tag{2.190}$$

Equation (2.190) has free-photon solutions $(j^\nu = 0)$ of the form

$$A^\mu = \epsilon^\mu \exp(-iq \cdot x) \tag{2.191}$$

Where q is the four-momentum of the photon satisfying $q^2 = 0$ and ϵ^μ is the four-vector polarization.

We will find that the Lagrangian term for the kinetic energy of a spin-1 particle is given by the general expression

$$\mathcal{L}_{\text{vector}} = -\frac{1}{4} F_{\mu\nu} F^{\mu\nu}. \tag{2.192}$$

For the specific case of the electromagnetic tensor (2.183), the Euler-Lagrange equation for A^ν applied to the kinetic energy term (2.192) can be shown to yield the equation of motion for the photon (2.190).

2.12 Relativistic Propagator Theory

Propagator theory is based on the Green's function method of solving inhomogeneous differential equations. The classical example of its use is from electrodynamics and the Poisson equation

$$\nabla^2 \phi(x) = -\rho(x) \tag{2.193}$$

for a known charge distribution $\rho(\boldsymbol{x})$. The Poisson equation comes from equation (2.179) for a static charge distribution $\partial\Phi/\partial t = 0$ and with the choice of the Coulomb gauge $\nabla \cdot \boldsymbol{A} = 0$. The approach is to first solve for the unit source

$$\nabla_x^2 G(\boldsymbol{x};\boldsymbol{x}') = -\delta(\boldsymbol{x}-\boldsymbol{x}') \tag{2.194}$$

where $G(\boldsymbol{x};\boldsymbol{x}')$ is the potential at $\boldsymbol{x}$ due to a unit source at $\boldsymbol{x}'$. We then construct the total potential at $\boldsymbol{x}$ by treating each point in the charge distribution as a source and integrating over the contribution from each volume element $d^3\boldsymbol{x}'$ solving the Poisson equation

$$\phi(\boldsymbol{x}) = \int G(\boldsymbol{x};\boldsymbol{x}')\rho(\boldsymbol{x}')d^3\boldsymbol{x}'. \tag{2.195}$$

In the case of the Dirac equation, we deviate from the classical example in a couple of ways: (1) the source depends on the Dirac wave function perturbed at some time t, not an externally specified charge distribution, and (2) the Green's function propagates in time as well as in space and consists of both positive- and negative-frequency solutions.

We can compute the free-particle propagator to the Dirac equation

$$(i\gamma^\mu\partial_\mu - m)S_F(x'-x) = \delta^{(4)}(x'-x) \tag{2.196}$$

by Fourier transforming to momentum space with

$$S_F(x'-x) = \int \frac{d^4p}{(2\pi)^4}\exp(-ip\cdot(x'-x))S_F(p). \tag{2.197}$$

Equation (2.196) becomes

$$(\not{p} - m)S_F(p) = I_4, \tag{2.198}$$

giving

$$S_F(p) = \frac{\not{p}+m}{p^2-m^2} \equiv \frac{1}{\not{p}-m} \quad \text{for } p^2 \neq m^2. \tag{2.199}$$

The interpretation given to the Green's function $S_F(x'-x)$ is that it represents the wave produced at the space-time point x' by a unit source located at the point x. In the *Feynman-Stückelberg formulation* of relativistic propagator theory, $S_F(x'-x)$ will only propagate positive-frequency solutions forward in time and negative-frequency solutions backward in time. This division is imposed via the boundary conditions of the p_0 integration of the $S_F(p)$ Fourier transform (2.197). In fact, the p_0 integration is the single most important step in constructing scattering calculations using Feynman diagrams, as will be described below.

The presence of simple poles at $(p_0 - E)$ and $(p_0 + E)$ in the Fourier-transformed propagator is common to any relativistic propagator. These poles correspond to when the particle state being used to propagate the disturbance is on-shell or, in other words, satisfies the free-particle energy-momentum relationship $E^2 = \boldsymbol{p}^2 + m^2$. However, a particle that begins at one point in space-time and propagates a distance to be annihilated or scattered at another point can never be a pure frequency. Such a particle is said to be virtual and, therefore, one has to integrate over all possible off-shell states to form the propagator. Thus, the p_0 integration is over all possible energy values from negative to positive infinity

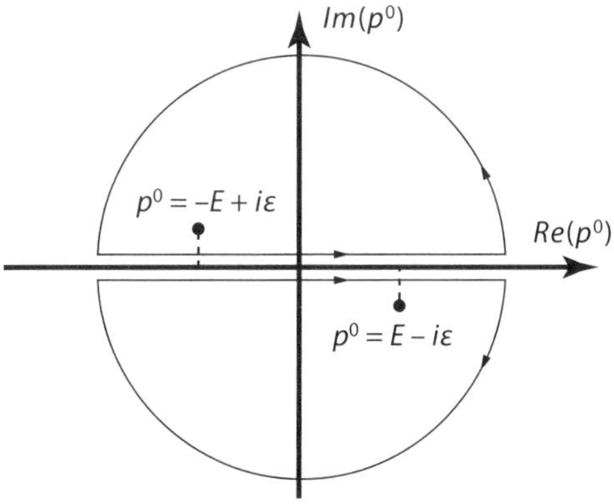

FIGURE 2.6. The p_0-integration for the Dirac $S_F(x'-x)$ relativistic propagator.

and encounters finite singularities only at the on-shell points $p_0 = \pm\sqrt{\boldsymbol{p}^2+m^2}$. Such an integration can be performed in the complex plane. For $t' > t$, the contour is closed in the lower half-plane and includes only the positive-frequency pole at $p_0 = E = \sqrt{\boldsymbol{p}^2+m^2}$. By adding a small positive imaginary part to the denominator (or, equivalently, $m^2 \to m^2 - i\epsilon$) and taking $\epsilon \to 0^+$, the positive (negative) poles are pushed into the appropriate negative (positive) $\operatorname{Im} p_0$ half-plane, as shown in figure 2.6. This gives

$$\begin{aligned} S_F^{(+)}(x'-x) &= \int \frac{d^3p}{(2\pi)^4}\exp(i\boldsymbol{p}\cdot(\boldsymbol{x}'-\boldsymbol{x}))\int_C dp_0 \exp(-ip_0(t'-t))\frac{\not{p}+m}{p^2-m^2} \\ &= (-2\pi i)\int\frac{d^3p}{(2\pi)^4}\exp(i\boldsymbol{p}\cdot(\boldsymbol{x}'-\boldsymbol{x})-iE(t'-t))\frac{E\gamma_0-\boldsymbol{p}\cdot\boldsymbol{\gamma}+m}{2E} \\ &= -i\int\frac{d^3p}{(2\pi)^3 2E}\exp(-ip\cdot(x'-x))(\not{p}+m) \\ &= -i\int\frac{d^3p}{(2\pi)^3 2E}\exp(-ip\cdot(x'-x))\sum_{s=1,2}u^s(p)\bar{u}^s(p) \qquad \text{for } t'>t \end{aligned} \tag{2.200}$$

using the *Cauchy residue theorem* for the simple pole $p^2 - m^2 = p_0^2 - (\boldsymbol{p}^2 + m^2) = (p_0 - E)(p_0 + E)$ with an overall minus sign coming from the clockwise contour integration. Note that by closing below, the exponential with p_0 negative imaginary goes to zero off the real axis. The exponential term $\exp(-ip\cdot(x'-x))$ can be thought of as arising from the unequal space-time locations of the $u\exp(-ip\cdot x')$ and $\bar{u}\exp(ip\cdot x)$ plane waves, and will functionally impose four-momentum conservation at the interaction vertex.

For $t' < t$, the contour can be closed above, encompassing the pole at $p_0 = -\sqrt{\boldsymbol{p}^2+m^2} = -E$ $-E$. For $t' < t$ and closing the counterclockwise contour in the positive $\operatorname{Im}(p_0)$ half-plane, we get

$$S_F^{(-)}(x'-x) = (2\pi i)\int\frac{d^3p}{(2\pi)^4}\exp(i\boldsymbol{p}\cdot(\boldsymbol{x}'-\boldsymbol{x})+iE(t'-t))\frac{-E\gamma_0-\boldsymbol{p}\cdot\boldsymbol{\gamma}+m}{-2E}, \tag{2.201}$$

which since the integration is over all three-momentum space, we can make the substitution $\boldsymbol{p} \to -\boldsymbol{p}$ (without changing the sign of the integral)

$$S_F^{(-)}(x'-x) = -i\int \frac{d^3p}{(2\pi)^3 2E} \exp(ip\cdot(x'-x))(-\not{p}+m)$$
$$= i\int \frac{d^3p}{(2\pi)^3 2E} \exp(ip\cdot(x'-x)) \sum_{r=1,2} v^r(p)\bar{v}^r(p) \qquad \text{for } t'<t. \tag{2.202}$$

Therefore, $S_F^{(-)}(x'-x)$ propagates negative-frequency waves backward in time. Any other choice for the two contour integrations corresponding to the $p_0 = \pm E$ simple poles would result in negative-frequency solutions propagating forward in time or positive-frequency solutions backward in time. The Feynman-Stückelberg prescription, therefore, gives us a unified treatment of the Dirac propagator (2.199) that makes no distinction between particle and antiparticle propagators. Hence, while the negative-frequency solutions from the Dirac equation were an unexpected consequence of relativity, they become a natural symmetry in relativistic propagator theory.

If we now return to the Green's function approach to solving the Dirac equation and introduce the minimal electromagnetic coupling on the right side, we now have

$$(\not{\pi} - m)\Psi = (\not{p} - e\not{A} - m)\Psi = 0, \tag{2.203}$$

which can be rewritten as

$$(\not{p} - m)\Psi = e\not{A}\Psi. \tag{2.204}$$

Comparing with the Green's function form of equation (2.196), the solution for the scattered wave $\Psi(x')$ is given by

$$\Psi(x') = e\int d^4x S_F(x'-x)\not{A}(x)\Psi(x). \tag{2.205}$$

As Ψ appears on both sides, an iterative perturbative series solution in powers of e is needed, where for n possible scatters of the incident wave, there are n orders of perturbation. Equations (2.200) and (2.202) for a single scatter can be used to factor off the final-state plane wave

$$\Psi_{\text{scat}}^{(+)}(x') = \int \frac{d^3p}{(2\pi)^3 2E} \sum_{s=1,2} \int d^4x\, u^s(p)\exp(-ip\cdot x')[-ie\bar{u}^s(p)\exp(ip\cdot x)\not{A}(x)\Psi(x)] \quad \text{as } t\to+\infty,$$
$$\Psi_{\text{scat}}^{(-)}(x') = \int \frac{d^3p}{(2\pi)^3 2E} \sum_{r=1,2} \int d^4x\, v^r(p)\exp(ip\cdot x')[+ie\bar{v}^r(p)\exp(-ip\cdot x)\not{A}(x)\Psi(x)] \quad \text{as } t\to-\infty. \tag{2.206}$$

The part in brackets is, therefore, the coefficient of the scattered wave. For multiple scatters, the propagator iteratively solves for each scattering step. For an initial state with an incoming electron with four-momentum p_i and spin s, the scattering amplitude for n scatters is given by

$$S_{fi}^{(n)} = -ie^n \int d^4y_1 \cdots d^4y_n \sum_{s'=1,2} \bar{u}^{s'}(p_f)\exp(ip_f\cdot y_n)\not{A}(y_n)S_F(y_n - y_{n-1})\not{A}(y_{n-1})\cdots$$
$$S_F(y_2-y_1)\not{A}(y_1)u^s(p_i)\exp(-ip_i\cdot y_1) \tag{2.207}$$

where both ordinary scattering and pair creation/annihilation are necessarily included in the series since the d^4y_i integrations also allow for a reverse-time ordering, $y_{n+1}^0 < y_n^0$, as shown in figure 2.7. Feynman showed that by summing over all time-ordered scattering

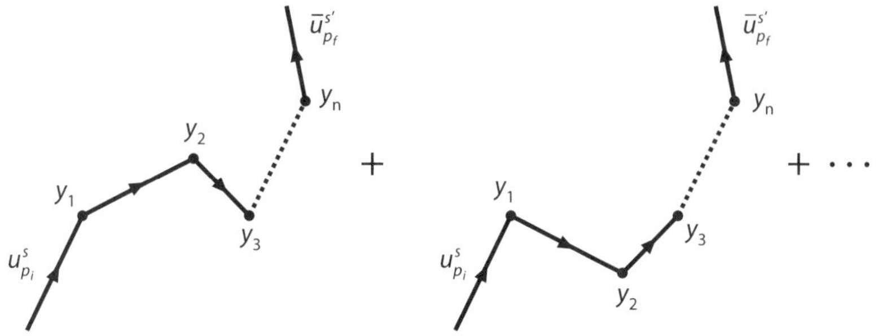

FIGURE 2.7. The invariant scattering amplitude for n scatters is a sum over all possible time orderings.

diagrams (including a minus sign for all odd permutations) the final scattering matrix was covariant, giving rise to what is known as the *invariant scattering amplitude.* Note the minus signs for odd permutations follows from equation (2.206) and is reminiscent of the *Pauli exclusion principle* in the time domain.

2.12.1 *Source Terms: Coulomb Scattering Potential*

The source term on the right-hand side of equation (2.205) is in terms of $A^\mu(x)$. The Fourier transform $A^\mu(q)$ of the electromagnetic four-vector potential is given by

$$A^\mu(q) = \int d^4x \exp(-iq \cdot x)\, A^\mu(x). \tag{2.208}$$

For a static source, A^μ is time independent, and therefore

$$\begin{aligned} A^\mu(q) &= \int dt \exp(-i(E_i - E_f)t) \int d^3x \exp(i\mathbf{q}\cdot\mathbf{x}) A^\mu(\mathbf{x}) \\ &= 2\pi\delta(E_f - E_i) A^\mu(\mathbf{q}). \end{aligned} \tag{2.209}$$

Notice that it is expected to get only the energy conservation δ-function when three-momentum is not conserved (the Coulomb potential is not allowed to move). If we start with the equation of motion (2.190) for A^μ and zero the time derivatives, we have

$$-\nabla^2 A^\mu(\mathbf{x}) = j^\mu. \tag{2.210}$$

The Fourier transform of the left-hand side can be done by partial integration, giving

$$\begin{aligned} -\int d^3x \left(\nabla^2 A^\mu(\mathbf{x})\right) \exp(i\mathbf{q}\cdot\mathbf{x}) &= -\int d^3x A^\mu(\mathbf{x}) \left(\nabla^2 \exp(i\mathbf{q}\cdot\mathbf{x})\right) \\ &= |\mathbf{q}|^2 A^\mu(\mathbf{q}). \end{aligned} \tag{2.211}$$

Combining this with the Fourier transform of j^μ gives

$$A^\mu(\mathbf{q}) = \frac{1}{|\mathbf{q}|^2} j^\mu(\mathbf{q}). \tag{2.212}$$

For the charge current

$$
\begin{aligned}
\boldsymbol{j}(x) &= 0, \\
j^0(x) &= \rho(x) = -Ze\delta(\boldsymbol{x}),
\end{aligned}
\tag{2.213}
$$

and the Fourier transform is

$$j^0(\boldsymbol{q}) = -Ze. \tag{2.214}$$

The four-vector potential $A^\mu(\boldsymbol{q})$ corresponding to the Coulomb potential is, therefore,

$$A^\mu(\boldsymbol{q}) = \frac{-Ze}{|\boldsymbol{q}|^2}\delta_0^\mu. \tag{2.215}$$

An alternative approach to calculating $A^\mu(\boldsymbol{q})$ for the Coulomb potential is to write the spatial dependence explicitly

$$A_0(x) = \frac{-Ze}{4\pi|\boldsymbol{x}|} \quad \text{and} \quad \boldsymbol{A}(x) = 0, \tag{2.216}$$

and then Fourier transforming the $1/|\boldsymbol{x}|$ potential

$$A^\mu(\boldsymbol{q}) = -Ze\int \frac{d^3x}{4\pi|\boldsymbol{x}|}\exp(-i\boldsymbol{q}\cdot\boldsymbol{x})\,\delta_0^\mu = \frac{-Ze}{|\boldsymbol{q}|^2}\delta_0^\mu \tag{2.217}$$

in agreement with the previous approach.

2.12.2 *Photon Propagator*

The free-photon equation of motion in Lorentz gauge is given by

$$g^{\mu\nu}\Box A_\nu = 0 \tag{2.218}$$

and can be cast into a Green's function equation as follows:

$$\Box' D_F^{\mu\nu}(x' - x) = \delta^{(4)}(x' - x)\,g^{\mu\nu}. \tag{2.219}$$

A Fourier transform to momentum space gives

$$D_F^{\mu\nu}(x' - x) = \int \frac{d^4q}{(2\pi)^4}\exp(-iq\cdot(x' - x))\,D_F^{\mu\nu}(q^2) \tag{2.220}$$

and thus, using equation (2.219), gives

$$D_F^{\mu\nu}(q^2) = -g^{\mu\nu}/q^2 \qquad \text{for } q^2 \neq 0. \tag{2.221}$$

The prescription for handling the pole at $q^2 = 0$ is to add a positive imaginary term to the denominator (or equivalently to give the photon a small negative imaginary mass), so that

$$D_F^{\mu\nu}(x' - x) = \int \frac{d^4q}{(2\pi)^4}\exp(-iq\cdot(x' - x))\left(\frac{-g^{\mu\nu}}{q^2 + i\epsilon}\right), \tag{2.222}$$

thereby splitting the singularity into two simple poles, one for the forward and the other for the backward light cone.

For an electron scattering off a positive static charge $-Ze$, the initial and final energies of the electron are equal and therefore the energy of the photon is zero $q_0 = 0$, giving

$$q^2 = -|\boldsymbol{q}|^2. \tag{2.223}$$

Therefore, the photon propagator $-g_{\mu\nu}/q^2$ reduces down to $1/|\boldsymbol{q}|^2$ for a Coulomb scattering potential (2.217), as expected.

2.12.3 *Massive Spin-1 Propagator*

A free massive vector particle is described by the *Proca equation*,

$$[g^{\mu\nu}(\Box + M^2) - \partial^\mu \partial^\nu] W_\nu = 0, \tag{2.224}$$

and the propagator satisfies the Green's function equation,

$$[g^{\mu\nu}(\Box' + M^2) - \partial'^\mu \partial'^\nu] D_{F\mu\nu}(x' - x) = \delta^{(4)}(x' - x). \tag{2.225}$$

The Fourier transform of (2.225) gives

$$[g^{\mu\nu}(-q^2 + M^2) + q^\mu q^\nu] D_{F\mu\nu}(q) = 1. \tag{2.226}$$

To solve for $D_F(q)$, we need to compute the inverse of the term in square brackets. In general, the inverse has the form

$$D_{F\nu\lambda}(q) = C_0 g_{\nu\lambda} + C_1 q_\nu q_\lambda \tag{2.227}$$

as required by Lorentz invariance. Applying the identity

$$[g^{\mu\nu}(-q^2 + M^2) + q^\mu q^\nu][C_0 g_{\nu\lambda} + C_1 q_\nu q_\lambda] = \delta^\mu_\lambda \tag{2.228}$$

gives

$$C_0(-q^2 + M^2)\delta^\mu_\lambda + q^\mu q_\lambda [C_0 + C_1(-q^2 + M^2) + C_1 q^2] = \delta^\mu_\lambda. \tag{2.229}$$

Equation (2.229) is solved with $C_0 = -1/(q_2 - M^2)$ and $C_1 = -C_0/M^2$. The propagator for a massive vector particle is therefore

$$D^{\mu\nu}_F(q) = (-g^{\mu\nu} + q^\mu q^\nu/M^2)/(q^2 - M^2). \tag{2.230}$$

2.13 S-Matrix and Feynman Rules for QED

If we consider a single scattering from an electromagnetic source term, then the amplitude of the scattered wave is given by

$$S_{fi} = -ie \int d^4x \bar{\Psi}_f(x) \not{A} \Psi_i(x) \qquad \text{for } (f \neq i), \tag{2.231}$$

where Ψ_i is an incoming plane wave state and $\bar{\Psi}_f$ an outgoing plane wave. We computed the form of $A^\mu(q)$ for a Coulomb potential. If the source electromagnetic four-vector potential originates from a charged current $j_\nu^{em}(x)$, then the Green's function for the photon can be used to solve for $A^\mu(x)$:

$$A^\mu(x_1) = \int d^4x_2 D_F^{\mu\nu}(x_1 - x_2) j_\nu^{em}(x_2). \tag{2.232}$$

Introducing $A^\mu(x_1)$ into S_{fi} gives, to lowest order ($f \neq i$),

$$S_{fi} = -i \int d^4x_1 d^4x_2 [e\bar{\Psi}_3(x_1)\gamma_\mu \Psi_1(x_1)] D_F^{\mu\nu}(x_1 - x_2) j_\nu^{em}(x_2). \tag{2.233}$$

Here we have separated the γ_μ from the $\not{A}$ term so that the $e\bar{\Psi}_3(x_1)\gamma_\mu\Psi_1(x_1)$ term representing the Dirac electron charged current is clearly visible. This term is "connected" to the source current $j_\nu^{em}(x_2)$ via the photon propagator $D_F^{\mu\nu}$. The current $j_\nu^{em}(x_2)$ can be, for example, a Dirac (pointlike) proton charged current ($e < 0$)

$$j_\nu^{em}(x_2) = -ej^\nu(x_2) = -e\bar{\Psi}_4(x_2)\gamma^\nu\Psi_2(x_2) \tag{2.234}$$

and, therefore, the scattering process is symmetric in structure with respect to the electron and proton Dirac charged currents, or, in other words, either current could have been called the source for the electromagnetic field according to our construction. The symmetry between the currents in S_{fi} parallels the symmetry that is restored when all possible time orderings are included in the computation of the invariant scattering amplitude.

The spatial integrals of equation (2.233) can be reduced from plane-wave factors to a δ-function representing four-momentum conservation and the Fourier transform of the propagator

$$\begin{aligned}
&\int d^4x_1 d^4x_2 \exp(ip_3x_1 - ip_1x_1 + ip_4x_2 - ip_2x_2) D_F^{\mu\nu}(x_1 - x_2) \\
&= \int d^4x_1 d^4x_2 \exp(ip_3x_1 - ip_1x_1 + ip_4x_2 - ip_2x_2) \int \frac{d^4q}{(2\pi)^4} \exp(-iq \cdot (x_1 - x_2)) D_F^{\mu\nu}(q^2) \\
&= (2\pi)^4 \delta^{(4)}(p_3 + p_4 - p_1 - p_2) D_F^{\mu\nu}(q^2).
\end{aligned} \tag{2.235}$$

We can therefore use (2.235) and the plane-wave normalization factors (2.66) to rewrite S_{fi} as

$$\begin{aligned}
S_{fi} &= \frac{(2\pi)^4\delta^{(4)}(p_3 + p_4 - p_1 - p_2)}{\sqrt{(2E_1V)(2E_2V)(2E_3V)(2E_4V)}} \\
&\quad \times \left[\bar{u}_3(-ie\gamma^\mu)u_1 \left(\frac{-ig_{\mu\nu}}{q^2} \right) \bar{u}_4(ie\gamma^\nu)u_2 \right]
\end{aligned} \tag{2.236}$$

where the *Feynman rules* for the scattering matrix, or S-matrix, element S_{fi} for a single scatter ($f \neq i$) can be summarized as follows:

1. the spinors $u_1(\bar{u}_3)$ and $u_2(\bar{u}_4)$ are the ingoing (outgoing) electron and proton, respectively, where the plane waves are normalized to a volume V and the spinors to $2E$,
2. vertex factors $-ie\gamma^\mu$ and $ie\gamma^\nu$ at the electron and proton vertices, respectively,
3. a photon propagator term $-ig_{\mu\nu}/(q^2)$, where $q^2 = (p_3 - p_1)^2 = (p_4 - p_2)^2$,

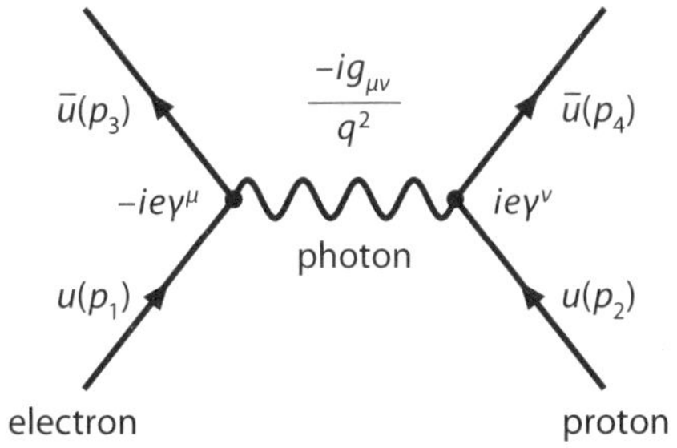

FIGURE 2.8. Feynman diagram for pointlike electron-proton scattering.

4. the term in brackets is $-i\mathcal{M}$, the *invariant amplitude*, and
5. the δ-function expressing overall four-momentum conservation.

The factors of i in the propagator and $-i$ in the vertex factors are such that the matrix elements for higher order diagrams can be written with the same rules.

The Feynman rules provide a method for constructing elements of the scattering matrix for a given number of interaction vertices, and a procedure for generating a perturbative expansion of the scattering process. Each order of the scattering process can be represented by a set of diagrams known as Feynman diagrams. The Feynman diagram for the term in brackets in equation (2.236) is shown in figure 2.8. Figure 2.9 lists the Feynman rules for external and internal lines of a Feynman diagram. Internal lines represent propagators. The spin-1/2 propagator S_F is related to the spin-0, or *Klein-Gordon*, propagator by

$$S_F(p) = (\not{p} + m)\Delta_F(p) \tag{2.237}$$

with the spin-0 propagator given by

$$\Delta_F(p) = \frac{1}{p^2 - m^2}. \tag{2.238}$$

The $p_0 = \pm E$ poles are common to S_F and Δ_F and give the same basic characteristics of forward propagation for positive-frequency solutions and backward propagation for negative-frequency solutions. The QED vertex factors for photons interacting with a charged spin-1/2 particle or a charged spin-0 particle are given in figure 2.10. The four-point interaction between photons and scalars is known as the *seagull* diagram, as shown in figure 2.10. Only the three-point interaction is possible for charged spin-1/2 particles.

2.13.1 *Cross Sections and Decay Rates*

The square of the scattering matrix element $|S_{fi}|^2$ is the probability to scatter into a particular final state from a given initial state in an unspecified time interval T. To allow comparisons with experimental measurements, the scattering probability has to be properly normalized. To do this, one first restricts the interaction to occur within a limited interval

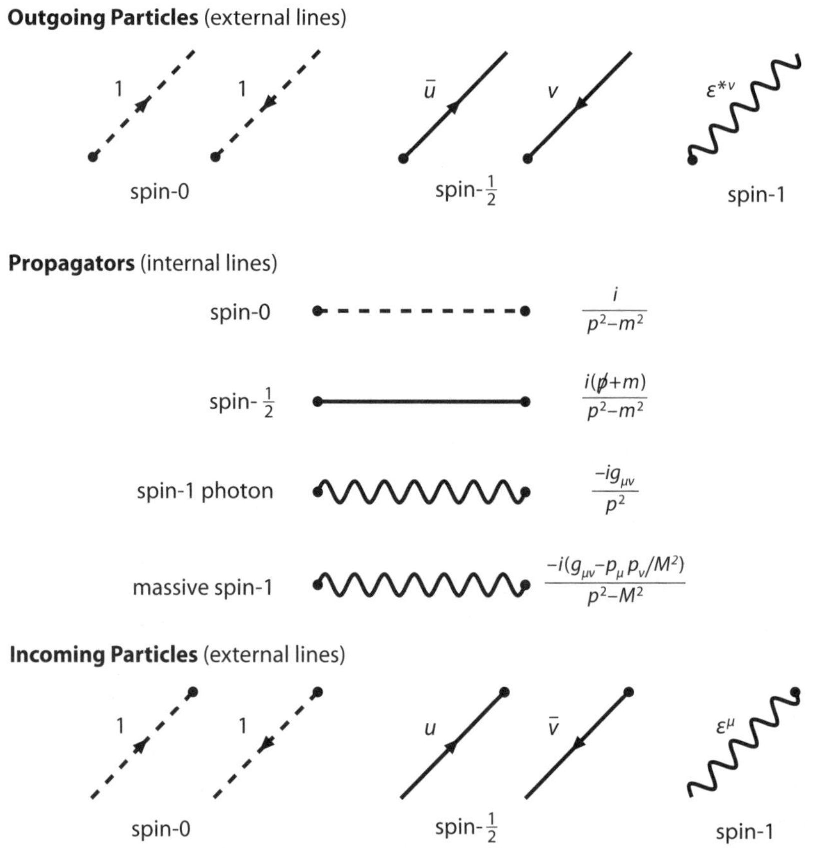

FIGURE 2.9. Feynman rules for external and internal lines.

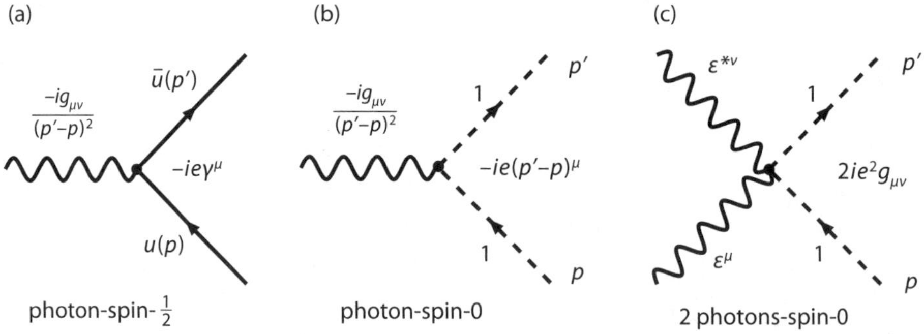

FIGURE 2.10. Feynman rules for QED vertex factors for fermions with change e, diagram (a), and for scalars with change e, diagrams (b) and (c).

of time and within a finite spatial volume. The next step is to include the number density of possible final states to count the number of scatterings that occur in a momentum volume around a particular final-state configuration. The final step in translating the scattering probability to an experimental measurement is to normalize to a known average flux of particles per unit area per unit time, a so-called particle beam. A flux of particles

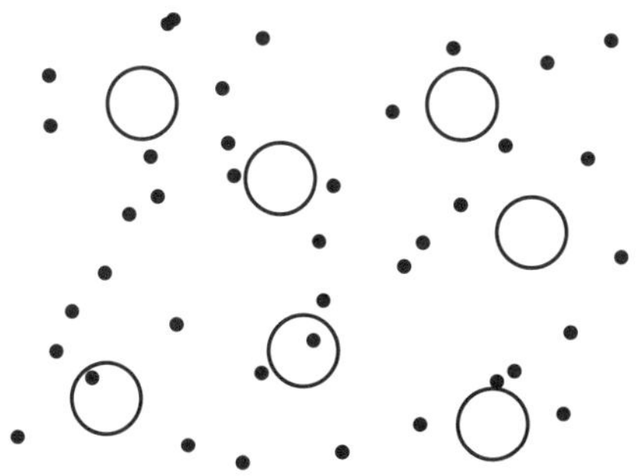

FIGURE 2.11. An integrated flux of particles incident on an ensemble of scattering centers will result in a number of interactions that is proportional to an effective area, or *cross section*.

incident on a scattering center will have a probability per unit time for an interaction. The fraction of particles in the beam that scatter in the target corresponds to an effective area, as sketched in figure 2.11. This effective area is known as a *cross section*. The total cross section is an integral over all possible kinematic configurations of the final-state particles.

To compute a probability per unit time, or transition rate, ω_{fi}, from the square of the scattering matrix element, we divide by an interval of time T and multiply by the number density $d\Phi$ of possible final states:

$$\begin{aligned}\omega_{fi} &= |S_{fi}|^2 d\Phi/T \\ &= \left(\frac{1}{V(2E_1)(2E_2)}\right)(2\pi)^4\delta^{(4)}(p_3+p_4-p_1-p_2)\,|\mathcal{M}_{fi}|^2\frac{d^3p_3d^3p_4}{(2\pi)^3 2E_3(2\pi)^3 2E_4}.\end{aligned} \tag{2.239}$$

Equation (2.239) is derived as follows. The phase-space factor $d\Phi$ is the product of the terms $Vd^3p_f/(2\pi)^3$ for each of the scattered particles and comes from counting the number of plane waves in a volume $V = L_xL_yL_z$ with

$$dp_xdp_ydp_z = \left(\frac{2\pi}{L_x}\right)\left(\frac{2\pi}{L_y}\right)\left(\frac{2\pi}{L_z}\right)dn. \tag{2.240}$$

The term $\mathcal{M}_{fi}$ in equation (2.239) is known as the *invariant amplitude* and for pointlike electron-proton scattering is given by

$$-i\mathcal{M}_{fi} = \bar{u}_3(-ie\gamma^\mu)u_1\left(\frac{-ig_{\mu\nu}}{q^2}\right)\bar{u}_4(ie\gamma^\nu)u_2. \tag{2.241}$$

In general, the invariant amplitude has a one-to-one correspondence to a Feynman diagram, which for equation (2.241) is given by figure 2.8.

A crucial and nonobvious step in going from S_{fi} to equation (2.239) is the handling of the square of the δ-function. Here we have to manually impose the constraint that the interaction is localized in space and time to the volume VT. The way this is done is to rewrite one of the two δ-functions from $|S_{fi}|^2$ in terms of its Fourier transform integral definition as

$$\begin{aligned}\left[\delta^{(4)}(p_f - p_i)\right]^2 &= \delta^{(4)}(p_f - p_i)\int_{VT}\frac{d^4x}{(2\pi)^4}\exp(i(p_f - p_i)\cdot x)\big|_{p_f = p_i} \\ &= \delta^{(4)}(p_f - p_i)\,VT/(2\pi)^4\end{aligned} \tag{2.242}$$

where p_f and p_i are the sum over the final- and initial-state particle four-momenta, respectively. We have imposed the condition of $p_f - p_i = 0$ from the first δ-function, rendering the integral representing the second δ-function a constant over d^4x, thus resulting in the finite space-time volume VT over which the interaction is localized.

The integral over momenta in equation (2.239) has the following general structure for n-particles in the final state:

$$\int d\Phi_n = \left(\prod_f \int \frac{d^3p_f}{(2\pi)^3 2E_f}\right)(2\pi)^4\delta^{(4)}(P_i - \sum p_f) \tag{2.243}$$

with P_i the total initial four-momentum. The integral (2.243) is manifestly Lorentz invariant since it is built up from invariant three-momentum integrals constrained by a four-momentum δ-function. To explicitly show that the three-momentum integrals are invariant, we can rewrite the term $d^3p/2E$ as

$$\begin{aligned}\frac{d^3p}{2E} &= \int_0^\infty dp_0\,\frac{\delta(p_0 - E)}{2p_0}\,d^3p \\ &= \int_0^\infty dp_0\,\delta(p_0^2 - E^2)\,d^3p \\ &= \int_0^\infty dp_0\,\delta(p^2 - m^2)\,d^3p \\ &= \int_{-\infty}^\infty d^4p\,\delta(p^2 - m^2)\,\theta(p_0)\end{aligned} \tag{2.244}$$

where $\theta(p_0)$ is the *Heaviside step function*,

$$\theta(p_0) = \int_{-\infty}^{p_0}\delta(t)\,dt = \begin{cases}0 & \text{if } p_0 < 0,\\ 1 & \text{if } p_0 \geq 0,\end{cases} \tag{2.245}$$

and where we have also used in the above the following property of the δ-function:

$$\delta(f(x)) = \sum_i \frac{\delta(x - a_i)}{|df/dx|_{x = a_i}}. \tag{2.246}$$

Expression (2.244) is, therefore, Lorentz invariant provided that p^μ is timelike, as is the case for a final-state particle. The integral (2.243) is known as the *Lorentz-invariant n-particle phase space.*

The differential cross section for an interaction is given by the transition rate divided by the incident flux:

$$d\sigma = \omega_{fi}/|J_{\text{inc}}|. \tag{2.247}$$

The incident flux $|J_{\text{inc}}|$ is the number of particles per unit area which pass by each other per unit time. By construction, the volume V contains the 2-particle interaction, and therefore we have

$$V|J_{\rm inc}| = |v_1 - v_2|. \tag{2.248}$$

Collecting the terms from the incident electron and proton spinor normalizations and the remaining volume term V from equation (2.239) and the incident flux, we get the resulting term in the denominator

$$2(E_1)2(E_2)V|J_{\rm inc}| = 4E_1E_2|v_1 - v_2| = 4E_1E_2\left|\frac{\mathbf{p}_1}{E_1} - \frac{\mathbf{p}_2}{E_2}\right| = 4|\mathbf{p}_1E_2 - \mathbf{p}_2E_1|. \tag{2.249}$$

This factor is frequently replaced with the expression

$$\begin{aligned}\mathcal{F} = 4\sqrt{(p_1\cdot p_2)^2 - m_1^2m_2^2} &= 4\sqrt{(E_1E_2 - \mathbf{p}_1\cdot\mathbf{p}_2)^2 - (E_1^2 - \mathbf{p}_1^2)(E_2^2 - \mathbf{p}_2^2)}\\ &= 4\sqrt{(\mathbf{p}_1E_2 - \mathbf{p}_2E_1)^2 + \mathbf{p}_1\cdot\mathbf{p}_2^2 - \mathbf{p}_1^2\mathbf{p}_2^2}\\ &= 4\sqrt{(\mathbf{p}_1E_2 - \mathbf{p}_2E_1)^2 - (\mathbf{p}_1\times\mathbf{p}_2)^2}\\ &= 4|\mathbf{p}_1E_2 - \mathbf{p}_2E_1| \quad \text{for } \mathbf{p}_1 \| \mathbf{p}_2,\end{aligned} \tag{2.250}$$

which is equal to term (2.249) if $\mathbf{p}_1$ and $\mathbf{p}_2$ are collinear. Effectively, cross sections are always computed in a frame where the incident particles are collinear. The final expression in equation (2.250) is Lorentz-invariant for boosts along the axis collinear with the incoming particles, giving this term the same transformation properties as a cross-sectional area.

Rewriting equation (2.247) in terms of the collinear Lorentz-invariant quantities computed above, the general expression for a differential cross section is given by

$$d\sigma = \frac{|\mathcal{M}|^2}{\mathcal{F}}\, d\Phi \tag{2.251}$$

where $\mathcal{F}$ is given by equation (2.250). Collecting the results of equations (2.239), (2.243), and (2.250), the differential cross section for interaction for particles p_1 and p_2 with masses m_1 and m_2, respectively, is given by

$$d\sigma = \frac{|\mathcal{M}|^2}{4\sqrt{(p_1\cdot p_2)^2 - m_1^2m_2^2}}\left(\prod_f \int \frac{d^3p_f}{(2\pi)^3 2E_f}\right)(2\pi)^4\delta^{(4)}\left(p_1 + p_2 - \sum p_f\right). \tag{2.252}$$

The total cross section σ for a particular set of initial-state and final-state spin configurations is computed by integrating equation (2.252) over the momenta of final-state particles. To compute the total cross section for unpolarized initial-state particles, the invariant amplitude $|\mathcal{M}|^2$ can be replaced by

$$\overline{|\mathcal{M}|^2} \equiv \frac{1}{(2s_1+1)(2s_2+1)}\sum_{s_1,s_2,s_3,s_4} |\mathcal{M}(s_1,s_2 \rightarrow s_3,s_4)|^2 \tag{2.253}$$

where an average is made over initial spins and a sum is taken over all possible final-state spins.

The total cross section allows one to predict the number of interactions for a given amount of integrated flux, known as *integrated luminosity*, where

$$N_{\rm events} = \sigma\mathcal{L}_{\rm int} \tag{2.254}$$

with

$$\mathcal{L}_{\text{int}} = \int \mathcal{L}(t)dt. \tag{2.255}$$

Here, $\mathcal{L}(t)$ is the *instantaneous luminosity* and is typically quoted in units of $[\text{cm}]^{-2}[\text{s}]^{-1}$. A particle accelerator produces collections of particles known as "bunches" and can pass two bunches through each other to produce collisions. The formula for the instantaneous luminosity of a collider with collinear intersecting particle beams is given by

$$\mathcal{L} = f\frac{N_1 N_2}{4\pi\sigma_x\sigma_y} \tag{2.256}$$

where N_1 and N_2 are the numbers of particles in each of the two bunches, f is the frequency in Hz at which the bunches are brought into collision, and σ_x and σ_y specify the one-dimensional transverse sizes of the beam in cm, typically given as Gaussian standard deviations along the horizontal and vertical axes defined relative to one of the two incoming beam directions. For modern colliders and planned colliders, the current maximum values for the instantaneous luminosity are given in table 2.1. The relation (2.254) is heavily used in particle physics to translate a cross section into a predicted number of events. The size of a dataset is generally quoted in terms of integrated luminosity in units of *inverse femtobarns* $[\text{fb}]^{-1} = 10^{39}\,\text{cm}^{-2}$, probably one of the most obscure units in physics. The history of this unit is that the *barn* is a term used to describe the hard-sphere cross section of a uranium nucleus ($A = 235$, $\sigma_{\text{hard}} = \pi(A^{1/3}\text{ fermi})^2 \approx 10^{-24}\,\text{cm}^2 \equiv 1$ barn, where compared to most processes the uranium hard-sphere scattering cross section is "as big as a barn door").

A decay rate is a simplified cross section in which there is only one particle at rest in the initial state. The only modifications relative to the cross-section calculation is to remove the $\mathcal{F}$ term (2.249), which included the flux and initial-state wave-function normalizations, and to replace it with a spinor normalization of $1/2m$ for a particle of mass m decaying at rest. Thus, the differential decay rate formula for a particle with mass m is

Table 2.1 The current maximum values for the instantaneous luminosity for a selection of modern colliders and planned colliders [3].

$\mathcal{L}_{\text{inst}}\,(\text{cm}^{-2}\text{s}^{-1})$	Collider (relevant cross section)
2×10^{31}	LEP100 e^+e^- (Z peak ~30 nb)
10^{32}	LEP200 e^+e^- (WW pair production ~16 pb)
3×10^{32}	Tevatron $p\bar{p}$ (top quark pair production ~7 pb)
10^{33}	HERA $e^{\pm}p$ (deep inelastic scattering)
2×10^{33}	planned LHC pp startup (supersymmetry? ~1–10 pb)
10^{34}	planned ILC e^+e^- and LHC pp high-luminosity (Higgs? ~0.1–10 pb)
2×10^{34}	PEP-II/KEK-B asymmetric e^+e^- (B meson factory ~4 nb)
10^{35}	planned super-LHC phase-II upgrade (~10–100 fb)

$$d\Gamma = \frac{1}{2m}|\mathcal{M}|^2 \left(\prod_f \int \frac{d^3p_f}{(2\pi)^3 2E_f}\right)(2\pi)^4\delta^{(4)}\left(p - \sum p_f\right) \tag{2.257}$$

where p is the initial-state four-momentum $p = (m, 0, 0, 0)$. As we will see, there are two primary categories of $\mathcal{M}$ in a decay rate calculation. The first is a decay coming from a single 3-particle vertex term, resulting in a $1 \to 2$ process. The second is a decay through a propagator, typically a $1 \to 3$ process such as muon decay $\mu^- \to \nu_\mu e^- \bar{\nu}_e$.

The two-body Lorentz-invariant phase space for a $1 \to 2$ decay is given by

$$d^6\Phi_2 = \frac{d^3p_2}{(2\pi)^3 2E_2}\frac{d^3p_3}{(2\pi)^3 2E_3}(2\pi)^4\delta^{(4)}(p_1 - p_2 - p_3). \tag{2.258}$$

In the rest frame of the particle of mass M, the integral of the $\delta^{(3)}$-function over d^3p_3 gives

$$d^3\Phi_2 = \frac{d^3p_2}{(2\pi)^2 4E_2E_3}\delta(M - E_2 - E_3). \tag{2.259}$$

The differential element d^3p_2 can be rewritten using $p_2dp_2 = E_2dE_2$

$$d^3p_2 = p_2^2dp_2d\Omega = p_2E_2dE_2d\Omega \tag{2.260}$$

for an element of solid angle $d\Omega$. Therefore, if we use the constraint that $\boldsymbol{p}_2 = -\boldsymbol{p}_3$ in the rest frame of M, then the δ-function in equation (2.259) can be written

$$\delta(M - E_2 - E_3) = \delta(M - E_2 - \sqrt{E_2^2 - m_2^2 + m_3^2}) \tag{2.261}$$

with

$$\left|\frac{d(M - E_2 - E_3)}{dE_2}\right| = 1 + E_2/E_3 = M/E_3. \tag{2.262}$$

Integrating over the last δ-function in equation (2.259) gives

$$d\Phi_2 = \frac{|\boldsymbol{p}|d\Omega}{16\pi^2 M}, \tag{2.263}$$

showing that the size of the two-body phase space scales linearly with the magnitude of the three-momentum, $|\boldsymbol{p}|$, of a final-state particle in the rest frame of the decaying particle.

The total decay rate Γ is computed from equation (2.257) by integrating over the momenta of final-state particles and is inversely related to the lifetime τ of the particle

$$\tau = 1/\Gamma. \tag{2.264}$$

As with a cross-section calculation, the square of the invariant amplitude $|\mathcal{M}|^2$ is replaced by an average over initial spins and a sum over final spins when calculating the total decay width of an unpolarized unstable particle, giving

$$\overline{|M|^2} \equiv \frac{1}{(2s+1)}\sum_{s,s_1,s_2}|\mathcal{M}(s \to s_1, s_2)|^2 \tag{2.265}$$

for a $1 \to 2$ decay.

In general, the evaluation of the square of the invariant amplitude $|\mathcal{M}|^2$ can be reduced down to a trace. Consider, for example, an invariant amplitude with a general bilinear Γ at the interaction vertex, which gives

$$
\begin{aligned}
|\bar{u}_f \Gamma u_i|^2 &= (\bar{u}_f \Gamma u_i)(u_i^\dagger \Gamma^\dagger \gamma^0 u_f) \\
&= \bar{u}_f \Gamma u_i \bar{u}_i \bar{\Gamma} u_f \\
&= (\bar{u}_f)_\alpha \Gamma_{\alpha\beta} (u_i \bar{u}_i)_{\beta\gamma} (\bar{\Gamma})_{\gamma\delta} (u_f)_\delta \\
&= (u_f \bar{u}_f)_{\delta\alpha} \Gamma_{\alpha\beta} (u_i \bar{u}_i)_{\beta\gamma} (\bar{\Gamma})_{\gamma\delta} \\
&= \sum_{\delta=1}^{4} \left[(u_f \bar{u}_f) \Gamma (u_i \bar{u}_i) \bar{\Gamma} \right]_{\delta\delta} \\
&= \mathrm{tr}\{(u_f \bar{u}_f) \Gamma (u_i \bar{u}_i) \bar{\Gamma}\}
\end{aligned}
\tag{2.266}
$$

where $\bar{\Gamma} = \gamma^0 \Gamma^\dagger \gamma^0$. The trace arises from the summation over the index δ, the final step in reducing down the 4×4 matrix terms to a real number. Note that for $\Gamma = \gamma^\mu$ or $\gamma^\mu \gamma^5$, then $\bar{\Gamma} = \Gamma$. If we now start with the result in equation (2.266), we see that if Γ has a Lorentz index, then the term in equation (2.266) is a 2-indexed tensor $V^{\mu\nu}$ associated with the interaction vertex. A Feynman diagram with an internal photon propagator will couple the $V^{\mu\nu}$ vertex to another vertex tensor $V'_{\mu\nu}$. For example,

$$
\begin{aligned}
|\mathcal{M}|^2 &= \left| \bar{u}_3 \gamma^\mu u_1 \frac{e^2}{q^2} \bar{u}_4 \gamma_\mu u_2 \right|^2 \\
&= \frac{e^4}{q^4} \mathrm{tr}\{(u_3 \bar{u}_3) \gamma^\mu (u_1 \bar{u}_1) \gamma^\nu\} \mathrm{tr}\{(u_4 \bar{u}_4) \gamma_\mu (u_2 \bar{u}_2) \gamma_\nu\} \\
&= \frac{e^4}{q^4} V^{\mu\nu} V'_{\mu\nu}.
\end{aligned}
\tag{2.267}
$$

If there is a sum over spins, as in (2.253), the spinor sums can be replaced by the completeness relations $\sum u\bar{u} = \not{p} + m$ and in the case of positrons $\sum v\bar{v} = \not{p} - m$.

The remaining task is the calculation of the trace of an expression which, in general, contains products of γ-matrices and four-momenta or polarization four-vectors.

Some useful trace properties are listed here:

1. The trace of an odd number of γ-matrices is zero.

2. $\mathrm{tr}\{\gamma^\mu \gamma^\nu\} = 4g^{\mu\nu}$,
   $\mathrm{tr}\{\not{p}_1 \not{p}_2\} = 4p_1 \cdot p_2$.

3. $\mathrm{tr}\{\gamma^\alpha \gamma^\mu \gamma^\beta \gamma^\nu\} = 4[g^{\alpha\mu} g^{\beta\nu} + g^{\alpha\nu} g^{\mu\beta} - g^{\alpha\beta} g^{\mu\nu}]$,
   $\mathrm{tr}\{\not{p}_1 \not{p}_2 \not{p}_3 \not{p}_4\} = 4[(p_1 \cdot p_2)(p_3 \cdot p_4) + (p_1 \cdot p_4)(p_2 \cdot p_3) - (p_1 \cdot p_3)(p_2 \cdot p_4)]$,
   $\mathrm{tr}\{\not{p}_1 \cdots \not{p}_n\} = (p_1 \cdot p_2) \mathrm{tr}\{\not{p}_3 \cdots \not{p}_n)\} - (p_1 \cdot p_3) \mathrm{tr}\{\not{p}_2 \not{p}_4 \cdots \not{p}_n\} + \cdots$
   $+ (p_1 \cdot p_n) \mathrm{tr}\{\not{p}_2 \cdot \not{p}_{n-1}\}$.

4. $\mathrm{tr}\{\gamma^5\} = 0$,
   $\mathrm{tr}\{\gamma^5 \not{p}_1 \not{p}_2\} = 0$,
   $\mathrm{tr}\{\gamma^\alpha \gamma^\mu \gamma^\beta \gamma^\nu \gamma^5\} = 4i\epsilon^{\alpha\mu\beta\nu}$,
   $\mathrm{tr}\{\not{p}_1 \not{p}_2 \not{p}_3 \not{p}_4 \gamma^5\} = 4i\epsilon^{\alpha\mu\beta\nu} p_{1\alpha} p_{2\mu} p_{3\beta} p_{4\nu}$.

In addition, these are frequently used relationships from the Dirac algebra (2.95):

1. $\gamma_\mu \gamma^\mu = 4,$
   $\gamma_\mu \not{p} \gamma^\mu = -2\not{p},$
   $\gamma_\mu \not{p}_1 \not{p}_2 \gamma^\mu = 4p_1 \cdot p_2,$
   $\gamma_\mu \not{p}_1 \not{p}_2 \not{p}_3 \gamma^\mu = -2\not{p}_3 \not{p}_2 \not{p}_1,$
   $\gamma_\mu \not{p}_1 \not{p}_2 \not{p}_3 \not{p}_4 \gamma^\mu = 2(\not{p}_4 \not{p}_1 \not{p}_2 \not{p}_3 + \not{p}_3 \not{p}_2 \not{p}_1 \not{p}_4).$
2. $\not{p}_1 \not{p}_2 + \not{p}_2 \not{p}_1 = 2p_1 \cdot p_2,$
   $\not{p}\not{p} = p^2.$

In the high-energy limit, it is often useful to express cross sections in terms of the invariant *Mandelstam variables*

$$
\begin{aligned}
s &= (p_1 + p_2)^2 = (p_3 + p_4)^2 \simeq 2p_1 \cdot p_2 \simeq 2p_3 \cdot p_4, \\
t &= (p_1 - p_3)^2 = (p_2 - p_4)^2 \simeq -2p_1 \cdot p_3 \simeq -2p_2 \cdot p_4, \\
u &= (p_1 - p_4)^2 = (p_2 - p_3)^2 \simeq -2p_1 \cdot p_4 \simeq -2p_2 \cdot p_3, \\
& s + t + u \simeq 0,
\end{aligned} \tag{2.268}
$$

where the approximate equalities correspond to neglecting particle masses.

2.13.2 *Worked Example: Mott Scattering*

Mott scattering is the scattering of relativistic electrons off of a Coulomb potential. In this example, we compute the differential cross section $d\sigma/d\Omega$ of this process as a function of the scattering angle θ of the electron.

We begin with the invariant amplitude $-i\mathcal{M}$ for an electron scattering off a Coulomb potential (2.215):

$$
\begin{aligned}
-i\mathcal{M} &= \bar{u}(p_f, s_f)(-ie\gamma^\mu)u(p_i, s_i)\left(\frac{-Ze}{|q|^2}\delta^0_\mu\right) \\
&= \frac{iZe^2}{|q|^2}\bar{u}(p_f, s_f)\gamma^0 u(p_i, s_i).
\end{aligned} \tag{2.269}
$$

We can apply equation (2.253) with one electron in the initial state and final state to compute the unpolarized total differential cross section:

$$
\begin{aligned}
\overline{|\mathcal{M}|^2} &= \frac{1}{(2s_1 + 1)}\sum_{s_i, s_f} |\mathcal{M}(s_i \to s_f)|^2 \\
&= \frac{Z^2 e^4}{2|q|^4}\sum_{s_i, s_f} \mathrm{tr}\{u(p_f, s_f)\bar{u}(p_f, s_f)\gamma^0 u(p_i, s_i)\bar{u}(p_i, s_i)\gamma^0\} \\
&= \frac{Z^2 e^4}{2|q|^4}\mathrm{tr}\{(\not{p}_f + m)\gamma^0(\not{p}_i + m)\gamma^0\} \\
&= \frac{Z^2 e^4}{2|q|^4}\left(\mathrm{tr}\{\not{p}_f \gamma^0 \not{p}_i \gamma^0\} + m^2 \mathrm{tr}\{(\gamma^0)^2\}\right) \\
&= \frac{2Z^2 e^4}{|q|^4}(p_i^0 p_f^0 + p_f^0 p_i^0 - \mathbf{p}_i \cdot \mathbf{p}_f + m^2)
\end{aligned}
$$

$$= \frac{2Z^2e^4}{|q|^4}\left(E_iE_f + p_i \cdot p_f + m^2\right)$$

$$= \frac{4Z^2e^4}{|q|^4}\left(|p|^2(1+\cos\theta)/2 + m^2\right)$$

$$= \frac{4Z^2e^4}{|q|^4}\left(|p|^2\cos^2(\theta/2) + m^2\right) \tag{2.270}$$

where $E = E_i = E_f$, $|p_i| = |p_f| = |p|$, and $|q| = 2|p|\sin(\theta/2)$ from the kinematics of Coulomb scattering. The flux factor $\mathcal{F}$ is given by

$$\mathcal{F} = |v|2E_i = 2|p_i|. \tag{2.271}$$

The Lorenz-invariant phase space $d\Phi$ is given by

$$d\Phi = d\Omega\int_0^\infty \frac{p_f^2\,dp_f}{(2\pi)^3 2E_f}2\pi\delta(E_f - E_i) = d\Omega\left(\frac{E_f}{|p_f|}\right)\frac{|p_f|^2}{(2\pi)^2 2E_f} = d\Omega\frac{|p_f|}{2(2\pi)^2}. \tag{2.272}$$

Therefore, applying equation (2.251) and dividing each side by $d\Omega$, we find

$$\begin{aligned}\frac{d\sigma}{d\Omega} &= \frac{1}{2|p_i|}\frac{4Z^2e^4}{(2|p|)^4\sin^4(\theta/2)}\left(|p|^2\cos^2(\theta/2) + m^2\right)\frac{|p_f|}{2(2\pi)^2}\\ &= \left(\frac{Z^2\alpha_{\text{QED}}^2}{4|p|^2}\right)\frac{\cos^2(\theta/2) + (m/|p|)^2}{\sin^4(\theta/2)}\end{aligned} \tag{2.273}$$

giving a low-angle $1/\theta^4$ divergence as expected from *Rutherford scattering.*

2.14 Spin Statistics

In relativistic quantum theory, particle and antiparticle states are created out of the vacuum. Therefore, Bose-Einstein and Fermi-Dirac spin statistics are expected to be inherent to the relativistic wave functions for bosons and fermions. In section 2.10, we learned that two successive time-reversal operations on a spin-1/2 Dirac particle resulted in an overall minus sign $TT = -1$. As this derivation was based on a 2π rotation of a Dirac spinor, the corresponding analysis applied to a spin-0 particle yields $TT = +1$. The connection between the sign difference for two successive time reversals and spin statistics is presented in Feynman's 1986 Dirac memorial lecture [5]. A basic summary of the arguments is repeated here.

If one starts with a particle in an initial state $u(p_i)$ and constructs the S-matrix expansion of the scattered wave interacting with a real potential U, then the sum of the probabilities to end up in all possible final states at fixed order in U will give unity. This is the *unitarity* property of the S-matrix. In nonrelativistic scattering, two scattering diagrams contribute to the probability of a null transition $u(p_i) \to u(p_i)$ up to first order in U, the zeroth-order transition shown in figure 2.12(a) and the interference term between the zeroth-order diagram and the second-order scattering diagram shown in figure 2.12(b). The interference

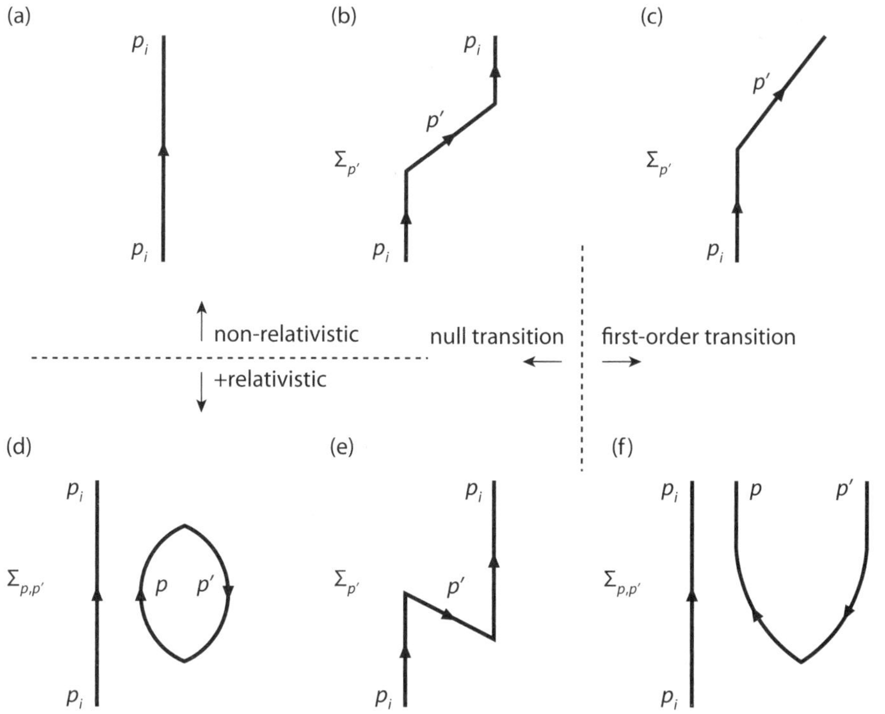

FIGURE 2.12. S-matrix expansion with non-relativistic transition diagrams given by (a), (b), and (c) and additional uniquely relativistic transitions given by (d), (e), and (f). The null transitions are the four leftmost diagrams and the first-order transitions are the two rightmost diagrams.

term between diagrams 2.12(a) and 2.12(b) is negative and, therefore, reduces the probability of a null transition by twice the real component of the interference amplitude

$$|\mathcal{M}_a + \mathcal{M}_b|^2 = |\mathcal{M}_a|^2 + \mathcal{M}_a\mathcal{M}_b^* + \mathcal{M}_b\mathcal{M}_a^* + \mathcal{O}(U^4) = 1 + 2\,\mathrm{Re}\{\mathcal{M}_b\} + \mathcal{O}(U^4), \quad (2.274)$$

where the amplitude of the corresponding diagram in figure 2.12 is indicated with the subscript. The probability of the first-order transition $u(p_i) \to u(p')$ shown in figure 2.12(c) is positive and exactly compensates for the interference term

$$-|\mathcal{M}_c|^2 = 2\,\mathrm{Re}\{\mathcal{M}_b\}. \quad (2.275)$$

This relation is also seen diagrammatically by interpreting the square of diagram 2.12(c) as flipping the diagram vertically and then joining it with the existing diagram to reproduce the topology of diagram 2.12(b). The correspondence between the interference and first-order transition is expected from unitarity, and can also be seen from the identical sum that occurs in diagram 2.12(b) over intermediate states $u(p')$ and in diagram 2.12(c) over all possible final states.

In relativistic scattering, there are two diagrams, shown in figures 2.12(d) and 2.12(e), with negative-frequency propagators that also contribute along with the nonrelativistic terms to the probability of a null transition $u(p_i) \rightarrow u(p_i)$ to first order in U. In relativistic theory, the probability contributions to the null transition from the interference of the second-order terms with the zeroth-order term are not strictly negative, in contrast to the nonrelativistic summation. Furthermore, in the relativistic theory an additional diagram contributes to the probability of a first-order transition. This is the pair-production diagram, shown in figure 2.12(f). From the commonality of the factors in the loop diagram 2.12(d) and the squared pair-production diagram 2.12(f), we can conclude that the probability contribution of these diagrams to the S-matrix will cancel to first order in U. In other words, the interference of diagram 2.12(d) with 2.12(a) will drain the null transition probability by an amount that is equal to the probability for additional final states from 2.12(f), giving the diagrammatic relationship in figure 2.13(a). This cancellation, however, leaves a nonzero contribution from the interference of diagram 2.12(e) with the zeroth-order null transition that would violate the unitarity of the S-matrix, thus creating a paradox. The resolution of this paradox is the existence of one additional first-order scattering diagram. The spectator particle in diagram 2.12(f) can exchange with the pair-produced positive-frequency particle if the two are in the same momentum state, thereby making the pair-produced positive-frequency particle indistinguishable from the initial-state particle. The presence of the exchange diagram must, therefore, exactly compensate the probability contribution from diagram 2.12(e) to maintain the unitarity of the S-matrix.

The exchange diagram will be equal to the spectator diagram up to a relative sign, as shown in figure 2.13(b). If the relative sign in figure 2.13(b) is negative, then the destructive interference of the exchange and spectator diagrams will zero out the contribution on the left-hand side of figure 2.13(a) for $p = p_i$. However, the contribution from $p = p_i$ on the right-hand side of figure 2.13(a) still remains and, therefore, must compensate the probability contribution from the interference of diagram 2.13(e) with the zeroth-order null transition, which must be positive in this case.

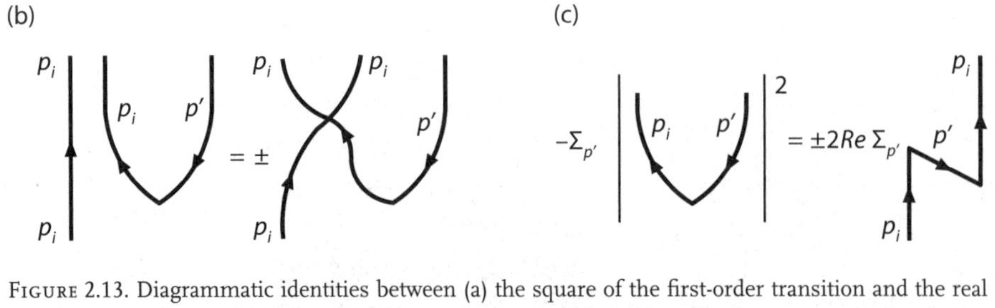

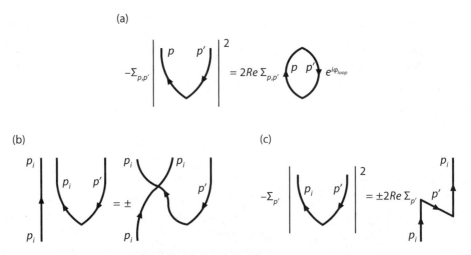

FIGURE 2.13. Diagrammatic identities between (a) the square of the first-order transition and the real part of the loop diagram including an additional phase factor $\exp(i\phi_{\text{loop}})$, (b) the spectator and exchange diagrams, and (c) the square of the exchange diagram and the real part of the relativistic null transition.

Now consider the case where the relative sign between the exchange and spectator diagrams in figure 2.13(b) is positive. This has two effects. Not only is the $p = p_i$ contribution on the left-hand side of figure 2.13(a) nonzero; the contribution would appear to be four times larger than the specatator diagram alone due to the constructive interference. However, this poses a problem as the interference terms cannot compensate for a fourfold increase in first-order scattering. What we need to introduce is a symmetrization factor $1/\sqrt{2}$ when summing diagrams whose final states differ by only an exchange of indistinguishable particles. Second, the interference term from diagram 2.12(e) must change sign and give a negative contribution to the probability for a null transition. Figure 2.13(c) shows the diagrammatic relationship summarizing the magnitude and relative sign of the probability contribution from diagram 2.12(e), where the sign in figure 2.13(c) is correlated to the sign in figure 2.13(b).

We are now left with one last unexplained diagrammatic relationship; namely, the summations in diagrams 2.12(d) and 2.12(e) are such that if the interference from diagram 2.12(e) with the zeroth-order null transition is positive, so should the interference term from diagram 2.12(d) be. However, we can see that the right-hand side of diagrammatic relationship 2.13(a) must be strictly negative. To satisfy 2.13(a) the loop diagram is multiplied by a phase factor, $\exp(i\phi_{\text{loop}})$. The phase factor is set by a Feynman rule. We know from spin statistics that the relative sign in the diagrammatic relationship 2.13(b) must be positive for bosons and negative for fermions. In other words, the amplitude for the creation of a boson into the same state as N existing bosons is amplified by the factor $N+1$. Similarly, the amplitude for the creation of a fermion into the same state as an existing fermion is zero by the *Pauli exclusion principle*. Therefore, to have the correct spin statistics, the Feynman rule for loop diagrams is that $\exp(i\phi_{\text{loop}}) = -1$ for fermions and $\exp(i\phi_{\text{loop}}) = +1$ for bosons, equivalent to the rule for adding a negative sign for an odd number of positrons in the diagram. Feynman pointed out that a particle traveling around a loop will undergo two time reversals before returning to its starting point, as indicated in figure 2.14. Therefore, a fermion loop diagram has the opposite sign relative to a boson loop, corresponding to $TT = -1$ for fermions and $TT = +1$ for bosons. The Feynman rule for loops thereby establishes the link between relativistic quantum theory and spin statistics. It should be noted that the spin-statistics connection is more explicit in the multiparticle techniques of quantum field theory where the field operators directly

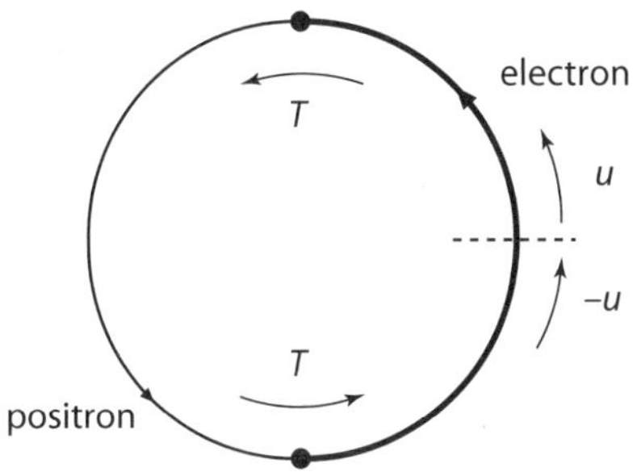

FIGURE 2.14. Particle-antiparticle loop with the two successive time reversals shown.

incorporate the corresponding commutation and anticommutation relations of bosons and fermions. A selection of QFT references can be found in section 2.16.

2.15 Exercises

1. Calculate $d\hat{x}/dt$ separately for the nonrelativistic Hamiltonian $\hat{E}=\hat{\mathbf{p}}^2/2m$ and for the Dirac Hamiltonian $\hat{E}=\boldsymbol{\alpha}\cdot\hat{\mathbf{p}}+\beta m$.

2. Show that $\boldsymbol{\pi}\times\boldsymbol{\pi}=(\hat{\mathbf{p}}-e\mathbf{A})\times(\hat{\mathbf{p}}-e\mathbf{A})=ie\mathbf{B}$.

3. Show that the Dirac equation with the minimal electromagnetic coupling substitution predicts that the mass in the Bohr magneton becomes relativistically heavy for non-zero electron momentum.

4. Show that $d\hat{\mathbf{L}}/dt=\boldsymbol{\alpha}\times\hat{\mathbf{p}}$ and that $\frac{1}{2}d\boldsymbol{\Sigma}/dt=-\boldsymbol{\alpha}\times\hat{\mathbf{p}}$ for a free Dirac particle.

5. Show that $d\boldsymbol{\alpha}/dt=-2(\boldsymbol{\Sigma}\times\hat{\mathbf{p}})-i2m\boldsymbol{\alpha}\beta$.

6. Show that $d\boldsymbol{\Sigma}\cdot\boldsymbol{\pi}/dt=0$ for a static $\mathbf{B}$ field and a vanishing $\mathbf{E}=0$ electric field.

7. Find a unitary transformation U in terms of $\boldsymbol{\alpha}$ and β matrices that interchanges α^2 and β in the Dirac equation. Show that this transform yields the correct exchange when applied to the Dirac equation for $\Psi'=U\Psi$.

8. Show that $v^{(2)}(p)=(i\beta\alpha^2)[u^{(2)}(p)]^*$.

9. **Off-axis neutrino beam.** A typical high-energy neutrino beam is made from the decay of pions that have been produced in proton interactions on a target. Assume that downstream of the target, a "horn" collects a pure π^+ beam into a decay pipe. The main source of neutrinos comes from the decay $\pi^+\to\mu^+\nu_\mu$. Facts: $m_{\pi^\pm}=139.6$ MeV/c^2, $m_\mu=105.7$ MeV/c^2. In this problem, neutrinos can be taken as massless. The spin and parity quantum numbers for pions are $J^P=0^-$.

   (a) If the decay pions have $E_\pi\gg m_\pi$, what is the characteristic angle θ_C of decay neutrinos with respect to the direction of the π^+?

   (b) Compute the energy of the neutrino E_ν^* in the pion rest frame and derive the relationship between $\cos\theta$, the neutrino decay angle relative to the pion direction of flight, and $\cos\theta^*$, the decay angle in the rest frame of the pion. Note that $\cos\theta^*$ can be written in terms of $\lambda\equiv E_\nu/E_\nu^*$ and $\cos\theta$.

   (c) Show that when observing the neutrino beam energy spectrum at a fixed lab angle θ with respect to the pion beam direction, there is an upper limit on the value of E_ν independent of E_π.

   (d) Start with an isotropic angular distribution of neutrino decay angles in the pion rest frame and boost into the lab frame, using the axis $\theta=0$ as the pion flight direction. There is a large off-axis enhancement in the observed neutrino flux in the lab frame. Assume that in the region of phase space corresponding to the enhancement, λ is a constant independent of the energy of the pion that produced it. At what lab angle θ_K is there an enhanced rate?

(f) What is the minimum π^+ energy E_π needed to produce an off-axis enhanced neutrino beam of $E_\nu = 1.5$ GeV? The MINOS detector is located in the Soudan mine 730 km from the decay pipe at an angle $\theta = 0$ with respect to the π^+ meson beam. How far away from the Soudan mine normal to the beam direction should a neutrino detector be built to observe the off-axis peak of $E_\nu = 1.5$ GeV neutrinos?

10. Show with an explicit example that the operators that generate Lorentz boosts, S^{0i}, are not Hermitian ($S^{0i} \neq (S^{0i})^\dagger$).

11. Show that the matrix representation of the Lorentz group proposed by Dirac, $S^{\mu\nu} = \frac{i}{4}[\gamma^\mu, \gamma^\nu]$, satisfies the commutation relations (2.94)

$$[S^{\mu\nu}, S^{\rho\sigma}] = i(g^{\nu\rho}S^{\mu\sigma} - g^{\mu\rho}S^{\nu\sigma} - g^{\nu\sigma}S^{\mu\rho} + g^{\mu\sigma}S^{\nu\rho}).$$

12. Show that the Dirac spin operator can be written $\frac{1}{2}\boldsymbol{\Sigma} = -\frac{i}{4}\boldsymbol{\alpha} \times \boldsymbol{\alpha}$ and that the components of the antisymmetric tensor S^{0j} can be written in terms of the $\boldsymbol{\alpha}$ matrices. Using these results, write the finite Lorentz transformation for a Dirac spinor (2.106) in terms of a three-vector of boosts $\vec{\eta}$, a three-vector of rotations $\vec{\theta}$, and the $\boldsymbol{\alpha}$ matrices.

13. Show that $\exp\left(\gamma^0\gamma^j\eta_j/2\right) = I_4\cosh(\eta_j/2) + (\gamma^0\gamma^j)\sinh(\eta_j/2)$.

14. Show that $\left[\exp\left(\gamma^0\gamma^j\eta_j/2\right)\right]^{-1}\gamma^0\exp\left(\gamma^0\gamma^j\eta_j/2\right) = \gamma^0\cosh\eta_j + \gamma^j\sinh\eta_j$.

15. Show that $\bar{\Psi} \to \bar{\Psi}\Lambda_{\frac{1}{2}}^{-1}$.

16. Consider the quantity $T^{\mu\nu}$, the Dirac stress-energy tensor, defined as (neglecting possible *improvement terms*)

$$T^{\mu\nu} = i\bar{\Psi}\gamma^\mu\partial^\nu\Psi - g^{\mu\nu}(i\bar{\Psi}\gamma^\sigma\partial_\sigma\Psi - m\bar{\Psi}\Psi) \tag{2.276}$$

for a Dirac wave function Ψ. What kinematic quantity of Ψ is equal to $T^{0\nu}$? If $T^{0\nu}$ is a conserved quantity, what is the Lorentz-invariant continuity equation for the corresponding conserved current?

17. Show that in the Weyl representation $\bar{\Psi}(i\gamma^\mu\partial_\mu - m)\Psi = \xi^\dagger i\bar{\sigma}^\mu\partial_\mu\xi + \eta^\dagger i\sigma^\mu\partial_\mu\eta - m(\xi^\dagger\eta + \eta^\dagger\xi)$ where ξ and η are left- and right-handed chirality two-component spinors.

18. Given an explicit form for the charge-conjugation operator C in the Weyl representation, compute $C[u^{(1)}(p)]^*$ and $C[u^{(2)}(p)]^*$ explicitly using the matrix representation for C and the column-vector Weyl solutions for $u^{(1)}(p)$ and $u^{(2)}(p)$.

19. High-energy photon beams. High-energy photons can be produced through the inverse Compton scattering of laser light off a high-energy electron beam. These are the parameters for the laser pulse:

- Energy of pulse: $E_{\text{pulse}} = 1$ J
- Wavelength of light: $\lambda = 1\ \mu$m

The parameters of the electron beam are:

- Energy of electrons: $E_0 = 80$ GeV
- Total charge of electron bunch: $Q = 1$ nC
- Cross-sectional area of bunch: $A = 1\ \mu\text{m}^2$

(a) Give an order-of-magnitude estimate of the number of inverse Compton scatters that will occur if the entire laser pulse passes through the electron bunch. Use $\alpha_{\text{QED}} = 1/137$, $m = 0.511$ MeV/c^2, and $\hbar c = 0.2$ GeV fm.

(b) The laser is pulsed at the electron beam with an angle of incidence of 90 degrees with respect to the electron beam direction. Compute the energy of the photons scattered in the direction parallel to the electron beam. *Hint:* The parallel scattered photon energy is substantially lower than the incoming electron energy.)

(c) An upgrade increases the electron beam energy to $E_0 = 800$ GeV while keeping all other parameters fixed. In these operating conditions, it is found that very few high-energy photons arrive at the interaction region. What reaction is occurring with the laser beam to prevent photons from reaching the interaction region and what is the minimum electron beam energy E_{min} at which this process will begin to occur?

20. Electron-positron pair annihilation.

(a) Draw the two tree-level Feynman diagrams for pair annihilation into two photons, $e^+e^- \to \gamma\gamma$.

(b) Compute the total differential cross section for the process $e^+e^- \to \gamma\gamma$ in the center-of-mass system in the limit that the electron mass is negligible compared to its energy. Assume unpolarized electron-positron beams and sum over all possible photon polarizations in the final state.

21. Coulomb approximation.

(a) Solve the Dirac equation and obtain a spinor for a free electron of positive energy E and momentum $\boldsymbol{p}$. The spinor equation is

$$(\not{p} - m)u = 0$$

where one choice of matrices γ^μ in terms of the 2×2 matrices I_2 (identity matrix) and $\boldsymbol{\sigma}$ (Pauli spin matrices) is

$$\gamma^0 = \begin{pmatrix} I_2 & 0 \\ 0 & -I_2 \end{pmatrix}, \quad \boldsymbol{\gamma} = \begin{pmatrix} 0 & \boldsymbol{\sigma} \\ -\boldsymbol{\sigma} & 0 \end{pmatrix}.$$

(b) Consider scattering of unpolarized electrons with a massive pointlike proton. The interaction Lagrangian density of electrons with the electromagnetic field is

$$\mathcal{L}_{\text{int}} = -e\bar{\Psi}\gamma^{\mu}\Psi A_{\mu}.$$

Compute the full $\overline{|\mathcal{M}|^2}$ for electron-proton scattering and then show that when the energy of the electron E is small compared to the proton mass M, then we recover the scattering result from the static Coulomb potential. The Coulomb potential is given by

$$A_0 = \frac{-Ze}{4\pi|\mathbf{x}|}, \qquad \mathbf{A} = 0.$$

(c) Derive the dependence of the differential scattering cross section in terms of the electron deflection angle θ in the limit of Coulomb scattering.

22. Differential cross section for $e^+e^- \to \mu^+\mu^-$.

(a) Draw the lowest-order Feynman diagram for

$$e^+(p^+, s^+)e^-(p^-, s^-) \to \mu^+(k^+, r^+)\mu^-(k^-, r^-)$$

and write down the expression for the amplitude $\mathcal{M}$. The symbols $p^{\pm}$ and $k^{\pm}$ refer to the momenta of the particles and $s^{\pm}$ and $r^{\pm}$ to the spins.

(b) The expression you found in part (a) depends on the spin of the particles. Compute $\overline{|\mathcal{M}|^2}$ when the particles are unpolarized. The electron mass, but not the muon mass, can be neglected compared to the center-of-mass energy E. Write the answer in terms of center-of-mass scattering angle θ and E.

(c) Use the result of part (b) to calculate the center-of-mass differential cross section as a function of energy, scattering angle, and m_{μ}.

(d) In the high-energy limit where the mass of the electrons and muons can be neglected, show what combination of initial- and final-state electron and muon helicities are active in the amplitude.

23. It is well known that one can "bend" the external legs of a Feynman diagram in spacetime to convert a Compton scattering process to a two-photon annihilation process. Write down the matrix element for both processes and determine the substitution rules that are needed to go from one matrix element to the other. What are the corresponding rules for the squared spin-summed matrix element?

24. Prove the diagrammatic identity 2.13(c).

25. Catastrophic muon energy loss. Figure 2.15 shows that the bremsstrahlung energy loss per unit length of an electron passing through copper becomes comparable to ionization energy loss at electron energies of $E_e \approx 20$ MeV. Here, bremsstrahlung refers to the process of an electron scattering off a Coulomb potential and radiating a

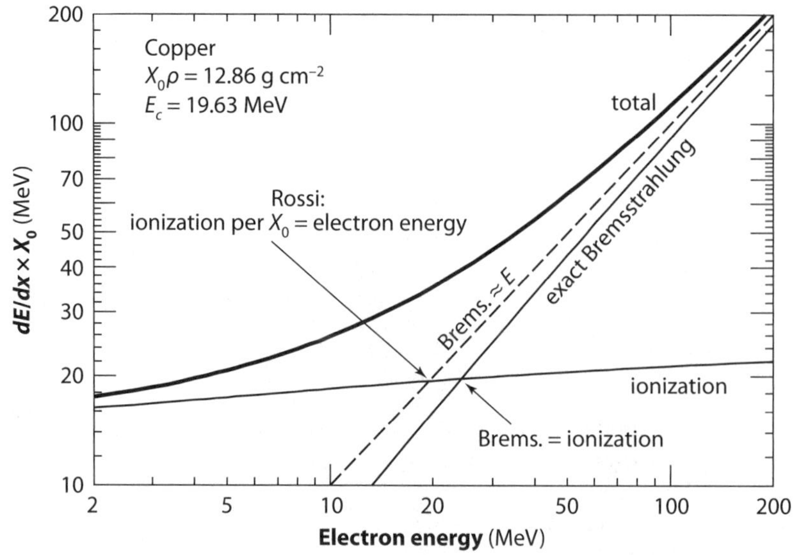

FIGURE 2.15. Energy dependence of bremsstrahlung energy loss compared with ionization for electrons (Credit: PDG) [3].

real photon either before or after the Coulomb scatter. A muon will also suffer bremsstrahlung energy loss through the same scattering diagram. Estimate the muon energy at which the energy loss from bremsstrahulung will begin to dominate the total energy loss of a muon passing through matter. Use $m_e = 0.511$ MeV/c^2 for the electron mass and $m_\mu = 106$ MeV/c^2 for the muon mass.

26. Two-photon collider. Two photon beams of equal energy are brought into collision. The energy of the beams is adjustable and therefore it is possible to scan the center-of-mass energy of the collisions. The reference process at the two-photon collider is dimuon production $\gamma\gamma \to \mu^+\mu^-$. The rate of hadron production $\gamma\gamma \to q\bar{q}$ is normalized to the dimuon rate. The muon mass is $m_\mu = 106$ MeV and the bare quark masses are approximately $m_u = 3$ MeV, $m_d = 5$ MeV, $m_s = 0.3$ GeV, $m_c = 1.3$ GeV, $m_b = 4.4$ GeV. The masses of the pions are $m_{\pi^0} = 135$ MeV and $m_{\pi^\pm} = 140$ MeV.

(a) What are the Feynman diagrams for the processes $\gamma\gamma \to \mu^+\mu^-$ and $\gamma\gamma \to q\bar{q}$?

(b) The relative rate of hadron production is measured relative to dimuon production for the scattering angle $\theta = \pi/2$ measured relative to one of the incoming photon beam directions. We define the ratio $R_{\gamma\gamma}^{\theta=\pi/2}$ to be

$$R_{\gamma\gamma}^{\theta=\pi/2} = \frac{\sum_i \frac{d\sigma}{d\Omega}(\theta = \pi/2)(\gamma\gamma \to q_i\bar{q}_i)}{\frac{d\sigma}{d\Omega}(\theta = \pi/2)(\gamma\gamma \to \mu^+\mu^-)}. \tag{2.277}$$

Compute the ratio $R_{\gamma\gamma}^{\theta=\pi/2}$ of the $\theta = \pi/2$ scattering cross section for the process $\gamma\gamma \to q\bar{q}$ to the process $\gamma\gamma \to \mu^+\mu^-$ for all kinematically accessible quark flavors at $\sqrt{s} = 60$ GeV. Neglect the quark and lepton masses and treat quark-antiquark production as a leading-order diagram in perturbative QCD.

(c) Now assume the b-quark is massive while keeping the other quarks and leptons massless. Neglect the b-quark mass in the matrix element, but include it in the phase space calculation. Compute the ratio $R_{\gamma\gamma}^{\theta=\pi/2}$ at $\sqrt{s} = 4m_b$.

(d) Sketch $R_{\gamma\gamma}^{\theta=\pi/2}$ as a function center-of-mass energy in the range $\sqrt{s} = 2m_\mu$–20 GeV and show the general behavior expected as the center-of-mass energy crosses the quark-antiquark production thresholds.

Try to estimate the shape based on nonzero quark and lepton masses and the need to produce hadrons in the final state. Indicate in the plot where you expect to see resonances and indicate what hadron is being produced and the spin and quark content of the resonance. There is no need to write down explicit hadron wave functions, only the valence quark content and spin.

27. **Polarized $\mu^+\mu^-$ linear collider.** You decide to construct a collider to collide left-handed helicity μ^+ with left-handed helicity μ^- in the center of mass. With this helicity configuration, the sum of the components of the μ^+ and μ^- spins along the collision axis is zero. The source of muons that you have is a single high-momentum source of charged pions. The charged pion is a spin-0 particle and decays through the weak charged current interaction to a muon and a muon neutrino or, specifically, $\pi^+ \to \mu^+ + \nu_\mu$ and $\pi^- \to \mu^- + \bar{\nu}_\mu$ where the bar indicates an antineutrino. Recall that the weak interaction only couples to left-handed neutrinos and right-handed antineutrinos.

(a) Consider the charged pion two-body decay in the rest frame of the pion. What is the polarization of the μ^+ from the decay and what is the polarization of the μ^-?

(b) In the lab frame, there is some loss of polarization, but it's possible to select out the most energetic of the muons to select dominantly the (pion) rest-frame polarizations of the μ^+ and μ^-. You decide that you need to flip the helicity of the μ^- before colliding the beams. The way you do this is to rotate the helicity of the μ^- by bending the μ^- in a half-circular orbit with a constant magnetic field. The angular frequency at which the polarization vector of the muon rotates with respect to the momentum direction of the muon is given in the lab frame by

$$\omega_a = a_\mu \frac{|e|B}{m_\mu c}$$

where $a_\mu \approx \alpha_{\text{QED}}/(2\pi)$ is the anomalous magnetic moment of the muon. Note the magnetic field B needed to bend a particle of momentum p is given by $B = (pc)/(|e|R)$. If needed, it is convenient to use units with B in tesla, p in GeV, c is unity, $|e| = 0.3$, R in meters, and $m_\mu = 0.106$ GeV. What is the momentum of muon p in GeV that will simultaneously bend the μ^- beam by 180° and reverse the direction of the μ^- helicity? See figure 2.16 for a schematic of the muon linear collider.

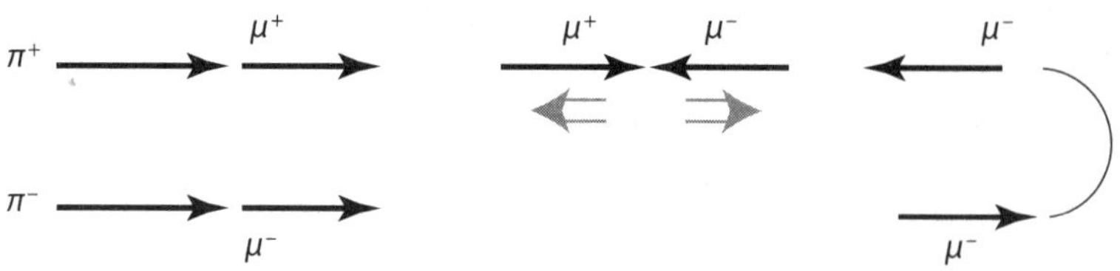

FIGURE 2.16. Schematic of the polarized $\mu^+\mu^-$ linear collider with open arrows indicating the spin polarizations of the colliding muons.

(c) Assume that that the collider can scan a wide range of center-of-mass energies with this helicity configuration. Consider only the high-energy limit, where the mass of the muon can be ignored when projecting out helicity states. What are possible final states from polarized $\mu^+\mu^-$ annihilation within the Standard Model? No calculations needed other than to check that the incoming helicities will yield nonzero matrix elements.

28. Heavy stable charged particles. Suppose a high-energy collider were able to produce a heavy stable charged spin-1/2 particle, denoted χ^+, with mass M and charge $+1$, at a nonrelativistic velocity β. If the χ^+ tranverses matter, it can scatter with atomic electrons producing what are known as δ-rays, energetic scattered electrons.

(a) Draw the Feynman diagram for lowest-order QED scattering in the process $\chi^+(p_1)e^-(p_2) \to \chi^+(p_3)e^-(p_4)$.

(b) Write down the invariant amplitude $-i\mathcal{M}$ corresponding to the scattering diagram in part (a). Assume the electrons are free particles.

(c) Compute the spin-averaged squared amplitude $\overline{|\mathcal{M}|^2}$ and write an expression in terms of the four-momenta p_1, p_2, p_3, p_4, and the parameters α_{QED}, M, and m_e, the electron mass.

(d) In the approximation that the electron is a free particle at rest, and that $m_e/M \ll 1$, and (as stated in the problem) that χ^+ has a nonrelativistic velocity $\beta = v/c$, compute the maximum total energy of a scattered electron up to and including order β^2 for a head-on collision, i.e., $\cos\theta = 1$.

(e) Compute the ratio of the differential cross section for forward scattering of electrons for two values of β for the χ^+, denote the two values β_1 and β_2. In this part, use the approximation for the scattered electron energy from part (d). Use a collinear scattering approximation where the χ^+ has no angular deflection from the scatter in the lab frame (continues forward) and the electron originally at rest scatters forward, collinear with the χ^+.

29. Muon Bremsstrahlung to an off-shell photon. A muon with three-momentum $\boldsymbol{p}$ is traveling past a very massive (assume infinitely heavy) pointlike charged nucleus and scatters off the Coulomb potential into an off-shell muon propagator that

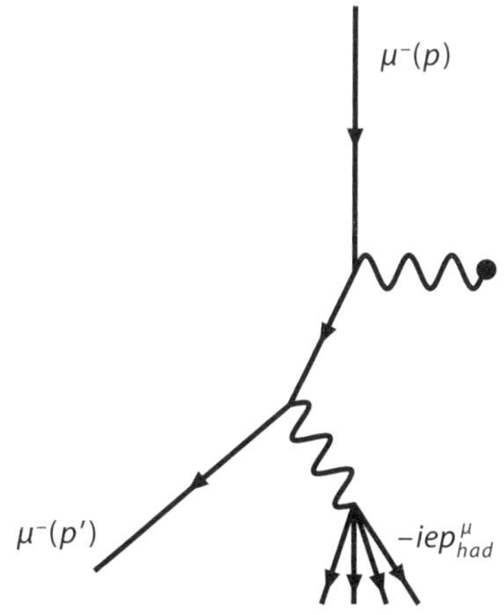

FIGURE 2.17. Inelastic scattering of a muon off of a Coulomb potential.

subsequently decays to an on-shell muon and an off-shell photon, as shown in figure 2.17. The on-shell muon has a three-momentum $\boldsymbol{p}'$. The off-shell photon decays to hadrons whose total invariant mass is given by m_{had} and whose total three-momentum is $\boldsymbol{p}_{\text{had}}$. Assume the vertex term of the off-shell photon coupling to hadrons is given by $-iep^{\mu}_{\text{had}}$ and that the hadrons are spin-0 and therefore contribute a factor of 1 to the invariant amplitude. The nucleus has a charge $-Ze$, where $e < 0$ is the elementary muon charge. Set the muon mass to zero in this problem.

(a) Write down the invariant amplitude $-i\mathcal{M}$ for this process. Assume there is only one diagram.

(b) Compute the squared invariant amplitude $\overline{|\mathcal{M}|^2}$ for unpolarized muons, summing over the two spin states of the scattered muon. Assume there is only one diagram.

(c) What is the minimum three-momentum transfer to the nucleus in this process, given the finite mass m_{had} of the off-shell photon?

30. High-energy cosmic rays. A 100 TeV/c energetic proton impinges on the upper atmosphere of Earth colliding with a proton in a hydrogen molecule. Treat the hydrogen nucleus as being at rest in the lab frame. A high-energy proton-proton circular accelerator is colliding two proton beams with proton momenta 7 TeV/c and -7 TeV/c such that the center of mass of the colliding protons is at rest in the lab frame. Compare the center-of-mass energy of the proton-proton collider with that of the high-energy cosmic ray described above.

2.16 References and Further Reading

A selection of texts on quantum electrodynamics can be found here: [2, 4, 5, 6, 7, 8, 9, 10, 11, 12].

A selection of introductory particle physics texts can be found here: [13, 14].

A selection of references for quantum field theory can be found here: [1, 15, 16, 17, 18, 19].

[1] Michael E. Peskin and Daniel V. Schroeder. *An Introduction to Quantum Field Theory*. Westview Press, 1995. ISBN 0-201-50397-2.

[2] Julian Schwinger, editor. *Selected Papers on Quantum Electrodynamics*. Dover Publications, 1958. ISBN 0-486-60444-6.

[3] K. Nakamura et al. (Particle Data Group). *J. Phys. G* **37**, (2010). http://pdg.lbl.gov.

[4] Raymond F. Streater and Arthur S. Wightman. *PCT, Spin and Statistics, and All That*. Princeton University Press, 1989. ISBN 0-691-07062-8.

[5] Richard P. Feynman and Steven Weinberg. *Elementary Particles and the Laws of Physics*. Cambridge University Press, 1987. ISBN 0-521-65862-4.

[6] Florian Scheck. *Quantum Physics*. Springer, 2007. ISBN 978-3-540-25645-8.

[7] Otto Nachtmann. *Elementary Particle Physics*. Springer-Verlag, 1990. ISBN 0-387-51647-6.

[8] V. B. Berestetskii, E. M. Lifshitz, and L. P. Pitaevskii. *Quantum Electrodynamics*. Pergamon Press, 1980. ISBN 0-08-026504-9.

[9] J. J. Sakurai. *Advanced Quantum Mechanics*. Addison-Wesley, 1967. ISBN 0-201-06710-2.

[10] James Bjorken and Sidney Drell. *Relativistic Quantum Mechanics*. McGraw-Hill, 1964. ISBN 07-005493-2.

[11] Richard P. Feynman. *Quantum Electrodynamics*. Westview Press, 1961. ISBN 0-201-36075-6.

[12] Hermann Weyl. *The Theory of Groups and Quantum Mechanics*. Dover Publications, 1950. ISBN 0-486-60269-9.

[13] David Griffiths. *Introduction to Elementary Particles*. Wiley-VCH, 2008. ISBN 978-3-527-40601-2.

[14] Gordon Kane. *Modern Elementary Particle Physics*. Westview Press, 1993. ISBN 0-201-62460-5.

[15] A. Zee. *Quantum Field Theory in a Nutshell*. Princeton University Press, 2010. ISBN 978-0-691-14034-6.

[16] Mark Srednicki. *Quantum Field Theory*. Cambridge University Press, 2007. ISBN 978-0-521-86449-7.

[17] Steven Weinberg. *The Quantum Theory of Fields*. Cambridge University Press, 2000. Vol. 1, ISBN 0-521-55001-7; vol. 2, ISBN 0-521-55002-5; vol. 3, ISBN 0-521-66000-9.

[18] J. Zinn-Justin. *Quantum Field Theory and Critical Phenomena*. Oxford University Press, 1993. ISBN 0-19-852053-0.

[19] Claude Itzykson and Jean-Bernard Zuber. *Quantum Field Theory*. McGraw-Hill, 1980. ISBN 0-07-032071-3.

3 Gauge Principle

Gauge theories originate from the existence of degrees of freedom in the description of elementary particle states that are indeterminate and have no effect on the predicted outcomes of any experiment. An example from classical physics is the four-vector potential A^μ of the electromagnetic interaction. From Maxwell's equations, the electric and magnetic fields are the physical origin of the electromagnetic interaction. In terms of A^μ, the electromagnetic fields are given by the elements of the tensor $F_{\mu\nu}(x) = \partial_\mu A_\nu(x) - \partial_\nu A_\mu(x)$. The values of the components of A^μ are arbitrary up to the addition of a real, differentiable function χ such that $A'^\mu \to A^\mu - \partial^\mu \chi$ gives an equivalent tensor $F'_{\mu\nu} = F_{\mu\nu}$, and, therefore, the addition of $\partial^\mu \chi$ has no effect on Maxwell's equations. The specification of definite values for A^μ is called *gauging*. If, on the other hand, the electron wave function changes its phase such that it varies at each point in space, then the relative phase changes would normally lead to a physical observable. However, a locally varying phase of the electron wave function can be made indeterminate (no physically observed consequence) if this local phase change is compensated in the equation of motion by a corresponding change in A^μ of the type described above. This is an example of a gauge interaction known previously as the minimal coupling substitution for incorporating electromagnetism. In gauge theories, the gauge freedom is extended to be a dynamical principle. Namely, the *gauge principle* states that the existence and form of an interaction may be deduced from the existence of physically indeterminate, gaugable quantities. This is explained in more detail below.

3.1 Global Internal Symmetries

Global internal symmetries may be divided into *discrete* and *continuous* symmetries. Parity conservation in QED is an example of a global discrete symmetry. An example of a continuous global symmetry is a constant phase transformation of a wave function

$$\Psi'(x) = \exp(i\phi)\,\Psi(x) \tag{3.1}$$

where global means specifically that ϕ does not depend on space and time. The invariance of the equation of motion due to this transformation $U = \exp(i\phi)$ is a consequence of the commutation with the Hamiltonian $[U,H] = 0$. This invariance property is associated with a conserved quantity, the norm of the wave function

$$\partial_0 \left[\int \sum_{r=1}^{4} \Psi_r^*(x) \Psi_r(x) d^3x \right] = 0. \tag{3.2}$$

Noether's theorem states that for every global transformation under which the Lagrangian density is invariant there exists a conserved quantity. In classical mechanics, the conservation of linear momentum, angular momentum, and energy follows from translational invariance, rotational invariance, and invariance under translations in time, respectively. These quantities are external invariances related to space-time properties. Internal invariances associated with particle properties can also give rise to conserved quantities.

Noether's theorem. If $\mathcal{L}(\Psi(x), \partial_\mu \Psi(x))$ is invariant under the transformation of the wave function $\Psi(x) \to \Psi'(x)$, where

$$\Psi'(x) = \Psi(x) + \delta\Psi(x), \tag{3.3}$$

then there exists a conserved current

$$\partial_\mu \left(\frac{\partial \mathcal{L}(x)}{\partial(\partial_\mu \Psi(x))} \delta\Psi(x) \right) = 0 \tag{3.4}$$

where $\delta\Psi$ denotes the variation of the wave function Ψ.

The Noether result can be shown to be an extremum property of the Lagrangian. If the Lagrangian density is invariant under the transformation (3.3), then the variation $\delta\mathcal{L}$ is equal to zero, where $\delta\mathcal{L}(x)$ is given by

$$\delta\mathcal{L}(x) = \frac{\partial \mathcal{L}(x)}{\partial \Psi(x)} \delta\Psi(x) + \frac{\partial \mathcal{L}(x)}{\partial(\partial_\mu \Psi(x))} \delta(\partial_\mu \Psi(x)). \tag{3.5}$$

In the second term, we have $\delta(\partial_\mu \Psi(x))$ where

$$\delta(\partial_\mu \Psi(x)) = \partial_\mu \Psi'(x) - \partial_\mu \Psi(x) = \partial_\mu (\Psi'(x) - \Psi(x)) = \partial_\mu (\delta\Psi). \tag{3.6}$$

The first term of equation (3.5) can be rewritten using the Euler-Lagrange equations (2.118), repeated here:

$$\partial_\mu \left(\frac{\partial \mathcal{L}}{\partial(\partial_\mu \Psi)} \right) - \frac{\partial \mathcal{L}}{\partial \Psi} = 0.$$

Substituting equations (2.118) and (3.6) into (3.5) gives

$$\begin{aligned} \delta\mathcal{L}(x) &= \left(\partial_\mu \frac{\partial \mathcal{L}(x)}{\partial(\partial_\mu \Psi(x))} \right) \delta\Psi(x) + \frac{\partial \mathcal{L}(x)}{\partial(\partial_\mu \Psi(x))} \partial_\mu (\delta\Psi(x)) \\ &= \partial_\mu \left(\frac{\partial \mathcal{L}(x)}{\partial(\partial_\mu \Psi(x))} \delta\Psi(x) \right) = 0, \end{aligned} \tag{3.7}$$

reproducing the Noether result (3.4).

For a phase transformation given by

$$\Psi'(x) = \exp(ie\chi)\,\Psi(x) \tag{3.8}$$

for small χ we have

$$\Psi'(x) = \Psi(x) + ie\chi\Psi(x) + \cdots \tag{3.9}$$

identifying

$$\delta\Psi = ie\chi\Psi. \tag{3.10}$$

Applying Noether's theorem (3.4), we have

$$\partial_\mu\left(\frac{\partial\mathcal{L}}{\partial(\partial_\mu\Psi)}(ie\chi\Psi)\right) = 0, \tag{3.11}$$

which for $\chi = constant$ and the Dirac Lagrangian density $\mathcal{L}_{\text{Dirac}} = \bar{\Psi}(i\gamma^\mu\partial_\mu - m)\,\Psi$ gives

$$\partial_\mu(\bar{\Psi}i\gamma^\mu(ie\Psi)) = 0. \tag{3.12}$$

This is the simply the Dirac electric current continuity equation $\partial_\mu j^\mu = 0$ with $j^\mu = e\bar{\Psi}\gamma^\mu\Psi$. Charge conservation is a result of Green's theorem applied to the continuity equation, namely,

$$\begin{aligned}\partial_0 Q &= \int \partial_0 j^0(x)d^3x = -\int \nabla\cdot\boldsymbol{j}(x)d^3x \\ &= -\int \boldsymbol{j}(x)\cdot d\boldsymbol{\Omega} = 0\end{aligned} \tag{3.13}$$

since the surface integral is taken to be zero at infinity.

3.2 Local Gauge Symmetries

Now we go one step beyond having a conserved internal property, such as a charge, to having an interaction that couples to this charge. This is done by allowing the phase of equation (3.8) to be a function of space and time $\chi(x)$ such that

$$\Psi'(x) = \exp(ie\chi(x))\,\Psi(x). \tag{3.14}$$

It is easy to see that the Dirac equation for free particles is *not* invariant under (3.14). If $\Psi(x)$ is a solution to the Dirac equation

$$(i\gamma^\mu\partial_\mu - m)\,\Psi(x) = 0, \tag{3.15}$$

then we have

$$\begin{aligned}(i\gamma^\mu\partial_\mu - m)\,\Psi' &= (i\gamma^\mu\partial_\mu - m)\exp(ie\chi(x))\Psi(x) \\ &= \exp(ie\chi(x))\left[(i\gamma^\mu\partial_\mu - m)\Psi(x) - e(\partial_\mu\chi(x))\,\gamma^\mu\Psi(x)\right] \\ &= -e(\partial_\mu\chi(x))\,\gamma^\mu\Psi'(x) \neq 0.\end{aligned} \tag{3.16}$$

Thus, $\Psi'(x)$ is not a solution of the free-particle Dirac equation. This result comes from a fundamental principle of quantum mechanics whereby the absolute phase of a wave function cannot be measured, but phase differences can.

Equation (3.6) tells us that the kinetic energy term $\partial_\mu \Psi(x)$ does not transform in the same way as $\Psi(x)$ under the local gauge transformation (3.14). To achieve covariance of the derivative term under (3.14) we define a new operation known as the *covariant derivative*, a concept from the theory of general relativity.

A derivative is a comparison of a vector field $V(x)$, or in this case a Dirac wave function, at two displaced space-time points. In translating from x to $x+\Delta x$, any contribution to the derivative that comes from a curvilinear coordinate system is removed by defining a covariant derivative $\mathcal{D}_\mu$ such that $V_{||}(x+\Delta x)$ is a parallel transport back to x, with

$$\mathcal{D}_\mu \equiv \lim_{\Delta x^\mu \to 0} \frac{V_{||}(x+\Delta x) - V(x)}{\Delta x^\mu}, \tag{3.17}$$

as shown in figure 3.1. The parallel transport of $V(x+\Delta x) \to V_{||}(x+\Delta x)$ at x is achieved with a unitary operator $U(x, x+\Delta x)$ such that

$$V_{||}(x+\Delta x) = U(x, x+\Delta x)\, V(x+\Delta x). \tag{3.18}$$

The local expansion of $U(x, x+\Delta x)$ for small Δx^μ is given by

$$U(x, x+\Delta x) = 1 + ie\Delta x^\mu A_\mu(x) + \mathcal{O}(\Delta x^2) \tag{3.19}$$

where $A_\mu(x)$ is a four-vector potential. The operator $U(x, x+\Delta x)$ transforms under the local gauge transformation (3.14) according to

$$U(x, x+\Delta x) \to \exp(ie\chi(x))\, U(x, x+\Delta x) \exp(-ie\chi(x+\Delta x)) \tag{3.20}$$

and therefore connects the local gauge transformations at displaced space-time points. Furthermore, applying equation (3.20) to the local expansion of $U(x, x+\Delta x)$ in equation (3.19) yields the transformation law for $A_\mu(x)$, that is,

$$A_\mu(x) \to A_\mu(x) - \partial_\mu \chi(x) \tag{3.21}$$

in the limit that $\Delta x^\mu \to 0$.

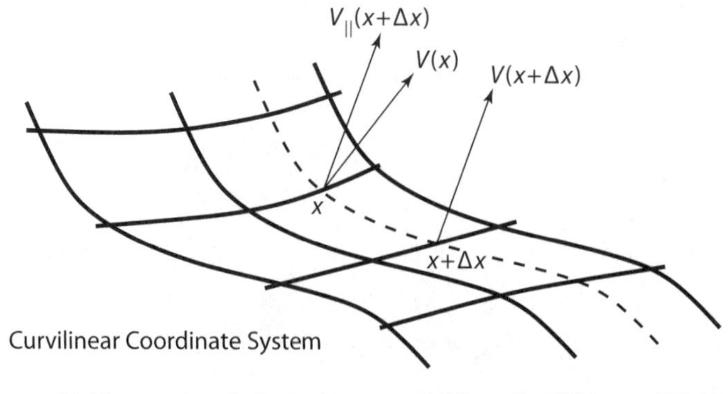

FIGURE 3.1. The covariant derivative is constructed from the difference of $V(x)$ and the parallel transport of $V(x+\Delta x)$ to x, denoted $V_{||}(x+\Delta x)$.

The gauge principle of particle interactions is that gauge degrees of freedom of four-vector potentials are related to the indeterminate space-time varying phases of particle wave functions. To retain invariance under (3.14), we must simultaneously transform the four-vector potentials so as to cancel the terms given in equation (3.16). Namely, $\Psi'(x)$ is a solution to the modified Dirac equation

$$i\gamma^{\mu}(\partial_{\mu} - ie\partial_{\mu}\chi(x))\Psi'(x) = m\Psi'. \tag{3.22}$$

To form a correspondence between the transformation (3.14) and the gauge degrees of freedom of the four-vector potential, we use a covariant derivative $\mathcal{D}_{\mu}$ according to equations (3.17), (3.18), and (3.20)

$$D_{\mu} = \partial_{\mu} + ieA_{\mu}. \tag{3.23}$$

Therefore, the simultaneous gauge transformation $A'_{\mu}(x) = A_{\mu}(x) - \partial_{\mu}\chi(x)$ and wave function transformation $\Psi'(x) = \exp(ie\chi(x))\Psi(x)$ leave the covariant-derivative substituted form of the Dirac equation invariant

$$(i\gamma^{\mu}\mathcal{D}_{\mu} - m)\Psi(x) = 0. \tag{3.24}$$

One can better understand the deep connection between the covariant derivative and the manifestation of a physical interaction by computing the commutator of two covariant derivatives $[D^{\mu}, D^{\nu}]$

$$\begin{aligned}
[D^{\mu}, D^{\nu}]\Psi &= (\partial^{\mu} + ieA^{\mu})(\partial^{\nu}\Psi + ieA^{\nu}\Psi) \\
&\quad -(\partial^{\nu} + ieA^{\nu})(\partial^{\mu}\Psi + ieA^{\mu}\Psi) \\
&= \partial^{\mu}\partial^{\nu}\Psi + ie(\partial^{\mu}A^{\nu})\Psi + ieA^{\nu}\partial^{\mu}\Psi + ieA^{\mu}\partial^{\nu}\Psi - e^2A^{\mu}A^{\nu}\Psi \\
&\quad -\partial^{\mu}\partial^{\nu}\Psi - ie(\partial^{\nu}A^{\mu})\Psi - ieA^{\mu}\partial^{\nu}\Psi - ieA^{\nu}\partial^{\mu}\Psi + e^2A^{\nu}A^{\mu}\Psi \\
&= ie(\partial^{\mu}A^{\nu} - \partial^{\nu}A^{\mu})\Psi = ieF^{\mu\nu}\Psi
\end{aligned} \tag{3.25}$$

where $F^{\mu\nu}$ is the electromagnetic field tensor. In general, the commutator of covariant derivatives will generate the corresponding field tensor of an interaction.

One of the remarkable parallels of local gauge invariance is the Aharonov-Bohm effect. Aharonov and Bohm predicted that a long, thin solenoid placed between the two slits of a single-electron interference experiment would induce a phase shift of the resulting interference pattern even though the electrons never enter the magnetic field of the solenoid. The presence of the solenoid forces a difference in the electron phase for trajectories that encircle the solenoid, which is equivalent to piecing together the partial trajectories of one single electron passing through one of the two slits and another single electron passing through the other slit, as shown in figure 3.2. The phase shift is given by

$$\Delta\phi = ie\oint \mathbf{A}\cdot dx. \tag{3.26}$$

This phase increment, albeit from a topological (nonlocal) constraint, imposes the same type of relationship between the phase of the electron wave function and the four-vector potential A^{μ} as does local gauge invariance. The fact that a difference in the phase is detected by the experiments means that the link between the phase transformation of the

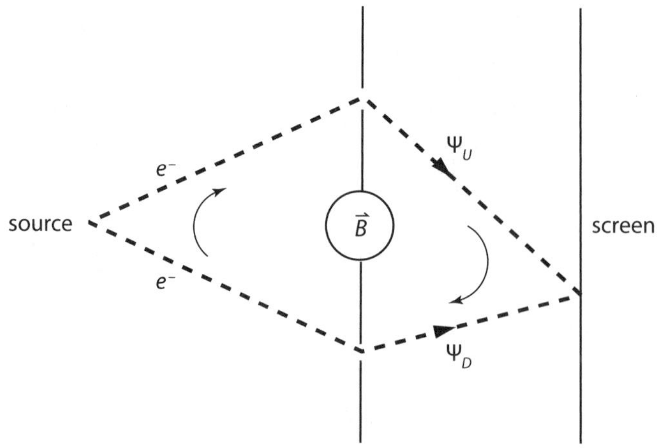

FIGURE 3.2. Single-electron interference in the Aharonov-Bohm effect. Single electrons are emitted from a source and encounter a double-slit barrier with openings on either side of a long solenoid before being detected at the screen.

electron and the gauge degree of freedom of the electromagnetic field is fundamental and goes beyond the predictions of the classical Lorentz-force interaction. In 1985, the single-electron interference experiment was repeated by Tonomura and collaborators using a toroidal ferromagnet covered with a superconducting layer to prevent magnetic field leakage and with the superconductor covered with a copper conducting layer to prevent the electron wave functions from penetrating into the magnetic field region [1]. The magnetic flux dependence on the phase shift of the interference pattern was confirmed.

In the case of the electromagnetic interaction, the gauge invariance of A^μ was a known property of Maxwell's equations. If particles carry a nonelectromagnetic charge, then we can use the gauge principle to specify the interaction that must exist for the Lagrangian density to be locally gauge invariant. Assume that the theory is invariant under the transformation of the particle states as follows:

$$\Psi' = U\Psi \tag{3.27}$$

for some unitary transformation U. We want to define

$$D^\mu = \partial^\mu + igB^\mu \tag{3.28}$$

where B^μ represents the interacting four-potential that has to be added to keep the theory invariant under (3.27). The question is then, how does B^μ transform? The condition of invariance is equivalent to requiring that $D^\mu\Psi$ transform as Ψ. Namely, the gauge transformation requirements came from derivatives in the Lagrangrian and, therefore, we can generalize the invariance requirement for any Lagrangian by specifying

$$D'^\mu\Psi' = U(D^\mu\Psi). \tag{3.29}$$

This allows us to solve for B'^μ:

$$(\partial^\mu + igB'^\mu)\, U\Psi = U(\partial^\mu + igB^\mu)\,\Psi, \tag{3.30}$$

and, therefore, we get

$$igB'^{\mu}U\Psi = -\partial^{\mu}(U\Psi) + U\partial^{\mu}\Psi + igUB^{\mu}\Psi$$
$$= -(\partial^{\mu}U)\Psi + igUB^{\mu}\Psi. \tag{3.31}$$

Since Ψ is arbitrary, for the purpose of this derivation, we can drop Ψ and multiply from the right by U^{-1}. This gives

$$B'^{\mu} = UB^{\mu}U^{-1} + \frac{i}{g}(\partial^{\mu}U)U^{-1}, \tag{3.32}$$

which for $U = \exp(ig\chi(x))$ and $UB^{\mu}U^{-1} = B^{\mu}$ reproduces the result for the electromagnetic four-potential $B'^{\mu} = B^{\mu} - \partial^{\mu}\chi$.

3.3 *SU*(2) and the Weak Interaction

The gauge transformation associated with the electromagnetic interaction is a one-dimensional unitary transformation $U(1)$. A $U(1)$ phase transformation cannot alter the identity of an electron wave function, i.e., there are no additional internal states. In the weak interaction, a neutrino can be transformed into an electron through the charged-current interaction. This process can be viewed as a generalization of a simple phase transformation, if the electron and the neutrino are combined into a single object. The first step is to generalize the electron wave function into a doublet

$$\Psi_D(x) = \begin{pmatrix} \Psi_\nu(x) \\ \Psi_e(x) \end{pmatrix} \quad \text{or simply} \quad \begin{pmatrix} \nu \\ e \end{pmatrix}. \tag{3.33}$$

The particle doublet $\Psi_D(x)$ is a vector with the neutrino $\Psi_\nu(x)$ and the electron $\Psi_e(x)$ forming the basis components. Now we consider transformations $\Psi'_D(x) = U\Psi_D(x)$ that would, in particular, convert an electron into a neutrino and vice versa. Clearly, a generalized phase transformation cannot interchange states with differing quantum numbers, not related to U, and differing masses. Most notably, the electron is electrically charged while the neutrino is neutral. Therefore, for this simple example, we will switch off the electromagnetic interaction and assume equal masses. By doing so, we can make the assumption that the neutrino and the electron are only distinguishable relative to each other, and that it is not *absolutely* possible to determine which is an electron and which is a neutrino. This concept is a crucial underpinning to multidimensional symmetry groups. Therefore, under this assumption, the transformed doublet Ψ'_D

$$\Psi'_D = \begin{pmatrix} e \\ \nu \end{pmatrix} \tag{3.34}$$

is equivalent to

$$\Psi_D = \begin{pmatrix} \nu \\ e \end{pmatrix} \tag{3.35}$$

in that all physical results of the new theory under this invariance are unaltered by an exchange $e \leftrightarrow \nu$. For example, the interaction cross section of ν-ν scattering would be exactly the same as that of e-e scattering.

The $U(1)$ gauge transformation of QED involved a real, continuous parameter $\chi(x)$ whose derivative $\partial^\mu \chi$ was part of the gauge transformation of the electromagnetic four-potential $A'^\mu = A^\mu - \partial^\mu \chi$. A general phase transformation depending on real, continuous parameters results from a representation of a *Lie group*. The smallest Lie group that will act on the doublet structure of $\Psi_D(x)$ is $SU(2)$. The $SU(2)$ group is the group of all unitary 2×2 matrices U with determinant $+1$. The need for positive determinant, hence the "S" or "Special" of $SU(2)$, should not be surprising as we saw a similar restriction when considering proper Lorentz transformations. Therefore, we have

$$\begin{pmatrix} \nu \\ e \end{pmatrix}' = U \begin{pmatrix} \nu \\ e \end{pmatrix}, \qquad U^\dagger = U^{-1}, \qquad \det(U) = +1. \tag{3.36}$$

Three real parameters $(\alpha_1, \alpha_2, \alpha_3)$ are required to parameterize the matrix U. Every 2×2 unitary matrix may be expressed in terms of the Pauli matrices $\boldsymbol{\sigma}$, but we will relabel $\boldsymbol{\sigma} = \boldsymbol{\tau}$ so as not to confuse this $SU(2)$ with the construction of the γ^μ matrices,

$$\begin{aligned} U(\alpha_1, \alpha_2, \alpha_3) &= \exp\left(ig(\alpha_1 \tau_1 + \alpha_2 \tau_2 + \alpha_3 \tau_3)/2\right) \\ &= \exp\left(ig\boldsymbol{\alpha} \cdot \boldsymbol{\tau}/2\right) \end{aligned} \tag{3.37}$$

where the $\boldsymbol{\tau}/2$ matrices are the *generators* of the transformations (3.37). The U matrices together with the particle doublet $\binom{\nu}{e}$ form the fundamental representation of the $SU(2)$ group.

In order to obtain an interaction through the gauge principle, we must admit local transformations, i.e., the parameters $\boldsymbol{\alpha}(x)$ are functions of x. Thus, the analog of $\Psi'(x) = \exp(ie\chi)\Psi(x)$ for $SU(2)$ is

$$\Psi'_D(x) = \begin{pmatrix} \Psi'_\nu(x) \\ \Psi'_e(x) \end{pmatrix} = \exp\left(ig\boldsymbol{\alpha}(x) \cdot \boldsymbol{\tau}/2\right) \begin{pmatrix} \Psi_\nu(x) \\ \Psi_e(x) \end{pmatrix}. \tag{3.38}$$

We also require that the equation of motion for $\Psi_D(x)$ should be invariant under transformation (3.38). To determine this we combine the Dirac equation for the free neutrino with that for the free electron using the matrix equation

$$\begin{pmatrix} i\gamma_\mu \partial^\mu & 0 \\ 0 & i\gamma_\mu \partial^\mu \end{pmatrix} \begin{pmatrix} \Psi_\nu \\ \Psi_e \end{pmatrix} - \begin{pmatrix} m_\nu & 0 \\ 0 & m_e \end{pmatrix} \begin{pmatrix} \Psi_\nu \\ \Psi_e \end{pmatrix} = \begin{pmatrix} 0 \\ 0 \end{pmatrix}, \tag{3.39}$$

which may be written symbolically as

$$i\gamma_\mu \partial^\mu \Psi_D - \boldsymbol{m}\Psi_D = 0. \tag{3.40}$$

Here $\boldsymbol{m}$ is a 2×2 matrix of diagonal 4×4 components

$$\boldsymbol{m} = \begin{pmatrix} m_\nu I_4 & 0 \\ 0 & m_e I_4 \end{pmatrix}. \tag{3.41}$$

Equation (3.40) is not invariant under (3.38). This can be seen by applying the infinitesimal transformation corresponding to (3.38),

$$\Psi'_D(x) = (1 + ig\boldsymbol{\alpha}(x)\cdot\boldsymbol{\tau}/2)\Psi_D(x), \qquad \alpha_j \ll 1, \tag{3.42}$$

to equation (3.40). This gives

$$\begin{aligned} &i\gamma_\mu\partial^\mu\Psi'_D(x) - \boldsymbol{m}\Psi'_D(x) \\ &\quad= i\gamma_\mu\partial^\mu(1 + ig\boldsymbol{\alpha}(x)\cdot\boldsymbol{\tau}/2)\Psi_D(x) - \boldsymbol{m}(1 + ig\boldsymbol{\alpha}(x)\cdot\boldsymbol{\tau}/2)\Psi_D(x) \\ &\quad= (1 + ig\boldsymbol{\alpha}(x)\cdot\boldsymbol{\tau}/2)(i\gamma_\mu\partial^\mu - \boldsymbol{m})\Psi_D(x) - g\gamma_\mu(\partial^\mu\boldsymbol{\alpha}(x))\cdot(\boldsymbol{\tau}/2)\Psi_D(x) \\ &\quad= -g\gamma_\mu(\partial^\mu\boldsymbol{\alpha}(x))\cdot(\boldsymbol{\tau}/2)\Psi_D(x) \neq 0 \end{aligned} \tag{3.43}$$

where $\boldsymbol{m}$ was assumed diagonal ($m_\nu = m_e$) in order to commute with $ig\boldsymbol{\alpha}(x)\cdot\boldsymbol{\tau}/2$. Here, as with the local gauge transformation of electromagnetism, we find that the Dirac equation must be modified to account for changes in $\boldsymbol{\alpha}(x)$ when comparing $\Psi_D(x+\Delta x)$ with $\Psi_D(x)$ in forming the derivative

$$i\gamma^\mu\left\{\partial_\mu - ig(\partial_\mu\boldsymbol{\alpha}(x))\cdot\boldsymbol{\tau}/2\right\}\Psi'_D(x) - \boldsymbol{m}\Psi'_D(x) = 0. \tag{3.44}$$

As with the electromagnetic interaction, we attempt to associate the additional term $ig(\partial^\mu\boldsymbol{\alpha}(x))\cdot\boldsymbol{\tau}/2$ in equation (3.44) with gauge degrees of freedom of the weak interaction. We can construct a covariant derivative with respect to an $SU(2)$ triplet of four-potentials,

$$\mathbf{W}_\mu(x) = \left(W^1_\mu(x), W^2_\mu(x), W^3_\mu(x)\right), \tag{3.45}$$

as follows:

$$\mathcal{D}_\mu = \partial_\mu - ig\mathbf{W}_\mu\cdot\boldsymbol{\tau}/2. \tag{3.46}$$

However, it is straightforward to check that the $U(1)$-analogous gauge transformation on the $\mathbf{W}_\mu$ four-potentials does not leave the covariant derivative form of the Dirac equation invariant under (3.38). Following the procedure for equations (3.28)–(3.32), the correct gauge transformation is given by

$$\mathbf{W}'_\mu(x) = \mathbf{W}_\mu(x) + \partial_\mu\boldsymbol{\alpha}(x) + g\mathbf{W}_\mu(x)\times\boldsymbol{\alpha}(x). \tag{3.47}$$

The simultaneous transform of the $\mathbf{W}_\mu$ four-potentials by (3.47) and the Dirac wave functions by (3.38) leaves the covariant form the Dirac equation invariant

$$\left(i\gamma^\mu\mathcal{D}'_\mu - m\right)\Psi'_D(x) = 0 \qquad \text{with } \mathcal{D}'_\mu = \partial_\mu - ig\mathbf{W}'_\mu(x)\cdot\boldsymbol{\tau}/2. \tag{3.48}$$

The covariant derivative (3.48) introduces an $SU(2)$ gauge interaction to Dirac particles analogous to the minimal coupling substitution of the electromagnetic interaction. However, for this gauge interaction to exist, the gauge transform (3.47) must correspond to a gauge degree of freedom of the $\mathbf{W}_\mu$ four-potentials. Note that it is also common to refer to the $\mathbf{W}_\mu$ as *fields* due to the quantum field theory (QFT) treatment of elementary interactions.

3.3.1 *Gauge Transformations of Massive Spin-1 Four-Potentials*

We can investigate $U(1)$ gauge interactions with a massive vector particle by going back to the Proca equation (2.224) and adding a source term on the right-hand side:

$$(\Box + M^2)W^\nu - \partial^\nu(\partial_\mu W^\mu) = -gJ^\nu \tag{3.49}$$

where J^ν represents a new type of current, one that couples to the W^ν in analogy to the electromagnetic interaction $\mathcal{H}_{\rm int} = -gJ_\nu W^\nu$. If we then apply ∂_ν to equation (3.49), we get

$$(\Box + M^2)\partial_\nu W^\nu - \partial_\nu\partial^\nu(\partial_\mu W^\mu) = -g\partial_\nu J^\nu, \tag{3.50}$$

which reduces to

$$M^2\partial_\nu W^\nu = -g\partial_\nu J^\nu. \tag{3.51}$$

For free particles ($J^\nu = 0$) and for conserved currents ($\partial_\nu J^\nu = 0$), this gives

$$\partial_\nu W^\nu = 0 \tag{3.52}$$

and, therefore, the free-particle equation (2.224) becomes

$$(\Box + M^2)W^\nu = 0. \tag{3.53}$$

The solutions to equation (3.53) have the same form as a free photon

$$W^\nu = \epsilon^\nu \exp(-iq\cdot x) \tag{3.54}$$

with two important exceptions. The first is simply $q^2 = M^2$, but the second has to do with the gauge transformation properties. For both the photon in the Lorentz gauge $\partial_\mu A^\mu = 0$ and the free massive vector particle $\partial_\mu W^\mu = 0$, the polarization vector satisfies the property

$$\epsilon^\mu q_\mu = 0. \tag{3.55}$$

However, in the case of the massless photon, the gauge condition has an additional degree of freedom. We can choose χ subject to

$$\Box\chi = 0 \tag{3.56}$$

and by local gauge invariance add $\partial^\mu\chi$ to A^μ where χ has the time dependence $\chi \sim \exp(-iq\cdot x)$ with $q^2 = 0$ from equation (3.56). This is equivalent to having the freedom to alter the photon polarization by a term proportional to q^μ:

$$\epsilon'^\mu = \epsilon^\mu + \lambda q^\mu \tag{3.57}$$

where we are free to choose λ without affecting the predictions of the electromagnetic interaction. A convenient choice is to choose λ such that $\epsilon^0 = 0$ and, therefore, the three-vector product $\boldsymbol{\epsilon}\cdot\boldsymbol{q} = 0$. This condition means that only two independent polarizations are available to a free photon and given the condition $\epsilon^\mu q_\mu = 0$, we know these are precisely the two helicity eigenstates where the photon polarization is transverse to the photon momentum. There are no "longitudinal" states of a free photon. Along this line, an invariant amplitude $\mathcal{M}$ involving free photons can be factorized

$$\mathcal{M} = \epsilon^\mu_\lambda \mathcal{M}_\mu \tag{3.58}$$

in terms of any one of the free-photon polarizations ϵ^μ_λ. As a consequence of gauge invariance, the substitution of the photon polarization with the photon four-momentum q^μ will yield zero:

$$q^\mu \mathcal{M}_\mu = 0. \tag{3.59}$$

Equation (3.59) is known as the *Ward identity*.

In the case of the massive vector particles, the $U(1)$ local gauge transformation

$$W'^{\mu} = W^{\mu} + \partial^{\mu}\chi \tag{3.60}$$

applied to equation (3.49) gives

$$(\Box + M^2)W^{\nu} - \partial^{\nu}(\partial_{\mu}W^{\mu}) + M^2\partial^{\nu}\chi = J^{\nu}. \tag{3.61}$$

Therefore, for $M \neq 0$ equation (3.61) differs from equation (3.49), indicating that the gauge degree of freedom is no longer present for massive vector particles. Indeed, a massive vector particle has three independent polarization states, including a longitudinal polarization state.

The three independent polarizations of a massive vector particle are evidenced in the corresponding completeness relation given by

$$\sum_{\lambda} \epsilon_{\lambda}^{\mu *} \epsilon_{\lambda}^{\nu} = -g^{\mu\nu} + q^{\mu}q^{\nu}/M^2 \tag{3.62}$$

with polarization vectors ($L \equiv$ Left, $R \equiv$ Right, $S \equiv$ Scalar), defined for $\boldsymbol{q}$ pointing along the z-axis,

$$\begin{aligned} \epsilon_{L,R}^{\mu} &= \frac{1}{\sqrt{2}}(0, 1, \mp i, 0), \\ \epsilon_{S}^{\mu} &= \frac{1}{M}(|\boldsymbol{q}|, 0, 0, E) = (\beta\gamma, 0, 0, \gamma). \end{aligned} \tag{3.63}$$

One can, therefore, identify the mass term with the scalar polarization. A free (massless) photon only has left- and right-handed polarizations.

The lack of gauge degrees of freedom in massive vector particles implies that an interaction involving a massive vector particle coupling to a current is not the result of local gauge invariance. On face value, the presence of mass in a vector-particle–mediated interaction is in direct conflict with the interaction being a consequence of local gauge invariance. We will return to this topic in the discussion of spontaneous symmetry breaking.

We can apply the same analysis to a massive vector particle with $SU(2)$ gauge interactions. The Proca equation (3.49) comes from the Euler-Lagrange equation applied to the Lagrangian:

$$\mathcal{L} = -\frac{1}{4}\boldsymbol{F}_{\mu\nu}\boldsymbol{F}^{\mu\nu} + \frac{1}{2}M^2\boldsymbol{W}_{\mu}\boldsymbol{W}^{\mu} + g\boldsymbol{J}^{\mu}\boldsymbol{W}_{\mu}. \tag{3.64}$$

The form of $\boldsymbol{F}^{\mu\nu}$ is different for $U(1)$ and $SU(2)$ local gauge symmetries. The $SU(2)$ generalization of $\boldsymbol{F}_{\mu\nu}$ requires that we define

$$\boldsymbol{F}^{\mu\nu} = \partial^{\mu}\boldsymbol{W}^{\nu} - \partial^{\nu}\boldsymbol{W}^{\mu} + g\boldsymbol{W}^{\mu} \times \boldsymbol{W}^{\nu}, \tag{3.65}$$

resulting from the commutation of covariant derivatives computed as in (3.25)

$$-ig\boldsymbol{F}^{\mu\nu} \cdot \boldsymbol{\tau}/2 = [D^{\mu}, D^{\nu}]. \tag{3.66}$$

The $SU(2)$ field strength $\boldsymbol{F}^{\mu\nu}$ is no longer gauge invariant as it is an $SU(2)$ triplet and can be rotated by the gauge transformation. The gauge transformation of the combination

$F_{\mu\nu} \cdot \boldsymbol{\tau}/2$ can be evaluated from the transformation properties of the commutator of covariant derivatives on the right-hand side of equation (3.66) and yields

$$F'_{\mu\nu} \cdot \frac{\boldsymbol{\tau}}{2} = U F_{\mu\nu} \cdot \frac{\boldsymbol{\tau}}{2} U^{-1} = F_{\mu\nu} \cdot \frac{\boldsymbol{\tau}}{2} - g\epsilon_{klm}\alpha^l F^m_{\mu\nu}\tau^k/2. \tag{3.67}$$

However, the kinetic energy term $-F_{\mu\nu}F^{\mu\nu}/4$ for $\mathbf{W}_\mu$ is gauge invariant under $SU(2)$, which can be seen by evaluating the trace

$$\mathcal{L}_{\text{kinetic}} = -\frac{1}{4}F_{\mu\nu}F^{\mu\nu} = -\frac{1}{2}\,\text{tr}\left\{\left(F_{\mu\nu} \cdot \boldsymbol{\tau}/2\right)^2\right\}. \tag{3.68}$$

Applying the gauge transformation to $\mathcal{L}_{\text{kinetic}}$ gives

$$\begin{aligned}
\mathcal{L}'_{\text{kinetic}} &= -\frac{1}{2}\,\text{tr}\left\{\left(F'_{\mu\nu} \cdot \boldsymbol{\tau}/2\right)^2\right\} \\
&= -\frac{1}{2}\,\text{tr}\left\{\left(U\left(F_{\mu\nu} \cdot \boldsymbol{\tau}/2\right)U^{-1}\right)^2\right\} \\
&= -\frac{1}{2}\,\text{tr}\left\{\left(F^i_{\mu\nu}\tau^i/2\right)U^{-1}U\left(F^{\mu\nu}_j\tau_j/2\right)U^{-1}U\right\} \\
&= \mathcal{L}_{\text{kinetic}}.
\end{aligned} \tag{3.69}$$

The corresponding Euler-Lagrange equation applied to the Lagrangian (3.64) with $F^{\mu\nu}$ defined in (3.65) gives

$$\partial_\mu F^{\mu\nu} + M^2\mathbf{W}^\nu + g\mathbf{W}_\mu \times F^{\mu\nu} = D_\mu F^{\mu\nu} + M^2\mathbf{W}^\nu = -gJ^\nu \tag{3.70}$$

where here we have used the form of the covariant derivative for an $SU(2)$ triplet in the adjoint representation

$$(\mathcal{D}_\mu)_{lm} = \partial_\mu\delta_{lm} + g\epsilon_{klm}W^k_\mu \tag{3.71}$$

obtained by replacing $\boldsymbol{\tau}/2$ in equation (3.46) with $i\epsilon_{klm}$. The mass term in equation (3.64) is not $SU(2)$ gauge invariant, and therefore, as with the $U(1)$ equation of motion, we must set $M=0$ to retain the gauge degrees of freedom in the free-particle ($J^\nu=0$) equation of motion for the $\mathbf{W}_\mu$ four-potentials.

3.3.2 *Non-Abelian Four-Potentials*

The gauge transform (3.47) of the triplet of $SU(2)$ $\mathbf{W}_\mu$ fields is *non-Abelian*, owing to the commutation relations of the group generators. Examining one component of (3.47) gives

$$W'^1_\mu(x) = W^1_\mu(x) + \partial_\mu\alpha_1(x) + g\left(W^2_\mu(x)\alpha_3(x) - W^3_\mu(x)\alpha_2(x)\right), \tag{3.72}$$

yielding, on the right, terms with W^2 and W^3. These terms act as additional source terms for W^1, showing that the $\mathbf{W}_\mu$ fields interact with each other. In other words, the $\mathbf{W}_\mu$ fields are in a triplet of the $SU(2)$ gauge group and are therefore charged under the $SU(2)$ gauge interaction. Gauge theories based on self-interacting non-Abelian fields are known as *Yang-Mills* theories.

It is not so surprising that the $\mathbf{W}_\mu$ fields carry the weak charge as the choice of the term $ig\mathbf{W}_\mu \cdot \boldsymbol{\tau}/2$ in the covariant derivative had to satisfy two constraints. The first is that the term has to transform as a Lorentz vector, as required by the Lorentz transformation properties of the covariant derivative. Hence, the $\mathbf{W}_\mu$ fields all correspond to spin-1 particles. The second requirement is that the Lagrangian density must be an $SU(2)$ scalar for $SU(2)$ gauge invariance, in the same way that it is a Lorentz scalar for Lorentz invariance. The $SU(2)$ scalar is formed by the dot product of the bilinear $\Psi_D^\dagger \boldsymbol{\tau}\Psi_D$, which acts like a vector in $SU(2)$, and the $SU(2)$ vector of $\mathbf{W}_\mu$ fields.

The vector behavior of the bilinear can be seen as follows. Consider a general spin-1/2 doublet state ψ in $SU(2)$ expanded into up and down components:

$$\psi = C_1|\uparrow\rangle + C_2|\downarrow\rangle. \tag{3.73}$$

The complex numbers C_i can be parameterized generally as

$$\begin{aligned} C_1 &= \exp(i\delta)\exp(-i\phi/2)\cos\left(\frac{\theta}{2}\right), \\ C_2 &= \exp(i\delta)\exp(i\phi/2)\sin\left(\frac{\theta}{2}\right). \end{aligned} \tag{3.74}$$

Then, recalling the form of the Pauli matrices

$$\begin{aligned} \psi^\dagger\tau_1\psi &= \left(\exp(-i\delta)\exp(i\phi/2)\cos\left(\tfrac{\theta}{2}\right) \quad \exp(-i\delta)\exp(-i\phi/2)\sin\left(\tfrac{\theta}{2}\right)\right) \\ &\quad \times \begin{pmatrix} 0 & 1 \\ 1 & 0 \end{pmatrix} \begin{pmatrix} \exp(i\delta)\exp(-i\phi/2)\cos\left(\tfrac{\theta}{2}\right) \\ \exp(i\delta)\exp(i\phi/2)\sin\left(\tfrac{\theta}{2}\right) \end{pmatrix} \\ &= \exp(i\phi)\cos\left(\frac{\theta}{2}\right)\sin\left(\frac{\theta}{2}\right) + \exp(-i\phi)\sin\left(\frac{\theta}{2}\right)\cos\left(\frac{\theta}{2}\right) \\ &= \sin\theta\cos\phi. \end{aligned} \tag{3.75}$$

Similarly,

$$\begin{aligned} \psi^\dagger\tau_2\psi &= \sin\theta\sin\phi, \\ \psi^\dagger\tau_3\psi &= \cos\theta. \end{aligned} \tag{3.76}$$

Thus, the angles θ and ϕ describe the orientation of a vector in spherical coordinates. A familiar example of this is the helicity operator involving $SU(2)$ spin angular momentum, where $\boldsymbol{\sigma}\cdot\mathbf{p}$ means $(\psi^\dagger\boldsymbol{\sigma}\psi)\cdot\mathbf{p}$. In general, for ψ in the fundamental representation and $\boldsymbol{\lambda}$ the generators of the group, the term $\psi^\dagger\boldsymbol{\lambda}\psi$ will behave like a vector in the adjoint representation and can be dotted into a multiplet of fields in the adjoint representation to form a scalar. Thus, there is a spin-1 gauge field for each of the generators of the gauge symmetry group, where the spin-1 property of the force-carrier particles is required by the Lorentz transformation properties of the covariant derivative.

The covariant form of the Dirac Lagrangian can be expanded to isolate the interaction Lagrangian for the gauge theory:

$$\begin{aligned} \mathcal{L} &= \bar{\Psi}_D(i\gamma^\mu\mathcal{D}_\mu - m)\Psi_D \\ &= \bar{\Psi}_D(i\gamma^\mu\partial_\mu - m)\Psi_D + g\bar{\Psi}_D\gamma^\mu\mathbf{W}_\mu\cdot\frac{\boldsymbol{\tau}}{2}\Psi_D \\ &= \mathcal{L}_{\text{Dirac}} + \mathcal{L}_{\text{int}}, \end{aligned} \tag{3.77}$$

where the second term is the interaction Lagrangian. In matrix form, $\mathcal{L}_{\text{int}}$ is given by

$$\begin{aligned}\mathcal{L}_{\text{int}} &= \frac{g}{2}(\bar{\nu} \quad \bar{e})\gamma^{\mu}\left\{W_{\mu}^{1}\begin{pmatrix}0 & 1\\ 1 & 0\end{pmatrix} + W_{\mu}^{2}\begin{pmatrix}0 & -i\\ i & 0\end{pmatrix}\right.\\ &\quad \left. + W_{\mu}^{3}\begin{pmatrix}1 & 0\\ 0 & -1\end{pmatrix}\right\}\begin{pmatrix}\nu\\ e\end{pmatrix}\\ &= \frac{g}{2}(\bar{\nu} \quad \bar{e})\gamma^{\mu}\begin{pmatrix}W_{\mu}^{3} & W_{\mu}^{1} - iW_{\mu}^{2}\\ W_{\mu}^{1} + iW_{\mu}^{2} & -W_{\mu}^{3}\end{pmatrix}\begin{pmatrix}\nu\\ e\end{pmatrix}\\ &= \frac{g}{2}(\bar{\nu} \quad \bar{e})\gamma^{\mu}\begin{pmatrix}W_{\mu}^{3} & \sqrt{2}\,W_{\mu}^{+}\\ \sqrt{2}\,W_{\mu}^{-} & -W_{\mu}^{3}\end{pmatrix}\begin{pmatrix}\nu\\ e\end{pmatrix}\end{aligned} \tag{3.78}$$

with

$$\begin{aligned}W_{\mu}^{+} &= \frac{1}{\sqrt{2}}(W_{\mu}^{1} - iW_{\mu}^{2}),\\ W_{\mu}^{-} &= \frac{1}{\sqrt{2}}(W_{\mu}^{1} + iW_{\mu}^{2}).\end{aligned} \tag{3.79}$$

The interaction Lagrangian from equation (3.78) can be decomposed into four terms:

$$\mathcal{L}_{\text{int}} = \frac{g}{2}(\bar{\nu}\gamma^{\mu}W_{\mu}^{3}\nu - \bar{e}\gamma^{\mu}W_{\mu}^{3}e) + \frac{g}{\sqrt{2}}(\bar{\nu}\gamma^{\mu}W_{\mu}^{+}e + \bar{e}\gamma^{\mu}W_{\mu}^{-}\nu). \tag{3.80}$$

The charged weak bosons are given by $W_{\mu}^{\pm}$ and the neutral weak boson is W_{μ}^{3}.

3.3.3 *Weak and Electromagnetic Interactions*

While our $SU(2)$ model for the weak interaction contains a full set of charged-current and neutral-current interactions between electrons and neutrinos, it fails to describe the physical weak interaction in many respects:

1. Parity violation is missing from the theory.
2. The $\mathbf{W}_{\mu}$ fields are required to be massless, which contradicts the observed short-range properties of the weak interaction.
3. The observed inequalities of mass and charge for the electron and neutrino are not allowed in the theory.

These three points all involve the origin of mass in some form. In particular, we could have attempted to transform only the left-handed fermion components by making use of the chirality projection operator $P_L = (1 - \gamma_5)/2$ in the U matrices:

$$\Psi'_D(x) = \left\{1 + ig\boldsymbol{\alpha}(x)\cdot\boldsymbol{\tau}/2\left(\frac{1-\gamma_5}{2}\right)\right\}\Psi_D(x). \tag{3.81}$$

However, the covariant form of the Dirac equation is not invariant under (3.81). This is because the $(1 - \gamma_5)/2$ term does not commute with γ_{μ}, although it does commute with the scalar mass term

$$
\begin{aligned}
(i\gamma^\mu \mathcal{D}_\mu - m)\Psi'_D &= (i\gamma^\mu \mathcal{D}_\mu - m)\left\{1 + ig\boldsymbol{\alpha}(x)\cdot\boldsymbol{\tau}/2\left(\frac{1-\gamma_5}{2}\right)\right\}\Psi_D(x) \\
&= \left\{1 + ig\boldsymbol{\alpha}(x)\cdot\boldsymbol{\tau}/2\left(\frac{1+\gamma_5}{2}\right)\right\} i\gamma^\mu \mathcal{D}_\mu \Psi_D(x) \\
&\quad - \left\{1 + ig\boldsymbol{\alpha}(x)\cdot\boldsymbol{\tau}/2\left(\frac{1-\gamma_5}{2}\right)\right\} m\Psi_D(x) \neq 0.
\end{aligned}
\tag{3.82}
$$

This has the physical interpretation that the dynamic term $i\gamma^\mu \mathcal{D}_\mu$ conserves left-handed chirality, but the mass term does not. This is another formulation of the fact that only massless particles have a well-defined chirality. In the Weyl representation, we saw that the mass term could be interpreted as a coupling between left- and right-handed chiral components. Thus, it is possible to construct a chiral gauge theory with $SU(2)$ interactions for *massless* fermions only. Points 1–3 will be simultaneously addressed via the *Higgs mechanism*.

The weak and electromagnetic interactions cannot be treated as independent gauge theories. To see this, we look at local gauge transformations of the $SU(2)\times U(1)$ direct product group. If we label the $U(1)$ charges of the $SU(2)$ doublet as $|e|Q_u$ and $|e|Q_d$, respectively, then we can write the product of infinitesimal $SU(2)$ and $U(1)$ transformations as

$$
\begin{aligned}
&(1 + ig\boldsymbol{\alpha}\cdot\boldsymbol{\tau}/2)\begin{pmatrix}(1 + i|e|Q_u\chi)\nu \\ (1 + i|e|Q_d\chi)e\end{pmatrix} \\
&= (1 + ig\boldsymbol{\alpha}\cdot\boldsymbol{\tau}/2)\left\{1 + i|e|\chi\left[\frac{(Q_u + Q_d)}{2} + (Q_u - Q_d)\frac{\tau_3}{2}\right]\right\}\begin{pmatrix}\nu \\ e\end{pmatrix}.
\end{aligned}
\tag{3.83}
$$

Since the $\boldsymbol{\tau}$ matrices do not commute, commutativity of the $SU(2)$ and the $U(1)$ transformations for all $\boldsymbol{\alpha}$ and χ is possible only if the $U(1)$ charges for the up and down components of the $SU(2)$ doublet are equal and therefore indistinguishable with respect to the $U(1)$ gauge transformation. Therefore, we cannot identify the $U(1)$ group in the $SU(2)\times U(1)$ direct product group with the electromagnetic gauge group $U(1)_{\mathrm{EM}}$.

3.4 Electroweak Gauge Interactions

The Glashow-Weinberg-Salam theory postulates that the weak and electromagnetic interactions are a consequence of an *electroweak Lagrangian* that is invariant under local gauge transformations of the direct product group $SU(2)_L \times U(1)_Y$ and that as a result of *spontaneous symmetry breaking*, the physical vacuum has only the $U(1)_{\mathrm{EM}}$ symmetry, where $U(1)_{\mathrm{EM}}$ is a subgroup of the product group [2]. Thus, the theory describes the observed gauge interaction of $U(1)_{\mathrm{EM}}$ QED, while the observed weak interaction results from the properties of the spontaneously broken product group symmetries.

The organization of matter into multiplets of the left-handed $SU(2)_L$ symmetry group and the values of the *hypercharge* $U(1)_Y$ quantum numbers are assigned based on the observed properties of the known particle states. We will therefore consider arbitrary assignments initially and then fix the assignments based on constraints from observation.

The weak charged-current interactions couple solely to the left-handed (L) chiral components of Dirac fermions, and therefore the $SU(2)_L$ symmetry group transforms left-handed chiral components while leaving the right-handed components unchanged. We will therefore define the $SU(2)_L$ doublet $\boldsymbol{L}$ as

$$\boldsymbol{L} \equiv \begin{pmatrix} \nu_L \\ e_L \end{pmatrix} \quad \text{where} \quad \begin{matrix} \nu_L \equiv \frac{1}{2}(1-\gamma_5)\Psi_\nu \\ e_L \equiv \frac{1}{2}(1-\gamma_5)\Psi_e \end{matrix} \tag{3.84}$$

and the right-handed chirality components as $SU(2)_L$ singlets

$$\nu_R \equiv \frac{1}{2}(1+\gamma_5)\Psi_\nu \quad \text{and} \quad e_R \equiv \frac{1}{2}(1+\gamma_5)\Psi_e, \tag{3.85}$$

where Ψ_ν and Ψ_e are the Dirac wave functions for the neutrino and electron, respectively. The gauge transformations of the doublet $\boldsymbol{L}$ and the singlets ν_R and e_R are

$$\begin{aligned} L' &= \exp(ig'\chi(x)Y_L/2 + ig\boldsymbol{\alpha}(x)\cdot\boldsymbol{\tau}/2)L, \\ \nu'_R &= \exp(ig'\chi(x)Y_R^\nu/2)\nu_R, \\ e'_R &= \exp(ig'\chi(x)Y_R^e/2)e_R, \end{aligned} \tag{3.86}$$

where Y_L, Y_R^ν, and Y_R^e are the hypercharge values of the different matter fields with respect to the $U(1)_Y$ gauge group. The factor of 1/2 in the exponents is convention. Note that ν_L and e_L carry the *same* hypercharge (as required to have commuting gauge symmetries), but the hypercharges of the right-handed singlets may be different from each other and from Y_L. Invariance of the Lagrangian under the combined $SU(2)_L \times U(1)_Y$ gauge transformation is achieved by replacing the derivative with the covariant derivative under the combined action of the product group as follows:

$$\partial_\mu \to \mathcal{D}_\mu = \begin{cases} \partial_\mu - ig'B_\mu Y_L/2 - ig\mathbf{W}_\mu\cdot\boldsymbol{\tau}/2 & \text{for } \boldsymbol{L}, \\ \partial_\mu - ig'B_\mu Y_R^i/2 & \text{for } i = \nu_R, e_R, \end{cases} \tag{3.87}$$

where B_μ and $\mathbf{W}_\mu$ are the gauge four-potentials of the $U(1)_Y$ and the $SU(2)_L$ transformations, respectively, and are minimally coupled to matter according to the covariant derivative with coupling constants g′ and g, respectively. The Lagrangian density is therefore given by (neglecting mass terms)

$$\begin{aligned} \mathcal{L} = {} & i\bar{L}\gamma^\mu\left(\partial_\mu - ig'B_\mu Y_L/2 - ig\mathbf{W}_\mu\cdot\boldsymbol{\tau}/2\right)L \\ & + i\bar{\nu}_R\gamma^\mu\left(\partial_\mu - ig'B_\mu Y_R^\nu/2\right)\nu_R \\ & + i\bar{e}_R\gamma^\mu\left(\partial_\mu - ig'B_\mu Y_R^e/2\right)e_R. \end{aligned} \tag{3.88}$$

If we consider for the moment only the neutral currents of the $SU(2)_L \times U(1)_Y$ Lagrangian (3.88), we have

$$\begin{aligned} \mathcal{L}_0 &= \bar{L}\gamma^\mu\left(g'B_\mu Y_L/2 + gW_\mu^3\tau_3/2\right)L + \bar{\nu}_R\gamma^\mu\left(g'B_\mu Y_R^\nu/2\right)\nu_R + \bar{e}_R\gamma^\mu\left(g'B_\mu Y_R^e/2\right)e_R \\ &= \bar{\nu}_L\gamma^\mu\left(g'B_\mu Y_L/2 + gW_\mu^3/2\right)\nu_L + \bar{\nu}_R\gamma^\mu\left(g'B_\mu Y_R^\nu/2\right)\nu_R \\ &+ \bar{e}_L\gamma^\mu\left(g'B_\mu Y_L/2 - gW_\mu^3/2\right)e_L + \bar{e}_R\gamma^\mu\left(g'B_\mu Y_R^e/2\right)e_R. \end{aligned} \tag{3.89}$$

We saw from QED that the electromagnetic interaction has the form

$$\mathcal{L}_{\text{EM}} = eQA_\mu\left(\bar{e}_L\gamma^\mu e_L + \bar{e}_R\gamma^\mu e_R\right) \tag{3.90}$$

and as the neutrino has no electric charge ($Q = 0$), the term in parentheses

$$\mathcal{L}_0^\nu = \bar{\nu}_L \gamma^\mu \nu_L \left(g' B_\mu Y_L / 2 + g W_\mu^3 / 2 \right) \tag{3.91}$$

must represent a physical gauge field which is orthogonal to the electromagnetic field. Therefore, we define the physical fields A^μ for the electromagnetic interaction and Z^μ for the neutral electroweak interaction to be orthogonal

$$A_\mu \propto g B_\mu - g' Y_L W_\mu^3, \qquad Z_\mu \propto g' Y_L B_\mu + g W_\mu^3, \tag{3.92}$$

where in analogy to a rotation, g is acting like a cosine and $g' Y_L$ like a sine. We normalize A_μ and Z_μ, treating W_μ^i and B_μ as normalized to unity,

$$A_\mu = \frac{g B_\mu - g' Y_L W_\mu^3}{\sqrt{g^2 + g'^2 Y_L^2}}, \qquad Z_\mu = \frac{g' Y_L B_\mu + g W_\mu^3}{\sqrt{g^2 + g'^2 Y_L^2}}, \tag{3.93}$$

with the corresponding reciprocal relationships

$$B_\mu = \frac{g A_\mu + g' Y_L Z_\mu}{\sqrt{g^2 + g'^2 Y_L^2}}, \qquad W_\mu^3 = \frac{-g' Y_L A_\mu + g Z_\mu}{\sqrt{g^2 + g'^2 Y_L^2}}. \tag{3.94}$$

Substituting (3.94) into $\mathcal{L}_0$ in equation (3.89) gives

$$\begin{aligned} \mathcal{L}_0 = \frac{1}{2\sqrt{g^2 + g'^2 Y_L^2}} \big[& (g'^2 Y_L^2 + g^2) \bar{\nu}_L \gamma^\mu Z_\mu \nu_L \\ & + 2 g g' Y_L \bar{e}_L \gamma^\mu A_\mu e_L + (g'^2 Y_L^2 - g^2) \bar{e}_L \gamma^\mu Z_\mu e_L + g g' Y_R^\nu \bar{\nu}_R \gamma^\mu A_\mu \nu_R \\ & + g'^2 Y_L Y_R^\nu \bar{\nu}_R \gamma^\mu Z_\mu \nu_R + g g' Y_R^e \bar{e}_R \gamma^\mu A_\mu e_R + g'^2 Y_L Y_R^e \bar{e}_R \gamma^\mu Z_\mu e_R \big]. \end{aligned} \tag{3.95}$$

The left-handed neutrino is prevented from having a coupling to A_μ by choosing the new neutral-current interaction Z_μ to be orthogonal to A_μ. However, the right-handed neutrino hypercharge must be set to zero $Y_R^\nu = 0$ to prevent a nonzero coupling to A_μ. This has the additional consequence of shutting off interactions with Z_μ and, indeed, all known gauge interactions with the right-handed neutrino. For the electron, in order for the coupling of the two chiral components e_R and e_L to have equal electric charge, we find that $Y_R^e = 2Y_L$.

Ultimately, A_μ is chosen to be the four-potential of a gauge interaction that has zero net coupling to one of the degrees of freedom of an $SU(2)$ complex-doublet scalar field $\mathbf{\Phi}$, known as the *Higgs field*. This defines the electric charge operator Q to be

$$Q = Y_\Phi \frac{\tau_3}{2} - \frac{\tau_3^{\langle\Phi\rangle}}{2} Y \tag{3.96}$$

where Y_Φ and $\tau_3^{\langle\Phi\rangle}/2$ are the hypercharge and third component of the isospin, respectively, of the electrically neutral component of $\mathbf{\Phi}$. We can choose the electrically neutral component of $\mathbf{\Phi}$ to be the lower component of the $SU(2)$ doublet, $\tau_3^{\langle\Phi\rangle}/2 = -1/2$. Applying the electric charge operator in equation (3.96) to the lepton doublet $\boldsymbol{L}$ gives for the neutrino and electron components, respectively,

$$\begin{aligned} 0 &= \frac{Y_\Phi}{2} + \frac{Y_L}{2}, \\ -1 &= -\frac{Y_\Phi}{2} + \frac{Y_L}{2}. \end{aligned} \tag{3.97}$$

This fixes $Y_L = -1$ and $Y_\Phi = +1$.

We can now define

$$\frac{g}{\sqrt{g^2 + g'^2}} = \cos\theta_W \quad \text{and} \quad \frac{g'}{\sqrt{g^2 + g'^2}} = \sin\theta_W \tag{3.98}$$

where θ_W is known as the *weak mixing angle*, or the *Weinberg angle*, and write the neutral gauge four-potentials of $SU(2)_L \times U(1)_Y$ from equation (3.94) in terms of the physical A_μ and Z_μ as follows:

$$\begin{aligned} B_\mu &= A_\mu \cos\theta_W - Z_\mu \sin\theta_W, \\ W_\mu^3 &= A_\mu \sin\theta_W + Z_\mu \cos\theta_W. \end{aligned} \tag{3.99}$$

As the coupling strength between A_μ and the electron is given by the elementary electronic charge e from (3.95), we can set

$$e = \frac{gg'}{\sqrt{g^2 + g'^2}} = g\sin\theta_W = g'\cos\theta_W. \tag{3.100}$$

3.5 Gauge Interaction of QCD

The elementary strong interaction, also known as quantum chromodynamics (QCD), is a gauge interaction between *colored* quarks [2]. The symmetry group of the gauge transformations is given by $SU(3)_C$, where C stands for color. The $SU(3)$ group is the group of all unitary 3×3 matrices U with determinant $+1$. The three basis states of the $SU(3)_C$ symmetry group are labeled by the three colors red, green, and blue. The choice of nomenclature reflects the known color neutral bound states of quarks, the *baryon* or three-quark bound state (hence the analogy to a three-color combination that yields neutral) and the *meson* or quark-antiquark bound state, where color and anticolor is also a neutral combination. The precise construction of wave functions that form stable dynamical states will be discussed in the chapter on hadrons.

As far as we know, $SU(3)_C$ is a symmetry of the Lagrangian of elementary particle interactions and the physical vacuum. As a result, the gauge bosons, called the *gluons*, are massless. At the same time, the electroweak interactions are "color-blind" and are not affected by the quark color charges. This means the $SU(3)_C$ transformations commute with the $SU(2)_L \times U(1)_Y$ transformations. Similar to the argument in the construction of the $SU(2)_L$ doublets, the members of an $SU(3)_C$ multiplet must differ only in their color charge, and not in any other quantum number; otherwise, members would be distinguishable by their properties under the electroweak interaction, or by their masses, and the $SU(3)_C$ symmetry would be broken.

The known *fundamental* $SU(3)_C$ *triplets* are

$$\begin{pmatrix} u_r \\ u_g \\ u_b \end{pmatrix}, \begin{pmatrix} d_r \\ d_g \\ d_b \end{pmatrix}, \begin{pmatrix} c_r \\ c_g \\ c_b \end{pmatrix}, \begin{pmatrix} s_r \\ s_g \\ s_b \end{pmatrix}, \begin{pmatrix} t_r \\ t_g \\ t_b \end{pmatrix}, \begin{pmatrix} b_r \\ b_g \\ b_b \end{pmatrix}, \tag{3.101}$$

where $r = red$, $g = green$, and $b = blue$. The six known flavors of quarks listed in (3.101) are called (from left to right) up, down, charm, strange, top, and bottom.

The $SU(3)_C$ gauge transformations are given by

$$\begin{pmatrix} q_r \\ q_g \\ q_b \end{pmatrix}' = \exp\left(ig_s \sum_{k=1}^{8} \eta_k(x) \frac{\lambda_k}{2} \right) \begin{pmatrix} q_r \\ q_g \\ q_b \end{pmatrix} \tag{3.102}$$

where the λ_k correspond to the τ_i in the $SU(2)_L$ case, and $\lambda_k/2$ are the generators of the $SU(3)_C$ transformations. One choice of representation for the eight linearly independent, unitary 3×3 matrices is known as the *Gell-Mann matrix representation*

$$\lambda_1 = \begin{pmatrix} 0 & 1 & 0 \\ 1 & 0 & 0 \\ 0 & 0 & 0 \end{pmatrix}, \lambda_2 = \begin{pmatrix} 0 & -i & 0 \\ i & 0 & 0 \\ 0 & 0 & 0 \end{pmatrix}, \lambda_3 = \begin{pmatrix} 1 & 0 & 0 \\ 0 & -1 & 0 \\ 0 & 0 & 0 \end{pmatrix},$$
$$\lambda_4 = \begin{pmatrix} 0 & 0 & 1 \\ 0 & 0 & 0 \\ 1 & 0 & 0 \end{pmatrix}, \lambda_5 = \begin{pmatrix} 0 & 0 & -i \\ 0 & 0 & 0 \\ i & 0 & 0 \end{pmatrix}, \lambda_6 = \begin{pmatrix} 0 & 0 & 0 \\ 0 & 0 & 1 \\ 0 & 1 & 0 \end{pmatrix}, \tag{3.103}$$
$$\lambda_7 = \begin{pmatrix} 0 & 0 & 0 \\ 0 & 0 & -i \\ 0 & i & 0 \end{pmatrix}, \lambda_8 = \frac{1}{\sqrt{3}} \begin{pmatrix} 1 & 0 & 0 \\ 0 & 1 & 0 \\ 0 & 0 & -2 \end{pmatrix}.$$

There are two diagonal matrices, $\lambda_3/2$ and $\lambda_8/2$, and their eigenvalues are used to label the components of an $SU(3)_C$ multiplet. A quark is in the fundamental representation of the $SU(3)_C$ color symmetry group, denoted as **3**. An antiquark is in the $\bar{\mathbf{3}}$ representation. These representations are plotted in terms of $(\lambda_3/2, \lambda_8/2)$ in figure 3.3.

The requirement for gauge invariance under the transformations (3.102) leads to eight gauge fields, the gluon fields G^k, $k = 1 \ldots 8$. These transform under infinitesimal transformations according to

$$G'^k_\mu = G^k_\mu + \partial_\mu \eta_k(x) + g_s f_{klm} \eta_l(x) G^m_\mu \tag{3.104}$$

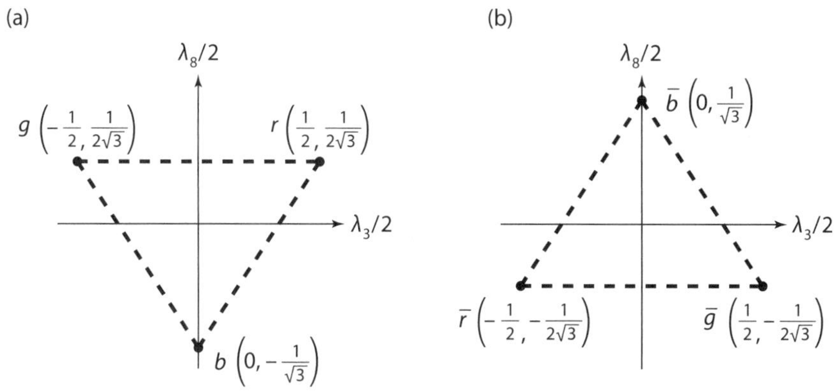

FIGURE 3.3. The $SU(3)_C$ quark representation (a) on the left and the antiquark representation (b) on the right.

so as to keep the Lagrangian invariant under (3.102). The term $f_{klm}\eta_l(x)G_\mu^m$ corresponds to the term $-\mathbf{W}_\mu \times \boldsymbol{\alpha}(x)$ in the $SU(2)_L$ case. The values f_{klm} are called the *structure constants* of the $SU(3)_C$ group, and correspond for $SU(2)_L$ to the tensor ϵ_{ijk}. The f_{klm} is a totally antisymmetric tensor satisfying the equation

$$\left[\frac{\lambda_k}{2}, \frac{\lambda_l}{2}\right] = if_{klm}\frac{\lambda_m}{2}. \tag{3.105}$$

The nonzero independent elements of the $SU(3)$ structure constants are

$$\begin{aligned} &f_{123} = 1, \\ &f_{147} = f_{246} = f_{257} = f_{345} = 1/2, \\ &f_{156} = f_{367} = -1/2, \\ &f_{458} = f_{678} = \sqrt{3}/2. \end{aligned} \tag{3.106}$$

The covariant derivative corresponding to invariance under $SU(3)_C$ gauge transformations is given by

$$\mathcal{D}^\mu = \partial^\mu - ig_s G_k^\mu \frac{\lambda_k}{2} \tag{3.107}$$

in the fundamental representation, i.e., an $SU(3)_C$ triplet of color. For an $SU(3)_C$ octet, the covariant derivative has the form

$$(\mathcal{D}^\mu)_{lm} = \partial^\mu \delta_{lm} + g_s f_{klm} G_k^\mu \tag{3.108}$$

where $\boldsymbol{\lambda}/2$ is replaced with if_{klm}. The sum $g_s G_k^\mu \lambda_k/2$ can be arranged in a matrix

$$\sum_{k=1}^{8} g_s G_k^\mu \lambda_k/2 = \frac{g_s}{2}\begin{pmatrix} G_3^\mu + \frac{1}{\sqrt{3}}G_8^\mu & G_1^\mu - iG_2^\mu & G_4^\mu - iG_5^\mu \\ G_1^\mu + iG_2^\mu & -G_3^\mu + \frac{1}{\sqrt{3}}G_8^\mu & G_6^\mu - iG_7^\mu \\ G_4^\mu + iG_5^\mu & G_6^\mu + iG_7^\mu & -\frac{2}{\sqrt{3}}G_8^\mu \end{pmatrix} \tag{3.109}$$

where the notation of the off-diagonal terms can be simplified using the equivalent ladder operation definitions (as was done for $SU(2)_L$ with $W^\pm$)

$$\sum_{k=1}^{8} g_s G_k^\mu \lambda_k/2 = \frac{g_s}{\sqrt{2}}\begin{pmatrix} \frac{1}{\sqrt{2}}G_3^\mu + \frac{1}{\sqrt{6}}G_8^\mu & G_{r\bar{g}}^\mu & G_{r\bar{b}}^\mu \\ G_{g\bar{r}}^\mu & -\frac{1}{\sqrt{2}}G_3^\mu + \frac{1}{\sqrt{6}}G_8^\mu & G_{g\bar{b}}^\mu \\ G_{b\bar{r}}^\mu & G_{b\bar{g}}^\mu & -\frac{2}{\sqrt{6}}G_8^\mu \end{pmatrix}. \tag{3.110}$$

The gluons carry both color (c) and anticolor ($\bar{c}$) charges, as indicated by the subscripts.

The Feynman diagrams for the strong interaction are constructed by labeling quarks by a color and indicating the corresponding gluon fields that mediate the $SU(3)_C$ interaction. For example, the two diagonal fields, G_3^μ and G_8^μ, mediate the interaction between two quarks of the same color. The vertex coupling is given by the coefficient of the corresponding gluon four-potential for the process being computed as it appears in the gluon matrix (3.110). Therefore, for two red quarks G_3^μ has a vertex coupling factor of $g_s/2$ and G_8^μ has a factor $g_s/(2\sqrt{3})$, as shown in figures 3.4(a) and 3.4(b). Notice that the scattering of a red and a blue quark only involves G_8^μ with a vertex coupling factor $g_2/(2\sqrt{3})$ on the red quark and $-g_s/\sqrt{3}$ on the blue quark, as shown in figure 3.4(c). For diagrams involving antiquarks, the sign of the vertex term flips for anticolor vertices, as shown in figure 3.5(a). Figure 3.5(a) gives the gluon exchange diagram for quark-antiquark scattering in two forms; the top form is the familiar Feynman diagram and the bottom form is called the *color flow*. The color flow diagrams show how the color indices move from quark to gluon and back.

As with the electromagnetic interaction, the strength of the interaction is proportional to the product of the two vertex coupling factors, and we define the *color factor* C_F to be equal to the coefficient in front of g_s^2 from the product of vertex coupling factors. The sign of the color factor indicates whether the gluon exchange is attractive or replusive (positive indicates repulsive).

The commutator of two covariant derivatives is used to compute the gluon field tensor

$$G_k^{\mu\nu} = \partial^\mu G_k^\nu - \partial^\nu G_k^\mu + g_s f_{klm} G_l^\mu G_m^\nu, \tag{3.111}$$

and the complete QCD Lagrangian (with the exception of a possible θ term) is given by

$$\mathcal{L}_{\text{QCD}} = \bar{q}(i\gamma_\mu D^\mu - m_q)q - \frac{1}{4} G_k^{\mu\nu}(G_k)_{\mu\nu}. \tag{3.112}$$

Note that the kinetic energy term for the gluon fields $-\frac{1}{4}G_k^{\mu\nu}(G_k)_{\mu\nu}$, given the equation for $G_k^{\mu\nu}$ (which is an octet of $SU(3)$) in (3.111), contains both three-point and four-point

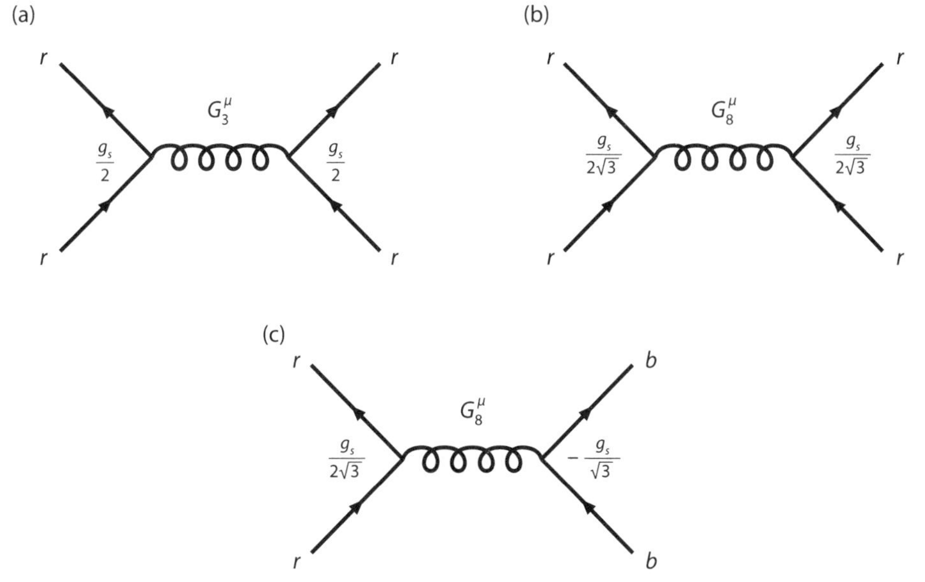

FIGURE 3.4. The QCD scattering of two quarks is shown above for two red quarks in diagrams (a) and (b) and for a red and a blue quark in diagram (c).

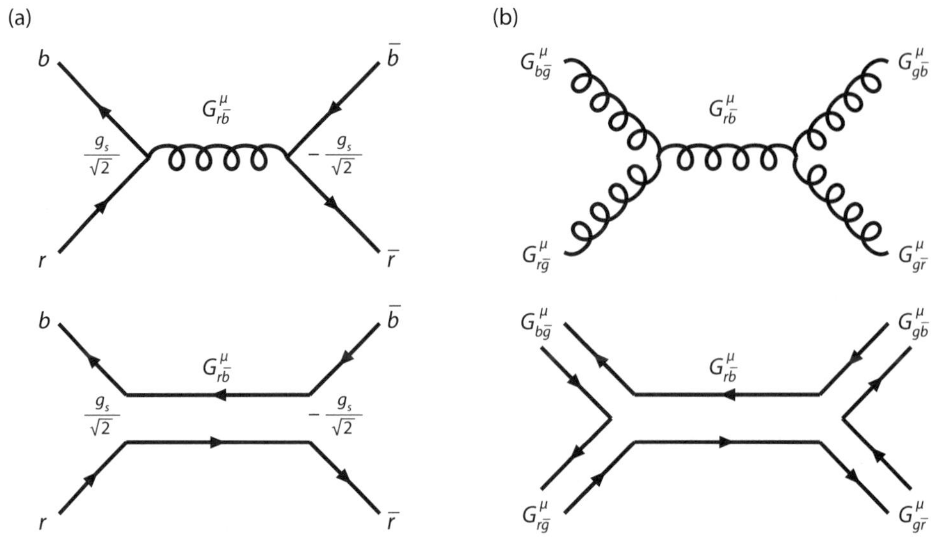

FIGURE 3.5. The QCD scattering of a quark and an antiquark (a) and the three-point gluon-gluon scattering diagram (b). The upper diagrams show the quark and gluon lines and the lower diagrams show the corresponding color flow.

self-couplings of the gluons fields, a consequence of the non-Abelian nature of the theory. These terms are identified by unpacking the compact $-\frac{1}{4}G_k^{\mu\nu}(G_k)_{\mu\nu}$ term, using the notation

$$\mathbf{G}^{\mu\nu} = \partial^{\mu}\mathbf{G}^{\nu} - \partial^{\nu}\mathbf{G}^{\mu} + g_s\mathbf{G}^{\mu}\times\mathbf{G}^{\nu} \tag{3.113}$$

where the cross-product is to be interpreted as the structure constants for $SU(3)_C$. This gives

$$\begin{aligned} -\frac{1}{4}G_k^{\mu\nu}(G_k)_{\mu\nu} &= -\frac{1}{2}(\partial_{\mu}\mathbf{G}_{\nu} - \partial_{\nu}\mathbf{G}_{\mu})\cdot(\partial^{\mu}\mathbf{G}^{\nu}) \\ &\quad - g_s(\mathbf{G}_{\mu}\times\mathbf{G}_{\nu})\cdot\partial^{\mu}\mathbf{G}^{\nu} \\ &\quad - \frac{1}{4}g_s^2\left[(\mathbf{G}^{\mu}\cdot\mathbf{G}_{\mu})^2 - (\mathbf{G}^{\mu}\cdot\mathbf{G}_{\nu})(\mathbf{G}^{\nu}\cdot\mathbf{G}_{\mu})\right]. \end{aligned} \tag{3.114}$$

The three-point vertex corresponds to the g_s term and the four-point vertex to the g_s^2 term. A color flow diagram involving the three-point gluon vertices is shown in figure 3.5(b).

Note that although in the absence of quark masses $m_q = 0$, the QCD Lagrangian appears to be scale invariant, the definition of physical observables will force the introduction of a finite mass scale, known as Λ_{QCD} ~0.1–0.3 GeV. This comes about from the antiscreening nature of the color force and, in particular, the antiscreening properties of the gluon fields. The consequences of Λ_{QCD} and further properties of the elementary strong interaction will be further discussed in the chapter on hadrons.

The gauge interaction of QCD is observed in experiment to conserve the discrete symmetries of parity, time reversal, and CP. However, an interesting consequence of the gauge principle is the prediction of the gauge-invariant θ term

$$\mathcal{L}_\theta = \frac{\theta g_s^2}{64\pi^2}\epsilon_{\mu\nu\rho\sigma} G_k^{\mu\nu} G_k^{\rho\sigma}. \tag{3.115}$$

This term violates P, T, and CP symmetries, and leads to a possible nonzero neutron electric dipole moment (EDM) for nonzero θ parameter of $d_n \sim 4 \times 10^{-16}\theta e \cdot cm$, which, compared to the current experimental limit of $d_n < 3 \times 10^{-26} e \cdot cm$ (90% CL) [3], implies $\theta \lesssim 10^{-10}$, or compatible with zero. The θ term has no physical relevance for Abelian theories such as QED as the term can be rewritten in terms of a total derivative whose surface integral at infinity is required to vanish. For non-Abelian theories and, in particular QCD, there are solutions to the QCD vacuum which are nonvanishing at infinity, known as *instantons*. No evidence for a nonzero θ term has been identified, and the strong interaction is believed to be invariant under P, T, and CP.

3.6 Structure of Elementary Matter

The gauge theory of $SU(3)_C \times SU(2)_L \times U(1)_Y$ is known as the *Standard Model of particle physics*. The matter content of the Standard Model, the leptons and quarks, is a regular array of fermions with fixed spacings in hypercharge quantum numbers whose interactions enter the Lagrangian through a gauge-covariant derivative given by

$$\mathcal{D}_\mu = \partial_\mu - ig' B_\mu \frac{Y}{2} - igW_\mu^a \frac{\tau_a}{2} - ig_s G_\mu^k \frac{\lambda_k}{2}. \tag{3.116}$$

The pattern of hypercharge splittings can be seen in figure 3.6 for the first generation of matter,

$$\{(e_R, L, \nu_R), (d_R, Q, u_R)\}. \tag{3.117}$$

The $|\Delta Y| = 1$ spacing corresponds to the $SU(2)_L$ singlet to doublet splitting. The $|\Delta Y| = 4/3$ spacing is the $SU(3)_C$ singlet to triplet spacing for corresponding $SU(2)_L$ doublets and up-type and down-type $SU(2)_L$ singlets. As will be seen in the following sections on the Higgs mechanism, the fixed hypercharge spacing between left-handed and right-handed chirality fermions of the same species allows one to write a set of Yukawa interactions in the presence of a spin-0 $SU(2)_L$ doublet $\mathbf{\Phi}$ with hypercharge $Y = 1$, namely:

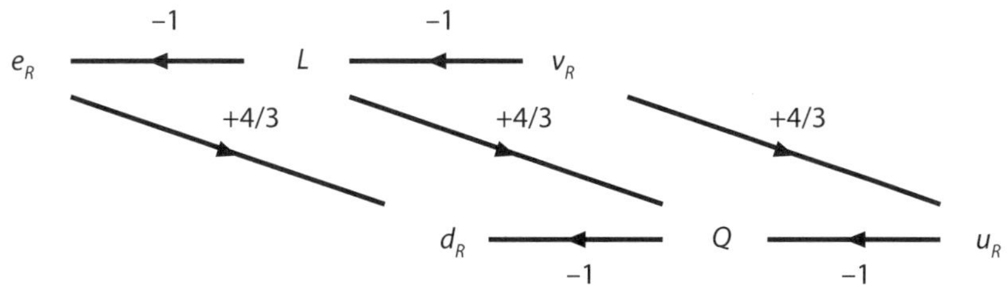

FIGURE 3.6. Two hypercharge splittings are observed in the leptons and quarks, a $|\Delta Y| = 1$ splitting along the $SU(2)_L$ singlet to doublet to singlet transformation and a $|\Delta Y| = 4/3$ splitting between corresponding $SU(3)_C$ singlets and triplets.

$$\begin{aligned}\Delta\mathcal{L}_{\Phi} &= (\mathcal{D}_{\mu}\Phi)^{\dagger}(\mathcal{D}^{\mu}\Phi)\\ &-\lambda_{\ell}^{ij}\bar{L}^{i}\cdot\Phi\ell_{R}^{j} - \lambda_{d}^{ij}\bar{Q}^{i}\cdot\Phi d_{R}^{j}\\ &-\lambda_{\nu}^{ij}\bar{L}^{i}\cdot\Phi_{c}\nu_{R}^{j} - \lambda_{u}^{ij}\bar{Q}^{i}\cdot\Phi_{c}u_{R}^{j} + h.c.\end{aligned} \tag{3.118}$$

including Hermitian conjugate (*h.c.*) terms and where the λ^{ij} are general complex-valued matrices, $\Phi_c = i\tau^2\Phi^*$ indicates charge conjugation, and

$$\begin{gathered}Q^{i} = \begin{pmatrix}u_L^i\\ d_L^i\end{pmatrix} = \left(\begin{pmatrix}u_L\\ d_L\end{pmatrix},\begin{pmatrix}c_L\\ s_L\end{pmatrix},\begin{pmatrix}t_L\\ b_L\end{pmatrix}\right),\ L^{i} = \begin{pmatrix}\nu_L^i\\ \ell_L^i\end{pmatrix} = \left(\begin{pmatrix}\nu_{eL}\\ e_L\end{pmatrix},\begin{pmatrix}\nu_{\mu L}\\ \mu_L\end{pmatrix},\begin{pmatrix}\nu_{\tau L}\\ \tau_L\end{pmatrix}\right),\\ u_R^i = (u_R, c_R, t_R),\ \nu_R^i = (\nu_{eR}, \nu_{\mu R}, \nu_{\tau R}),\ d_R^i = (d_R, s_R, b_R),\ \ell_R^i = (e_R, \mu_R, \tau_R)\end{gathered} \tag{3.119}$$

whose $(SU(3)_C,\ SU(2)_L,\ U(1)_Y)$ quantum numbers are given by

$$\begin{aligned} u_R &\ \text{transforms as}\ (\mathbf{3},\mathbf{1},4/3),\\ Q = \begin{pmatrix}u_L\\ d_L\end{pmatrix} &\ \text{transforms as}\ (\mathbf{3},\mathbf{2},1/3),\\ d_R &\ \text{transforms as}\ (\mathbf{3},\mathbf{1},-2/3),\\ \nu_R &\ \text{transforms as}\ (\mathbf{1},\mathbf{1},0),\\ L = \begin{pmatrix}\nu_L\\ e_L\end{pmatrix} &\ \text{transforms as}\ (\mathbf{1},\mathbf{2},-1),\\ e_R &\ \text{transforms as}\ (\mathbf{1},\mathbf{1},-2),\\ \Phi &\ \text{transforms as}\ (\mathbf{1},\mathbf{2},1).\end{aligned} \tag{3.120}$$

Here we note that the gauge quantum numbers of ν_R and $\bar{\nu}_R$ are vanishing, and, therefore, the Yukawa interaction terms (3.118) are ambiguous in the choice of the neutrino particle versus the antiparticle.

A scalar Yukawa coupling to either a $\bar{\Psi}\Psi$ or $\bar{\Psi}\gamma^5\Psi$ Dirac current must couple to opposite chirality fermions, as can be shown as follows:

$$\begin{aligned}\Psi &= P_R\Psi + P_L\Psi,\\ \bar{\Psi} &= \bar{\Psi}P_L + \bar{\Psi}P_R.\end{aligned} \tag{3.121}$$

And, therefore, only the combinations

$$\begin{aligned}\bar{\Psi}\Psi &= \bar{\Psi}P_RP_R\Psi + \bar{\Psi}P_LP_L\Psi = \bar{\Psi}P_R\Psi + \bar{\Psi}P_L\Psi,\\ \bar{\Psi}\gamma^5\Psi &= \bar{\Psi}P_R\gamma^5P_R\Psi + \bar{\Psi}P_L\gamma^5P_L\Psi = \bar{\Psi}\gamma^5P_R\Psi - \bar{\Psi}\gamma^5P_L\Psi\end{aligned} \tag{3.122}$$

are nonvanishing. This means, for instance, that a left-handed fermion that radiates or absorbs a scalar (or pseudoscalar) must change chirality at that vertex, as shown, for example, in figure 2.2.

The $|\Delta Y| = 1$ splitting is therefore the only splitting where one can introduce a scalar particle coupling to a $\bar{\Psi}\Psi$ or $\bar{\Psi}\gamma^5\Psi$ Dirac current. A second possibility arises for a set of Yukawa interactions between $SU(3)_C$ singlets and triplets, corresponding to the fixed hypercharge spacing of $|\Delta Y| = 4/3$ if one allows a *charge conjugation* operation at the interaction vertex. For instance, a *lepto-quark* interaction with a scalar particle $\mathbf{S}$ could be defined via

$$\mathcal{L}_S = -\lambda_L^{ij} \bar{L}_c^{i,k} \epsilon_{kl} Q_a^{j,l} S_a^* - \lambda^{R,ij} (\bar{\ell}_R)_c^i u_{R,a}^j S_a^* - \lambda_R^{ij} (\bar{\nu}_R)_c^i d_{R,a}^j S_a^* + h.c. \quad (3.123)$$

where $\Psi_c = C\Psi^*$ indicates charge conjugation with $C = i\sigma_2$ when acting on the Weyl two-component spinor representations and $\epsilon_{kl} = i\sigma_2$ is the $SU(2)_L$ contraction of the two fundamental representations, $\bar{L}_c$ and Q_a. The S boson is a strongly interacting, charge $-1e/3$, spin-0 boson whose $(SU(3)_C, SU(2)_L, U(1)_Y)$ quantum numbers are given by

$$S \text{ transforms as } (\mathbf{3}, \mathbf{1}, -2/3). \quad (3.124)$$

The λ_L^{ij}, $\lambda^{R,ij}$, and λ_R^{ij} are general complex-valued matrices with (ij) indices over the three mass generations. The appearance of the charge conjugation in equation (3.123) is needed to give nonvanishing vertex terms when coupling a scalar particle to a fermion pair with the same chirality. Note that if the fermions in equation (3.123) carry an additive quantum number not related to $SU(3)_C \times SU(2)_L \times U(1)_Y$, such as a lepton or baryon number, this quantum number would be violated at the interaction vertex. As there is no direct experimental evidence for lepton or baryon number violating interactions, we assume that these interactions are nonexistent in Nature or are suppressed at the energy scale of current experiments.

As we will see in the following sections, the couplings between left-handed and right-handed chiral components as described by the scalar $\mathbf{\Phi}$ Lagrangian (3.118) are the starting point for the theory of the origin of mass in the fermions and the electroweak gauge bosons and lead to the prediction of self-interactions of the $\mathbf{\Phi}$.

3.7 Spontaneous Symmetry Breaking

An implicit (but not mandatory) assumption of the previous discussion on gauge interactions is that symmetries of the Lagrangian are also symmetries of the vacuum ground state. However, in relativisitic quantum mechanics the vacuum is anything but a void. The vacuum consists of an ensemble of elementary particle *fields*, each of which is capable of creating and annihilating new particle states. In the Klein paradox, we saw that in the presence of a sharp potential, particle-antiparticle pairs stream directly out of the vacuum. One can consider the possibility that of the many elementary particle fields not all fields "vanish" in the vacuum. If a particle field takes on a nonzero expectation value in the vacuum, then the notion of free-particle equations of motion is called into question and must be revisited in the context of a nontrivial physical vacuum ground state. Depending on the quantum numbers of the nonzero fields, the physical vacuum will not be invariant under all symmetries of the Lagrangian.

There is no predictive theory for how symmetries are broken, and as a consequence the constraints for constructing the physical vacuum are understood in terms of symmetry-breaking mechanisms rather than equations of state. What we observe about the physical vacuum is that it obeys Lorentz invariance with translational and rotational symmetries in space-time, i.e., no preferred direction and no net angular momentum. It is also neutral with respect to the electromagnetic and strong interactions, as QED and QCD appear to be the only physically observed gauge symmetries. Therefore, if we are to choose a single nonzero field to explain the observed physical properties of the vacuum, we will choose a charge- and color-neutral scalar field.

In fact, as the gauge symmetries are symmetries of the Lagrangian, we have $[H,U]=0$, and therefore the vacuum ground state where $H|0\rangle = E_{\text{min}}|0\rangle$ can be transformed under U without changing the ground state energy

$$H(U|0\rangle) = UH|0\rangle = UE_{\text{min}}|0\rangle = E_{\text{min}}U|0\rangle. \tag{3.125}$$

If there is only one ground state $U|0\rangle=|0\rangle$, then the vacuum is invariant under U. If, however, there are several degenerate vacuum states $|0\rangle_i$, then the gauge transformation U has the potential of distinguishing one vacuum state from another

$$U|0\rangle_i = |0\rangle_j \qquad i \neq j. \tag{3.126}$$

Therefore, if only one $|0\rangle_i$ is the physical vacuum ground state, then the ground state will not necessarily be invariant under U.

The *Nambu-Goldstone theorem* states that for every symmetry of the Lagrangian that is not a symmetry of the vacuum ground state there exists a massless scalar boson, i.e., the nonrespected gauge symmetries of the vacuum manifest as physical particle states with interactions dictated by the gauge symmetries of the Lagrangian. These are known as *Goldstone bosons*. Therefore, whatever the symmetry-breaking mechanism is that allows for the ground-state vacuum to not respect the Lagrangian symmetries, it also must identify or explain the nonobservation of the corresponding Goldstone bosons as predicted from this theorem.

3.8 Higgs Mechanism

The choice of the physical states for the photon A^μ and the Z boson Z^μ must ultimately originate from the mass eigenstates, i.e., the free-particle solutions. However, we have seen that gauge invariance under $SU(2)_L$ requires massless fermions in order to have pure left-handed states. The force carriers of the interactions must also be massless for there to be gauge degrees of freedom as required by the gauge principle. The masslessness requirement is, therefore, based on enforcing the principle of local gauge invariance under $SU(3)_C \times SU(2)_L \times U(1)_Y$ of the Standard Model Lagrangian. If, however, the physical vacuum does not have all the symmetries of the Lagrangian, then the propagation of the gauge particles and of the fermions is modified [4]. If one assumes that the gauge principle is fundamental, then the symmetries of the physical vacuum must be spontaneously broken, i.e., gauge invariance must be maintained in the Lagrangian independent of the symmetries of the physical vacuum. The Higgs mechanism postulates the existence of a scalar field, or fields, with nonzero vacuum expectation values that reduce the gauge symmetries of the physical vacuum from $SU(3)_C \times SU(2)_L \times U(1)_Y$ down to $SU(3)_C \times U(1)_{EM}$. The Higgs vacuum expectation value is neutral under the color charge and electric charge. However, the nonzero scalar field component is necessarily nonzero (nonsinglet) for the quantum numbers of $SU(2)_L$ and $U(1)_Y$. The presence of nonzero fields, which by construction permeate all of space, impedes the propagation of the gauge bosons and fermions by causing them to interact with the nonzero components of the Higgs field.

The Higgs mechanism is not a prediction of a particular set of nonzero fields, but rather a minimum requirement on the elements of spontaneous symmetry breaking under a specific mechanism involving self-interacting scalar fields. The resulting broken symmetry of the Standard Model correctly predicts the mass ratio of the neutral and charged massive electroweak bosons and the relationship of the mass splitting to the coupling constants of the gauge interactions. One of the most important predictions of the Higgs mechanism is that at least one physical degree of freedom in the Higgs field is a particle degree of freedom. The free-particle degree of freedom is called the *Higgs boson*. The same mechanism that generates masses in the gauge particles also predicts a mass for the Higgs boson. However, while the location of the minimum of the Higgs potential can be constrained under some assumptions from previously measured parameters of the Standard Model, without the detection of the Higgs boson the parameters of the symmetry-breaking potential are only loosely constrained. The Higgs boson mass remains elusive, as it depends on the shape of the potential for small displacements away from the minimum and the shape is governed by an unpredicted self-coupling constant of the Higgs field.

3.8.1 *Minimum Single-Doublet of Complex Scalar Fields*

The minimal choice of Higgs field is determined by the following constraints. To induce spontaneous symmetry breaking of $SU(2)_L \times U(1)_Y$ down to $U(1)_{EM}$, the Higgs field should be charged under both subgroups of the direct product group. The smallest $SU(2)$ multiplet is the fundamental, or doublet. From equation (3.96), the hypercharge of an $SU(2)_L$ doublet, or multiplet, in general, will be equal to the sum of the electric charges of the components of the multiplet:

$$\sum_i Q_i = \sum_{T^3=\pm\frac{1}{2}} \left(T^3 + \frac{Y}{2}\right) = Y. \tag{3.127}$$

Therefore, the choice of the lower component of the Higgs multiplet to be electrically neutral forces a nonzero hypercharge value $Y_\Phi = 1$. We can also count the minimum number of field components by attributing one Higgs field degree of freedom to each of the massive weak bosons reponsible for the weak interaction; the rationale for this is described below. The basic requirement is that the additional longitudinal spin polarizations present in a massive spin-1 boson must originate from corresponding degrees of freedom in the Higgs sector. The short-range behavior of the charged-current interactions implies that the $W^\pm$ are massive particles. Similarly, the weak neutral current will be shown to originate from a massive Z boson. The requirement of three massive electroweak bosons forces the Higgs sector to have at least three degrees of freedom. This will be discussed in more detail in the section on the Goldstone boson equivalence theorem.

If we treat the scalar fields as elementary fields, the minimal single-doublet complex scalar Higgs field is given by

$$\Phi(x) = \begin{pmatrix} \phi^+(x) \\ \phi^0(x) \end{pmatrix} = \frac{1}{\sqrt{2}} \begin{pmatrix} \phi_1^+(x) + i\phi_2^+(x) \\ \phi_1^0(x) + i\phi_2^0(x) \end{pmatrix} \tag{3.128}$$

where $\phi_1^+, \phi_2^+, \phi_1^0$, and ϕ_2^0 are real, giving four degrees of freedom in the Higgs field. The charge-neutral nonzero vacuum expectation will be assigned by convention to the ϕ_1^0 component.

The kinetic energy of the complex-doublet scalar Higgs field is given by

$$T(\mathbf{\Phi}^\dagger, \mathbf{\Phi}) = (\mathcal{D}_\mu \mathbf{\Phi})^\dagger (\mathcal{D}^\mu \mathbf{\Phi}) \tag{3.129}$$

where $\mathcal{D}_\mu$ is the $SU(2)_L \times U(1)_Y$ gauge-covariant derivative. The self-interacting Higgs potential must contain at least the two-point and four-point interaction terms in order to form a nonzero vacuum expectation value that is bounded from below. Odd-powered terms are excluded if the potential is to be positive definite at small distances. The minimal self-interacting Higgs potential that is invariant under $SU(2)_L \times U(1)_Y$ is given by

$$V(\mathbf{\Phi}^\dagger \mathbf{\Phi}) = -\mu^2 \mathbf{\Phi}^\dagger \mathbf{\Phi} + \lambda (\mathbf{\Phi}^\dagger \mathbf{\Phi})^2, \qquad \mu^2 > 0, \lambda > 0, \tag{3.130}$$

where λ is the coupling strength of the four-point Higgs interaction. The minimum of the Higgs potential for this choice of μ and λ is not at $\mathbf{\Phi} = 0$ but at a finite value. The $SU(2)_L \times U(1)_Y$-invariant Lagrangian density $\mathcal{L}_\Phi$ for $\mathbf{\Phi}$ is then

$$\mathcal{L}_\Phi = (\mathcal{D}_\mu \mathbf{\Phi})^\dagger (\mathcal{D}^\mu \mathbf{\Phi}) + \mu^2 \mathbf{\Phi}^\dagger \mathbf{\Phi} - \lambda (\mathbf{\Phi}^\dagger \mathbf{\Phi})^2. \tag{3.131}$$

The physical vacuum is calculated to be the state of least energy, treating $\mathbf{\Phi}$ as a *classical* field. Thus, for vanishing kinetic energy (instanton solutions are not considered), the potential energy is minimal for

$$\mathbf{\Phi}_{\text{min}} = \frac{1}{\sqrt{2}} \begin{pmatrix} 0 \\ v \end{pmatrix} \quad \text{where} \quad v = \sqrt{\mu^2/\lambda}. \tag{3.132}$$

In the nonclassical interpretation, the vacuum expectation value (vev) of the Higgs field is finite $\langle \mathbf{\Phi} \rangle = \mathbf{\Phi}_{\text{min}}$. Notice the ground states of the physical vacuum given by $SU(2)_L \times U(1)_Y$ gauge transformations of $\mathbf{\Phi}_{\text{min}}$ can be written as

$$\mathbf{\Phi}' = \exp(ig'\chi(x) Y_\Phi/2 + ig\boldsymbol{\alpha}(x) \cdot \boldsymbol{\tau}/2) \frac{1}{\sqrt{2}} \begin{pmatrix} 0 \\ v \end{pmatrix} \neq \mathbf{\Phi}_{\text{min}}, \tag{3.133}$$

which have the same minimal energy and thus are also possible vacuum states. By choosing $\langle \mathbf{\Phi} \rangle$ to be a fixed orientation out of the possible continuous set of physical ground states, the $SU(2)_L \times U(1)_Y$ symmetry of the physical vacuum is spontaneously broken. The term *spontaneous* means that the symmetries of the Lagrangian remain intact, and the interactions are still governed by the gauge principle. It is only the dynamical selection of the vacuum from the self-interacting potential (3.130) that has reduced the symmetries of the vacuum. In fact, the only remaining symmetry of the vacuum is given by one linear combination of $SU(2)_L \times U(1)_Y$,

$$Q = T^3_{\langle \Phi \rangle} + Y_\Phi/2 = -\frac{1}{2} + (1)/2 = 0, \tag{3.134}$$

which is precisely the $Q = 0$ choice of the electrically neutral vacuum. Thus, the physical vacuum is broken from an $SU(2)_L \times U(1)_Y$-invariant vacuum in the absence of a Higgs field to a $U(1)_{\text{EM}}$ symmetry in the presence of a nonzero Higgs vev.

Excitations of the Higgs field will be with respect to the minimum $\mathbf{\Phi}_{\text{min}}$, and thus the field components can be parameterized by

$$\Phi(x) = \exp(i\xi(x)\cdot\tau)\,\frac{1}{\sqrt{2}}\begin{pmatrix}0\\ v+H(x)\end{pmatrix}. \tag{3.135}$$

The $\xi(x)$ are excitations of Φ_{min} along the potential minimum. The ξ are none other than the prediction from the Nambu-Goldstone theorem of an $SU(2)_L$ triplet of massless scalar particles corresponding to the broken symmetries of the vacuum. However, in the case of the Higgs mechanism, the additional terms generated by the mechanism couple these degrees of freedom directly to the electroweak bosons and become part of the *vacuum polarization* correction to the mass of these particles, as will be discussed in detail in the section on the Nambu-Goldstone theorem. Thus, this effect is often described as the gauge bosons "eating up" the Goldstone bosons to form the longitudinal polarizations of the massive spin-1 states. The $H(x)$ excitation is in the radial direction and corresponds to the prediction of a free-particle state, the Higgs boson. The Higgs boson, which would have been massless if it were not for the nonzero vev, also acquires a mass corresponding to the quadratic term from the expansion of the potential with respect to Φ_{min},

$$V(H) = -\frac{1}{4}\mu^2 v^2 + \mu^2 H^2 + \lambda v H^3 + \frac{1}{4}\lambda H^4, \tag{3.136}$$

and thus gives the Higgs mass m_H to be

$$m_H = \sqrt{2\mu^2} = \sqrt{2\lambda}\,v. \tag{3.137}$$

The current experimental limit corresponds to a value of $\lambda > 0.108$ (95% CL), where a value of $\lambda = 0.11$ corresponds to $m_H \approx 115$ GeV/c^2 (using the value of the Higgs vev determined below).

The value of the Higgs potential minimum can be determined from existing measurements and the Higgs mechanism prediction for the W mass. The W mass is determined from the Higgs kinetic energy term

$$\begin{aligned}(\mathcal{D}_\mu\Phi)^\dagger(\mathcal{D}^\mu\Phi) &= \left\{\left(\partial_\mu - ig'B_\mu Y_\Phi/2 - igW^i_\mu\tau_i/2\right)\Phi\right\}^\dagger\left(\partial^\mu - ig'B^T_\mu Y_\Phi/2 - igW^{j\mu}\tau_j/2\right)\Phi\\ &= \frac{g^2}{4}\sum_{ij}(W^i_\mu)^\dagger W^{j\mu}\Phi^\dagger\tau_i\tau_j\Phi + \cdots\end{aligned} \tag{3.138}$$

where for $i=j$ gives $\tau_i^2 = 1$. The charged W bosons states are given by

$$W^\pm_\mu = \frac{1}{\sqrt{2}}\left(W^1_\mu \mp iW^2_\mu\right) \tag{3.139}$$

and therefore continuing from (3.138) and evaluating for $\Phi = \Phi_{min} = \frac{1}{\sqrt{2}}\binom{0}{v}$ gives

$$\begin{aligned}&= \frac{g^2}{4}\sum_i^2 (W^i_\mu)^\dagger W^{i\mu}\Phi^\dagger_{min}\Phi_{min} + \cdots\\ &= \frac{g^2}{8}((W^-_\mu)^\dagger W^{-\mu} + (W^+_\mu)^\dagger W^{\pm\mu})(0 \quad v)\begin{pmatrix}0\\ v\end{pmatrix} + \cdots\\ &= \frac{g^2v^2}{8}((W^-_\mu)^\dagger W^{-\mu} + (W^+_\mu)^\dagger W^{\pm\mu}) + \cdots.\end{aligned} \tag{3.140}$$

The term $\frac{g^2v^2}{8}(W^\pm_\mu)^\dagger W^{\pm\mu}$ in the Lagrangian corresponds to a W boson mass ($M_{W^+} = M_{W^-}$)

$$M_W = \frac{gv}{2} \approx 80.4\ \mathrm{GeV}/c^2 \tag{3.141}$$

where the expression for the mass has the form of *a coupling times a vev* indicative of the Higgs mechanism. The value of the Higgs vev is then determined either by computing the $SU(2)_L$ coupling constant g using $g = e/\sin\theta_W$ or by extracting the ratio $(M_W/g)^2$ from the precisely measured muon lifetime. In Fermi weak interaction theory, the four-point interaction describing the low-energy weak decay of the muon is given by

$$\mathcal{M}(\mu^- \to e^- + \nu_\mu + \bar{\nu}_e) = (G_F/\sqrt{2})\,\bar{u}(\nu_\mu)\gamma_\mu(1-\gamma_5)\,u(\mu)\,\bar{u}(e)\gamma^\mu(1-\gamma_5)\,v(\nu_e) \tag{3.142}$$

resulting in $G_F \approx 1.166 \times 10^{-5}\ \mathrm{GeV}^{-2}$ from the measured value $\tau_\mu = 2.197 \times 10^{-6}$s. In electroweak theory, with a massive W boson, the corresponding matrix element is

$$\mathcal{M}(\mu^- \to e^- + \nu_\mu + \bar{\nu}_e) = \frac{g^2}{2}\bar{u}(\nu_\mu)\gamma_\mu\frac{(1-\gamma_5)}{2}u(\mu)\frac{-g^{\mu\nu} + q^\mu q^\nu/M_W^2}{q^2 - M_W^2}\bar{u}(e)\gamma^\mu\frac{(1-\gamma_5)}{2}v(\nu_e). \tag{3.143}$$

Comparing (3.142) with (3.143) in the limit $q^2 \to 0$ gives

$$\frac{G_F}{\sqrt{2}} = \frac{g^2}{8M_W^2}. \tag{3.144}$$

Using equations (3.141) and (3.144), we can solve for the Higgs vev

$$v = (\sqrt{2}\,G_F)^{-1/2} \approx 246\ \mathrm{GeV}. \tag{3.145}$$

Now we consider the coupling of the neutral gauge fields to the Higgs doublet. This is described by

$$\mathcal{L} = \frac{1}{4}\left\{(g' B_\mu Y_\Phi + g W_\mu^3 \tau_3)\boldsymbol{\Phi}\right\}^\dagger \left(g' B_\mu Y_\Phi + g W^{3\mu}\tau_3\right)\boldsymbol{\Phi}. \tag{3.146}$$

If we evaluate (3.146) at the vacuum expectation of $\boldsymbol{\Phi}$, we obtain

$$\begin{aligned}\mathcal{L}_m &= \frac{v^2}{8}\left(g' B_\mu^\dagger Y_\Phi + g W_\mu^{3\dagger}\tau_3^{\langle\Phi\rangle}\right)\left(g' B^\mu Y_\Phi + g W^{3\mu}\tau_3^{\langle\Phi\rangle}\right)\\ &= \frac{v^2}{8}\left[g^2 W_\mu^{3\dagger}W^{3\mu}(\tau_3^{\langle\Phi\rangle})^2 + g'^2 B_\mu^\dagger B^\mu Y_\Phi^2 + gg'(W_\mu^{3\dagger}B^\mu + B_\mu^\dagger W^{3\mu})Y_\Phi\tau_3^{\langle\Phi\rangle}\right]\\ &= \frac{v^2}{8}\begin{pmatrix} W_\mu^{3\dagger} & B_\mu^\dagger\end{pmatrix}\begin{pmatrix} g^2(\tau_3^{\langle\Phi\rangle})^2 & gg' Y_\Phi\tau_3^{\langle\Phi\rangle}\\ gg' Y_\Phi\tau_3^{\langle\Phi\rangle} & g'^2 Y_\Phi^2\end{pmatrix}\begin{pmatrix} W^{3\mu}\\ B^\mu\end{pmatrix}.\end{aligned} \tag{3.147}$$

The mass-squared matrix is

$$\boldsymbol{M}^2 = \frac{v^2}{4}\begin{pmatrix} g^2(\tau_3^{\langle\Phi\rangle})^2 & gg' Y_\Phi\tau_3^{\langle\Phi\rangle}\\ gg' Y_\Phi\tau_3^{\langle\Phi\rangle} & g'^2 Y_\Phi^2\end{pmatrix}, \tag{3.148}$$

which can be diagonalized with the unitary transformation

$$\boldsymbol{U} = \frac{1}{\sqrt{g^2(\tau_3^{\langle\Phi\rangle})^2 + g'^2 Y_\Phi^2}}\begin{pmatrix} g' Y_\Phi & -g\tau_3^{\langle\Phi\rangle}\\ g\tau_3^{\langle\Phi\rangle} & g' Y_\Phi\end{pmatrix}. \tag{3.149}$$

Setting the lower component of the Higgs field to be electrically neutral, $\tau_3^{\langle\Phi\rangle}/2=-1/2$ and $Y_\Phi=1$, the diagonalized mass-squared matrix is

$$M_D^2 = UM^2U^{-1} = \begin{pmatrix} 0 & 0 \\ 0 & \frac{v^2}{4}(g^2+g'^2) \end{pmatrix}. \tag{3.150}$$

The zero-mass eigenvalue corresponds to the four-potential resulting from the unitary transformation (3.149) of $(W^{3\mu}, B^\mu)$, whose gauge interaction coupling is given by

$$eQ \equiv \frac{gg'}{\sqrt{g^2(\tau_3^{\langle\Phi\rangle})^2 + g'^2 Y_\Phi^2}} \left(Y_\Phi \frac{\tau_3}{2} - \frac{\tau_3^{\langle\Phi\rangle}}{2} Y \right) \tag{3.151}$$

and where the coupling is defined by construction to be zero when acting on the Higgs vev. Thus, the electric charge operator Q and the massless four-potential, known as the photon and denoted A^μ, are a consequence of the choice of Higgs vev. The nonzero eigenvalue is the squared mass of the Z boson

$$M_Z^2 = \frac{v^2}{4}(g^2+g'^2) = \frac{M_W^2}{\cos^2\theta_W}. \tag{3.152}$$

3.9 Glashow-Weinberg-Salam Theory of the Electroweak Interactions

The electroweak sector is the minimal Lagrangian with a set of gauge interactions that can break down to $U(1)_{\text{EM}}$ through the Higgs mechanism and reproduce the low-energy Fermi weak interaction theory with good experimental agreement. The comparisons with experiment will be discussed later in the text.

The electroweak gauge-covariant derivative is

$$\mathcal{D}_\mu = \partial_\mu - ig'B_\mu Y/2 - igW_\mu^a T^a \tag{3.153}$$

where $T^a=\tau^a/2$ for isospin doublets (i.e., the left-handed fermions), while $T^a=0$ for isospin singlets (the right-handed fermions), and Y is arranged so that

$$Q = T^3 + \frac{Y}{2}. \tag{3.154}$$

Thus, we need the following $SU(2)_L \times U(1)_Y$ assignments for quarks and leptons:

$$\begin{aligned}
u_R \quad & \text{transforms as} \quad (\mathbf{1}, 4/3), \\
Q = \begin{pmatrix} u_L \\ d_L \end{pmatrix} \quad & \text{transforms as} \quad (\mathbf{2}, 1/3), \\
d_R \quad & \text{transforms as} \quad (\mathbf{1}, -2/3), \\
\nu_R \quad & \text{transforms as} \quad (\mathbf{1}, 0), \\
L = \begin{pmatrix} \nu_L \\ e_L \end{pmatrix} \quad & \text{transforms as} \quad (\mathbf{2}, -1), \\
e_R \quad & \text{transforms as} \quad (\mathbf{1}, -2).
\end{aligned} \tag{3.155}$$

The electroweak Lagrangian is a generalization of QED, given by

$$\mathcal{L} = -\frac{1}{4}(W^a_{\mu\nu})^2 - \frac{1}{4}B^2_{\mu\nu} + \sum_{\text{fermions } \psi} \bar{\psi} i\gamma^\mu \mathcal{D}_\mu \psi,$$
$$W^a_{\mu\nu} = \partial_\mu W^a_\nu - \partial_\nu W^a_\mu + g\epsilon_{abc} W^b_\mu W^c_\nu, \qquad B_{\mu\nu} = \partial_\mu B_\nu - \partial_\nu B_\mu, \tag{3.156}$$

only here *no gauge-invariant mass terms are possible.*

The minimal way to break $SU(2)_L \times U(1)_Y \rightarrow U(1)_{\text{EM}}$ through the Higgs mechanism is to have an isospin doublet scalar, the Higgs field $\mathbf{\Phi}$, with a nonzero vacuum expectation value (vev):

$$\mathbf{\Phi} \text{ transforms as } (\mathbf{2}, 1),$$
$$\langle \mathbf{\Phi} \rangle = \frac{1}{\sqrt{2}} \begin{pmatrix} 0 \\ v \end{pmatrix}. \tag{3.157}$$

We choose $Y_\Phi = 1$ so that $\langle \mathbf{\Phi} \rangle$ is $U(1)_{\text{EM}}$ invariant.

Gauge boson masses arise from $|D_\mu \langle \mathbf{\Phi} \rangle|^2$:

$$D_\mu \langle \mathbf{\Phi} \rangle = \left(\partial_\mu - ig' B_\mu/2 - igW^a_\mu \tau^a/2\right)\langle \mathbf{\Phi} \rangle = \frac{iv}{2\sqrt{2}} \begin{pmatrix} gW^1_\mu - igW^2_\mu \\ -gW^3_\mu + g'B_\mu \end{pmatrix},$$
$$|D_\mu \langle \mathbf{\Phi} \rangle|^2 = \frac{v^2}{8}\left[g^2(W^1_\mu)^2 + g^2(W^2_\mu)^2 + (-gW^3_\mu + g'B_\mu)^2\right]. \tag{3.158}$$

From the Higgs kinetic energy term, W^1_μ and W^2_μ get a mass of $gv/2$, and $Z_\mu \propto -gW^3_\mu + g'B_\mu$ acquires a similar mass, while the orthogonal combination $g'W_\mu + gB_\mu$ remains massless. Defining the cosine and sine of the weak mixing angle θ_W by

$$\cos\theta_W = \frac{g}{\sqrt{g^2 + g'^2}}, \quad \sin\theta_W = \frac{g'}{\sqrt{g^2 + g'^2}}, \tag{3.159}$$

we can write the physical electroweak fields as

$$W^\pm_\mu = \frac{1}{\sqrt{2}}(W^1_\mu \mp iW^2_\mu) \qquad \text{with } M_W = \frac{gv}{2},$$
$$Z_\mu = \cos\theta_W W^3_\mu - \sin\theta_W B_\mu \quad \text{with } M_Z = \frac{M_W}{\cos\theta_W},$$
$$A_\mu = \sin\theta_W W^3_\mu + \cos\theta_W B_\mu \quad \text{with } M_\gamma = 0. \tag{3.160}$$

The square of the sine of the *weak mixing angle* is given by

$$\sin^2\theta_W = \frac{g'^2}{g^2 + g'^2} = 1 - \left(\frac{M_W}{M_Z}\right)^2 \tag{3.161}$$

and thus predicts a relationship between the masses of the electroweak bosons and the gauge couplings. Note that these are tree-level relationships for $\sin^2\theta_W$. After higher-order corrections are included in the coupling strength, the tree-level mass relationship from (3.161) yields $\sin^2\theta_W \backsimeq 0.223$ while the effective weak mixing angle in the electroweak couplings, as determined in Z boson decays, is slightly larger, $\sin\theta_W^{\text{eff}}(M_Z^2) \approx 0.232$. The numerous experimental measurements in the electroweak sector have confirmed with no exceptions that the electroweak sector is governed by one unique value for the weak mixing angle.

3.10 Neutral-Current Feynman Rules

The interaction Lagrangian can be separated into charged-current and neutral-current components,

$$\begin{aligned}\mathcal{L}_{\text{int}} &= \left[g\left(\bar{L}\gamma_\mu \frac{\tau_i}{2} L\right) W_i^\mu + g'\left(\frac{Y_L}{2}\bar{L}\gamma_\mu L + \frac{Y_R}{2}\bar{e}_R\gamma_\mu e_R\right)B^\mu\right] \\ &= \left[g(j_i^t)_\mu W_i^\mu + g'\frac{(j^Y)_\mu}{2}B^\mu\right] \\ &= \mathcal{L}_{\text{CC}} + \mathcal{L}_{\text{NC}}\end{aligned} \tag{3.162}$$

with

$$\mathcal{L}_{\text{NC}} = \left[g(j_3^t)_\mu W_3^\mu + g'\frac{(j^Y)_\mu}{2}B^\mu\right]. \tag{3.163}$$

Writing W_3^μ and B^μ in terms of the physical states Z^μ and A^μ,

$$\begin{aligned}W_3^\mu &= \cos\theta_W Z^\mu + \sin\theta_W A^\mu, \\ B^\mu &= -\sin\theta_W Z^\mu + \cos\theta_W A^\mu,\end{aligned} \tag{3.164}$$

and substituting into the neutral-current interaction Lagrangian, gives

$$\mathcal{L}_{\text{NC}} = \left[g\sin\theta_W(j_3^t)^\mu + g'\cos\theta_W\frac{(j^Y)^\mu}{2}\right]A_\mu + \left[g\cos\theta_W(j_3^t)^\mu - g'\sin\theta_W\frac{(j^Y)^\mu}{2}\right]Z_\mu. \tag{3.165}$$

The coupling of A^μ to the electromagnetic current leads to the identification

$$g\sin\theta_W = g'\cos\theta_W = e \tag{3.166}$$

and the electromagnetic current then simply follows from $Q = T^3 + Y/2$

$$j_e^\mu = g\sin\theta_W(j_3^t)^\mu + g'\cos\theta_W\frac{(j^Y)^\mu}{2} = e\left[(j_3^t)^\mu + \frac{(j^Y)^\mu}{2}\right]. \tag{3.167}$$

Substituting $(j^Y)^\mu$ from (3.167) and $g' = g\sin\theta_W/\cos\theta_W$ from (3.166) back into equation (3.165) gives

$$\begin{aligned}\mathcal{L}_{\text{NC}} &= j_e^\mu A_\mu + \left[g\cos\theta_W(j_3^t)^\mu - \frac{g}{\cos\theta_W}\sin^2\theta_W(j_e^\mu/e - (j_3^t)^\mu)\right]Z_\mu \\ &= j_e^\mu A_\mu + \frac{g}{\cos\theta_W}\left[(j_3^t)^\mu - \sin^2\theta_W\frac{j_e^\mu}{e}\right]Z^\mu \\ &= j_e^\mu A_\mu + j_Z^\mu Z^\mu.\end{aligned} \tag{3.168}$$

The fermion couplings to the Z boson can be written in a couple of ways. One is to identify explicitly the left-handed C_L and right-handed C_R chiral couplings

$$\begin{aligned} j_Z^\mu &= \frac{g}{\cos\theta_W}\bar{\Psi}\gamma^\mu\left[(T^3 - Q\sin^2\theta_W)\frac{(1-\gamma^5)}{2} - Q\sin^2\theta_W\frac{(1+\gamma^5)}{2}\right]\Psi \\ &= \frac{g}{\cos\theta_W}\bar{\Psi}\gamma^\mu\left[C_L\frac{(1-\gamma^5)}{2} + C_R\frac{(1+\gamma^5)}{2}\right]\Psi \end{aligned} \tag{3.169}$$

with

$$C_L = T^3 - Q\sin^2\theta_W, \quad C_R = -Q\sin^2\theta_W. \tag{3.170}$$

In this form, it is clear that the coupling of the Z boson to right-handed matter would be zero if the weak mixing angle vanished. The creation of a weak boson from two weak-isospin scalars, as has been measured via $e^+e^- \to Z$ with right-handed chirality e^- in the initial state, would not be possible if the Z boson were part of a pure $SU(2)_L$ triplet. It is also common to write the neutral-current couplings in terms of the vector g_V and axial-vector g_A couplings:

$$j_Z^\mu = \frac{g}{\cos\theta_W}\bar{\Psi}\gamma^\mu\left[C_L\frac{(1-\gamma^5)}{2} + C_R\frac{(1+\gamma^5)}{2}\right]\Psi = \frac{g}{2\cos\theta_W}\bar{\Psi}\gamma^\mu(g_V - g_A\gamma^5)\Psi \tag{3.171}$$

with

$$g_V = C_L + C_R = T^3 - 2Q\sin^2\theta_W, \quad g_A = C_L - C_R = T^3. \tag{3.172}$$

As the axial-vector coupling has a particularly simple form $g_A = T^3$, one way of remembering the form of the vector coupling is that for $\sin^2\theta_W = 1/4$ (instead of ≈ 0.232), the vector coupling to a charged lepton would vanish, giving $g_V = -\frac{1}{2} + 2\sin^2\theta_W = 0$.

3.11 Fermion Masses and the CKM Mixing Matrix

In equation (3.156), left- and right-handed chiral components of the fermion fields are independent and can be explicitly separated:

$$\sum_{\text{fermions }\psi} \bar{\psi}i\not{D}\psi = \sum_{\text{left-fermions }\psi} \bar{\psi}_L i\not{D}\psi_L + \sum_{\text{right-fermions }\psi} \bar{\psi}_R i\not{D}\psi_R. \tag{3.173}$$

Explicit mass terms link the two chiral components and violate $SU(2)_L$ and $U(1)_Y$ gauge invariance

$$\Delta\mathcal{L}_e = -m_e(\bar{e}_L e_R + \bar{e}_R e_L). \tag{3.174}$$

However, gauge-invariant fermion mass terms appear naturally from spontaneous symmetry breaking

$$\Delta\mathcal{L}_e = -\lambda_e \bar{\mathbf{L}}\cdot\mathbf{\Phi} e_R + h.c. \tag{3.175}$$

where this general Yukawa interaction term is readily shown to be an $SU(2)_L$ singlet with zero net hypercharge.

The vacuum expectation of the Higgs field $\mathbf{\Phi}$ from equation (3.157) gives

$$\Delta\mathcal{L}_e = -\frac{1}{\sqrt{2}}\lambda_e v\bar{e}_L e_R + h.c. \tag{3.176}$$

with

$$m_e = \frac{1}{\sqrt{2}}\lambda_e v. \tag{3.177}$$

The largest Higgs-fermion Yukawa coupling is to the top quark. The current world average top quark mass is $m_t = 172.0 \pm 1.6\ \mathrm{GeV}/c^2$ [3], while the minimum of the Higgs potential is $v/\sqrt{2} \approx 174\ \mathrm{GeV}/c^2$, giving

$$\lambda_t = \sqrt{2}\, m_t/v = 0.989 \pm 0.009, \tag{3.178}$$

which is compatible with unity. The near unity value of the top quark-Higgs Yukawa coupling is unexplained in the Standard Model but is central to models of *radiative electroweak symmetry breaking.*

The Higgs mechanism for mass generation in the up-type quarks and neutrinos is not possible with a single complex-doublet Higgs field unless the charge-conjugate field is used asymmetrically with respect to $\mathbf{\Phi}$ in the Lagrangian

$$\mathbf{\Phi}_c = i\tau^2\mathbf{\Phi}^* = \frac{1}{\sqrt{2}}\begin{pmatrix} v+H \\ 0 \end{pmatrix} \tag{3.179}$$

which has $Y_{\Phi_c} = -1$.

Up- and down-type quark masses come from $\mathbf{\Phi}_c$ and $\mathbf{\Phi}$ Yukawa couplings, respectively,

$$\Delta\mathcal{L}_q = -\lambda_d^{ij}\bar{Q}^i\cdot\mathbf{\Phi} d_R^j - \lambda_u^{ij}\bar{Q}^i\cdot\mathbf{\Phi}_c u_R^j + h.c. \tag{3.180}$$

where λ_d^{ij} and λ_u^{ij} are general complex-valued matrices and

$$\begin{aligned} u_R^i &= (u_R, c_R, t_R), \qquad d_R^i = (d_R, s_R, b_R), \\ Q^i &= \begin{pmatrix} u_L^i \\ d_L^i \end{pmatrix} = \left(\begin{pmatrix} u_L \\ d_L \end{pmatrix}, \begin{pmatrix} c_L \\ s_L \end{pmatrix}, \begin{pmatrix} t_L \\ b_L \end{pmatrix}\right). \end{aligned} \tag{3.181}$$

As each successive mass generation is a replica of the lowest, with duplicate quantum numbers, the mass eigenstates of the fermions are determined by diagonalizing the 3×3 matrix of Higgs couplings over the three mass generations. Evaluating (3.180) at $\mathbf{\Phi}_{\mathrm{min}}$ and writing the weak interaction basis as u' and d' gives

$$\Delta\mathcal{L}_q = -\bar{d}'^i_L M'^{ij}_d d'^j_R - \bar{u}'^i_L M'^{ij}_u u'^j_R + h.c. \tag{3.182}$$

with

$$M'^{ij}_d = \lambda_d^{ij}\frac{v}{\sqrt{2}}, \qquad M'^{ij}_u = \lambda_u^{ij}\frac{v}{\sqrt{2}}. \tag{3.183}$$

We therefore define the following four unitary 3×3 transformations:

$$u_L^i \rightarrow (U_L^u)_{ij} u_L^j,\ \ d_L^i \rightarrow (U_L^d)_{ij} d_L^j,\ \ u_R^i \rightarrow (U_R^u)_{ij} u_R^j,\ \ d_R^i \rightarrow (U_R^d)_{ij} d_R^j \tag{3.184}$$

such that the mass matrices in equation (3.183) are diagonalized

$$\begin{aligned}\Delta\mathcal{L}_q &= -\bar{d}'_L M'_d d'_R - \bar{u}'_L M'_u u'_R \\ &= -\bar{d}'_L (U^d_L)^\dagger U^d_L M'_d (U^d_R)^\dagger U^d_R d'_R - \bar{u}'_L (U^u_L)^\dagger U^u_L M'_u (U^u_R)^\dagger U^u_R u'_R \\ &= -\bar{d}_L (U^d_L M'_d (U^d_R)^\dagger) d_R - \bar{u}_L (U^u_L M'_u (U^u_R)^\dagger) u_R \\ &= -\bar{d}^i_L (\delta_{ij} m^j_d) d^j_R - \bar{u}^i_L (\delta_{ij} m^j_u) u^j_R \\ &= -\sum_j m^j_d \bar{d}^j_L d^j_R - \sum_j m^j_u \bar{u}^j_L u^j_R \\ &= -m_d \bar{d}d - m_s \bar{s}s - m_b \bar{b}b - m_u \bar{u}u - m_c \bar{c}c - m_t \bar{t}t,\end{aligned} \tag{3.185}$$

giving the expected linear sum of independent Dirac masses for the six quark flavors. The matrices U_u and U_d commute with QCD couplings, but not necessarily with the $SU(2)_L \times U(1)_Y$ couplings.

The $SU(2)_L \times U(1)_Y$ Lagrangian for the quark sector (neglecting masses) can be written in terms of the weak interaction basis as follows:

$$\begin{aligned}\mathcal{L}_q &= i\bar{Q}'^j \gamma_\mu \left[\partial^\mu - igW^\mu_i \frac{\tau^i}{2} - ig'B^\mu \frac{Y_L}{2}\right] Q'^j \\ &+ i\bar{u}'^j_R \gamma_\mu \left[\partial^\mu - ig'B^\mu \frac{Y_R}{2}\right] u'^j_R + i\bar{d}'^j_R \gamma_\mu \left[\partial^\mu - ig'B^\mu \frac{Y_R}{2}\right] d'^j_R.\end{aligned} \tag{3.186}$$

One can see that the last two terms are invariant under the unitary transformations necessary to diagonalize the mass matrices, as $(U^u_R)^\dagger U^u_R = (U^d_R)^\dagger U^d_R = 1$. The neutral part of the first term, acting on the left-handed fermion doublet, is also invariant under the left-handed unitary transformations, U^u_L and U^d_L, as the neutral interactions are diagonal in quark flavor. The exception is the charged-current interaction, given by

$$\begin{aligned}\mathcal{L}_{CC} &= \frac{g}{\sqrt{2}} \bar{u}'^j_L \gamma_\mu \bar{d}'^j_L W^{+\mu} + h.c. \\ &= \frac{g}{\sqrt{2}} \left[\bar{u}^k_L \gamma_\mu (U^u_L)_{kj} (U^d_L)^\dagger_{jm} d^m_L\right] W^{+\mu} + h.c. \\ &= \frac{g}{\sqrt{2}} \left[\bar{u}^k_L \gamma_\mu V_{km} d^k_L\right] W^{+\mu} + h.c. \\ &= \frac{g}{\sqrt{2}} (\bar{u}_L \;\; \bar{c}_L \;\; \bar{t}_L) \gamma_\mu \begin{pmatrix} V_{ud} & V_{us} & V_{ub} \\ V_{cd} & V_{cs} & V_{cb} \\ V_{td} & V_{ts} & V_{tb} \end{pmatrix} \begin{pmatrix} d_L \\ s_L \\ b_L \end{pmatrix} W^{+\mu} + h.c.\end{aligned} \tag{3.187}$$

Thus, the three u^i_L quarks are linked with a unitary rotation of the triplet of d^i_L quarks

$$V = U^u_L (U^d_L)^\dagger. \tag{3.188}$$

The matrix V is known as the *Cabibbo-Kobayashi-Maskawa* (CKM) mixing matrix.

Chiral gauge theories naturally violate C and P. The CP transformation (and similarly with T) in (3.180) involves complex conjugation and therefore transforms the Higgs-fermion couplings as follows:

$$\lambda^{ij}_d \to (\lambda^{ij}_d)^*, \qquad \lambda^{ij}_u \to (\lambda^{ij}_u)^*. \tag{3.189}$$

CP would be a symmetry of (3.180) if the λ^{ij} were real-valued.

The matrix (3.188) is a general unitary $n \times n$ matrix. A unitary matrix is constrained relative to a general complex matrix in that the $2n^2$ parameters of a general complex matrix are reduced by n normalization and $n(n-1)$ orthogonality constraints, leaving n^2 free parameters of the unitary matrix. The $n^2 = 9$ parameters contain three rotation angles of an $O(3)$ rotation and $2n = 6$ quark phases. Therefore, there are $2n - 1 = 5$ parameters that are relative phases and can be chosen arbitrarily, but one phase remains as a free parameter. Thus, V contains $n^2 - (2n-1) = (n-1)^2 = 4$ free parameters, the three rotation angles, and one complex phase. The possibility of CP violation in the CKM mixing matrix from the presence of a nonzero complex phase requires a minimum of three generations, as a two-generation matrix has only one real rotation angle as a free parameter.

3.12 Neutrino Masses and the MNS Matrix

The Higgs-fermion Yukawa couplings provide a mechanism for mass generation in the fermions but provide little insight as to the origin and wide range of couplings constants, going from approximately unity strength at the top quark mass to 10^{-6} for the lowest mass generation of charged leptons and quarks. An even more profound mystery arises when one considers the finite mass of the neutrinos where the Yukawa coupling drops to less than 10^{-11}.

However, as the right-handed neutrino is a gauge singlet in the Standard Model, there are additional possibilities for mass terms beyond the Dirac mass terms originating from the Higgs mechanism

$$\Delta\mathcal{L}_{\text{Dirac}} = -\lambda_\ell^{ij} \bar{L}^i \cdot \Phi e_R^j - \lambda_\nu^{ij} \bar{L}^i \cdot \Phi_c \nu_R^j + h.c. \tag{3.190}$$

where λ_ℓ^{ij} and λ_ν^{ij} are general complex-valued matrices with the $\{ij\}$ indices running over the three generations of matter as defined in equation (3.119). The Dirac mass terms for neutrinos are constructed from the Φ_c interactions where $\Phi_c = i\tau^2\Phi^*$ indicates charge conjugation. In fact, the charge-conjugation terms involving the Higgs scalar field can be viewed as an unusual, though allowed within the Standard Model, addition to the possible Higgs-fermion interactions as required to generate masses in up-type fermions. Charge conjugation of the left- and right-handed two-component neutrino Weyl wave functions, Ψ_L and Ψ_R, is also a possibility when constructing mass terms. One can therefore consider the admixture of neutrino mass terms involving charge conjugation to include *Majorana mass terms*, $m_L \bar{\Psi}_L \Psi_R^c$ and $m_R \bar{\Psi}_L^c \Psi_R$, for the left- and right-handed chirality combinations of the Weyl wave functions, respectively,

$$\Delta\mathcal{L}_\nu = -\frac{1}{2}\begin{pmatrix} \bar{\Psi}_L & \bar{\Psi}_L^c \end{pmatrix} \begin{pmatrix} m_L & m_D \\ m_D & m_R \end{pmatrix} \begin{pmatrix} \Psi_R^c \\ \Psi_R \end{pmatrix} + h.c. \tag{3.191}$$

The diagonalization of the neutrino mass matrix (3.191) yields the mass eigenvalues

$$m_{1,2} = \frac{1}{2}\left[(m_L + m_R) \pm \sqrt{(m_L - m_R)^2 + 4m_D^2}\right]. \tag{3.192}$$

If we then hypothesize that the intrinsic Dirac mass terms of the neutrino are comparable to the charged leptons, $m_D \approx m_e$, and that the Majorana mass term m_R for the right-handed chirality combination is heavy compared to the left-handed chirality combination, $m_R \gg m_L$, and the Dirac mass terms, $m_R \gg m_D$, then we find the approximate solutions

$$m_1 \approx \frac{m_D^2}{m_R},$$
$$m_2 \approx m_R. \tag{3.193}$$

If m_R is of order 1000 TeV or heavier, then the light mass solution m_1 would be suppressed by the factor needed to explain the unusually light neutrino masses compared to the charged leptons and quarks. This effect is known as the *seesaw mechanism*. If the Majorana mass terms are indeed present as predicted by the seesaw mechanism, a heavy right-handed Majorana neutrino N would exist corresponding to the m_2 mass solution. In general, the existence of massive Majorana neutrinos would open up the possibility to observe lepton number violation. In a process known as neutrinoless double beta decay, a nucleus ${}^A_ZX$ with atomic number A and charge Z can transition to a lighter nucleus ${}^{A}_{Z+2}X$ through the emission of two electrons:

$${}^A_ZX \rightarrow {}^{A}_{Z+2}X + 2e^- \tag{3.194}$$

where the weak decay process proceeds through the internal emission and absorption of a virtual Majorana neutrino. A left-handed chiral weak charged current produces an electron and a virtual, right-handed neutrino at one vertex. Recall that there is no distinction between a neutrino and an antineutrino for Majorana particles. The chirality of the virtual, right-handed neutrino subsequently flips through a scalar chiral interaction with the physical vacuum and then is reabsorbed as a left-handed chirality neutrino in the second vertex to produce the second electron, thereby violating lepton number. Evidence for $0\nu\beta\beta$ processes in Nature still awaits experimental confirmation, and thus the existence of Majorana neutrinos is uncertain.

Dirac neutrinos produced promptly in weak charged-current interactions have a lepton flavor that corresponds to that of the charged leptons produced in association with the neutrino. Thus, according to the weak eigenstates of the neutrinos, there are three generations of Dirac neutrinos. However, neutrinos are stable in the Standard Model as their masses are below that of the charged leptons. Therefore, stable neutrinos will propagate through space according to the mass eigenstates, or admixture of mass eigenstates, into which the neutrino was produced as determined by the Heisenberg uncertainty principle. In general, the mass eigenstates ν_i are related to the weak eigenstates ν_α according to a 3×3 unitary matrix, called the *Maki-Nakagawa-Sakata* (MNS) matrix,

$$|\nu_\alpha\rangle = \sum_{i=1,3} U^*_{\alpha i}|\nu_i\rangle. \tag{3.195}$$

In the lepton sector, the mass mixing matrix gives rise to the phenomenon of neutrino oscillations. This effect and the corresponding neutrino mixing matrix are discussed in the chapter on neutrino oscillations.

3.13 Interaction Vertices in the Standard Model

We can now consider all possible interaction vertices in the Standard Model. These are constructed from the gauge interactions as they enter the covariant derivatives in the

Table 3.1 Mass dimensions of fields and coupling constants in gauge theories.

Scalar field	$\frac{1}{2}(\partial^\mu\phi)(\partial_\mu\phi)$	$\to [\phi]$	$= M^1$
Dirac field	$\bar{\Psi}M\Psi$	$\to [\Psi]$	$= M^{3/2}$
Vector field	$-\frac{1}{4}F_{\mu\nu}F^{\mu\nu}$	$\to [F_{\mu\nu}]$	$= M^2$
		$\to [A_\mu]$	$= M^1$
Minimal gauge coupling	$g\bar{\Psi}\gamma_\mu\Psi A^\mu$	$\to [g]$	$= M^0$
Triple-gauge boson coupling	$gA_\mu A^\mu \partial_\nu A^\nu$	$\to [g]$	$= M^0$
Quartic-gauge boson coupling	$g^2(A_\mu A^\mu)^2$	$\to [g]$	$= M^0$
Yukawa coupling	$g\bar{\Psi}\Psi\phi$	$\to [g]$	$= M^0$
Scalar couplings to gauge bosons	$\tilde{g}A_\mu A^\mu\phi$	$\to [\tilde{g}]$	$= M^1$
	$gA_\mu(\partial^\mu\phi)\phi$	$\to [g]$	$= M^0$
	$g^2\phi^2 A_\mu A^\mu$	$\to [g]$	$= M^0$
Scalar self-couplings	$\tilde{g}\phi\phi\phi$	$\to [\tilde{g}]$	$= M^1$
	$g\phi\phi\phi\phi$	$\to [g]$	$= M^0$

kinetic energy terms of the gauge bosons, the fermions, and the Higgs boson. There are additional diagrams from the Higgs self-interactions and the Higgs-fermion Yukawa interactions. A general construct for enumerating the interactions is to consider the *mass dimension* of the interaction vertex.

A Lagrangian density $\mathcal{L}$ has the dimension $[\mathcal{L}] = M^4$, following from the fact that the action $\int d^4x\mathcal{L}$ is dimensionless. Therefore, if one takes the list of kinetic energy and interaction terms in the Lagrangian density, it is possible to evaluate the mass dimension of fields and coupling constants. A list is given in table 3.1. With the exception of $\tilde{g}$, which has mass dimension from the nonzero Higgs (scalar) field vacuum expectation value, the gauge couplings have zero mass dimension.

A compilation of Standard Model interactions and their vertex terms is found in Appendix A.

3.14 Higgs Mechanism and the Nambu-Goldstone Theorem

In this section, we take a second look at what the Higgs mechanism is doing and touch on some of the theoretical considerations needed to make the Standard Model predictions finite at small length scales. One result, quoted here, is that a critical aspect of spontaneous symmetry breaking is the ability to generate masses in gauge bosons through vacuum polarization diagrams that obey the Ward identity. The Ward identity, shown in equation (3.59) and repeated here, is a consequence of gauge invariance; the substitution of the gauge boson polarization ϵ^μ with the photon four-momentum q^μ will yield zero when contracted against the rest of the vacuum polarization amplitude $\mathcal{M}$,

$$q^\mu \mathcal{M}_\mu = 0.$$

As the Ward identity follows from gauge invariance, we can infer that this property protects the finiteness of loop integrals that contribute to the gauge boson mass.

To see this property in action, consider QED with a $U(1)$ Higgs sector and no fermions,

$$\mathcal{L} = -\frac{1}{4}(F_{\mu\nu})^2 + |D_\mu \Phi|^2 + \mu^2 \Phi^* \Phi - \frac{\lambda}{2}(\Phi^* \Phi)^2, \tag{3.196}$$

with $D_\mu = \partial_\mu + ieA_\mu$ and the latter part identified as $-V(\Phi)$. The minimum of the Higgs potential occurs at (up to a $U(1)$ phase)

$$\langle \Phi \rangle = \Phi_{\text{min}} = \sqrt{\frac{\mu^2}{\lambda}}. \tag{3.197}$$

Expand the complex field $\Phi(x)$

$$\Phi(x) = \Phi_{\text{min}} + \frac{1}{\sqrt{2}}\big(\Phi_1(x) + i\Phi_2(x)\big) \tag{3.198}$$

and insert into the potential

$$V(\Phi) = -\frac{1}{2\lambda}\mu^4 + \frac{1}{2}\cdot 2\mu^2\Phi_1^2 + \mathcal{O}(\Phi_i^3) \tag{3.199}$$

to see that Φ_1 acquires a mass $m = \sqrt{2}\mu$ and Φ_2 remains massless. The massless Φ_2 contributes a $q^2 = 0$ pole in the vacuum polarization amplitude of A_μ in the following way. First, expanding the kinetic energy term gives

$$|D_\mu \Phi|^2 = \frac{1}{2}(\partial_\mu \Phi_1)^2 + \frac{1}{2}(\partial_\mu \Phi_2)^2 + \sqrt{2}e\Phi_{\text{min}} \cdot A_\mu \partial^\mu \Phi_2 + e^2 \Phi_{\text{min}}^2 A_\mu A^\mu + \cdots \tag{3.200}$$

where the last two terms both contribute to the vacuum polarization amplitude and the last term defines $m_A^2 = 2e^2\Phi_{\text{min}}^2$. Then, the vacuum polarization amplitude is pieced together from the fourth term from equation (3.200) and the third term acting twice, shown diagrammatically in figure 3.7, giving

$$\begin{aligned} \mathcal{M}^{\mu\nu}_{\text{pol}}(k) &= im_A^2 g^{\mu\nu} + (m_A k^\mu)\frac{i}{k^2}(-m_A k^\nu) \\ &= im_A^2\left(g^{\mu\nu} - \frac{k^\mu k^\nu}{k^2}\right), \end{aligned} \tag{3.201}$$

which has exactly the right form to make the amplitude satisfy the Ward identity

FIGURE 3.7. Diagrammatical relationship between the vacuum polarization amplitude and contributing terms from interactions with the Higgs field.

$$k_\mu \mathcal{M}^{\mu\nu}_{\text{pol}}(k) = 0. \tag{3.202}$$

This means that the mass of the gauge boson obtained via spontaneous symmetry breaking comes from an amplitude that is transverse and therefore well-behaved at small length scales. In the *unitary gauge*, the Goldstone boson $\mathbf{\Phi}_2$ does not appear as an independent physical particle but rather supplies the extra degree of freedom needed to give A_μ a longitudinal polarization. However, we will see in the following section on the Goldstone boson equivalence theorem that the Goldstone boson degree of freedom in a massive gauge boson can be treated as the original free scalar particle in the high-energy limit.

Now consider a non-Abelian gauge theory with a Higgs sector invariant under this group. If we choose $\mathbf{\Phi}$ to be in the vector representation of $SU(2)$, then the kinetic energy term can be written as

$$\frac{1}{2}(D_\mu \mathbf{\Phi})^2 = \frac{g^2}{2}\left(\epsilon_{abc} A^b_\mu (\mathbf{\Phi}_{\text{min}})_c\right)^2 + \cdots. \tag{3.203}$$

If we choose the vacuum expectation value $(\mathbf{\Phi}_{\text{min}})_c$ to point along the 3-direction

$$\langle \mathbf{\Phi}_c \rangle = (\mathbf{\Phi}_{\text{min}})_c = v\delta_{c3}, \tag{3.204}$$

then the mass term becomes

$$\frac{1}{2} m^2_{ab} A^a_\mu A^{b\mu} = \frac{g^2}{2} v^2 \left(\epsilon_{ab3} A^b_\mu\right)^2 = \frac{g^2}{2} v^2 \left((A^1_\mu)^2 + (A^2_\mu)^2\right) \tag{3.205}$$

where $m_1 = m_2 = gv$ and $m_3 = 0$ by the choice of pointing $\langle \mathbf{\Phi}_c \rangle$ along the 3-direction.

The choice of axis for the vacuum expectation of $\mathbf{\Phi}$ allows us to produce nonequal gauge boson masses and to preserve symmetries in the spontaneously broken theory. In larger symmetry groups, the direction of the vacuum expectation value is more easily seen in the adjoint representation

$$\mathbf{\Phi} = \mathbf{\Phi}_c t^c \tag{3.206}$$

where the t^c are the traceless $N \times N$ Hermitian matrices that represent the generators of the group. The commutation or noncommutation of $\mathbf{\Phi}$ with the generators of the group indicate the gauge bosons of the corresponding generators that remain massless or acquire mass, respectively.

For $SU(3)$, the choice of orientation along t^8, giving

$$\mathbf{\Phi}_{\text{min}} = v \begin{pmatrix} 1 & & \\ & 1 & \\ & & -2 \end{pmatrix}, \tag{3.207}$$

leaves the upper 2×2 diagonal, and therefore commutes with the generators

$$t^a = \frac{1}{2}\begin{pmatrix} \sigma^a & 0 \\ 0 & 0 \end{pmatrix}, \qquad t^8 = \frac{1}{2\sqrt{3}} \begin{pmatrix} 1 & & \\ & 1 & \\ & & -2 \end{pmatrix}, \tag{3.208}$$

breaking $SU(3)$ down to $SU(2) \times U(1)$.

For $SU(5)$, the choice of orientation

$$\mathbf{\Phi}_{\text{min}} = v \begin{pmatrix} 2 & & & & \\ & 2 & & & \\ & & 2 & & \\ & & & -3 & \\ & & & & -3 \end{pmatrix} \tag{3.209}$$

leaves the upper $SU(3)$ and the lower $SU(2)$ blocks intact in addition to one generator in diagonal representation. This breaks $SU(5)$ down to $SU(3) \times SU(2) \times U(1)$ and is an interesting consideration for a *grand unified theory* (GUT) model with spontaneous symmetry breaking. Note that the color triplet scalar as represented by the upper diagonal of equation (3.209) has the same quantum numbers under the broken $SU(3) \times SU(2) \times U(1)$ gauge interaction as the S boson described in equation (3.124).

3.15 Goldstone Boson Equivalence

When taken in the proper limit, the four degrees of freedom of the complex Higgs doublet in the Standard Model manifest themselves as real particles. Here we examine the behavior of the longitudinal polarization component of the massive W and Z bosons in the high-momentum limit as an example of a result known as the *Goldstone boson equivalence theorem.*

The longitudinal polarization of a massive vector boson grows with momentum $|\boldsymbol{k}|$ as

$$\epsilon_S^{\mu}(k) = \left(\frac{|\boldsymbol{k}|}{m}, 0, 0, \frac{E}{m}\right) \tag{3.210}$$

and becomes increasingly parallel to $k^{\mu} = (E, 0, 0, \boldsymbol{k})$ while maintaining $\epsilon_S \cdot k = 0$ and $k \cdot k = m^2$.

The enhanced longitudinal polarization behaves in high-energy interactions as the equivalent Goldstone boson with Higgs field couplings.

The mass of the top quark $m_t \approx 172$ GeV/c^2 is sufficiently larger than the mass of the W boson $m_W \approx 80.4$ GeV/c^2 to approximate the dominant $t \to W^+ b$ decay by $t \to \mathbf{\Phi}^+ b$ using Goldstone boson equivalence.

The Yukawa coupling of the top quark to the $\mathbf{\Phi}^+$ scalar field is

$$\Delta \mathcal{L} = \lambda_t \bar{b}_L \mathbf{\Phi}^+ t_R. \tag{3.211}$$

Computing the top decay amplitude

$$i\mathcal{M} = i\lambda_t \bar{u}(q)\left(\frac{1+\gamma^5}{2}\right) u(p) \tag{3.212}$$

gives the squared spin-averaged matrix element

$$\frac{1}{2}\sum_{\text{spins}} |\mathcal{M}|^2 = \lambda_t^2 q \cdot p. \tag{3.213}$$

The decay rate in this approximation is

$$\Gamma \approx \frac{\lambda_t^2}{32\pi} m_t = \frac{g^2}{64\pi} \frac{m_t^3}{m_W^2}, \tag{3.214}$$

correctly predicting the leading term of the full decay rate (approx. 70% of total rate).

An even more clear example of the Goldstone boson equivalence theorem arises when one considers the scattering of massive weak bosons off other weak bosons at high energy. In this limit, the longitudinal polarizations are dominant in the scattering matrices and one can drop the weak bosons altogether and compute the scattering in terms of the original ϕ^4-theory of the Higgs sector.

At high energies compared to m_W, the amplitude for $W^+W^- \to ZZ$ scattering in the gauge sector grows as E^2, as shown in figure 3.8,

$$\mathcal{M}_{\text{gauge}}(W_S^+ W_S^- \to Z_S Z_S) = \frac{s}{v^2}, \tag{3.215}$$

reminiscent of neutrino scattering in Fermi theory. The dominant contribution comes from longitudinal-longitudinal scattering, denoted by S as in equation (3.63), and can be computed with the Goldstone boson equivalence.

There is an additional scattering diagram involving the physical Higgs boson H, as shown in figure 3.9.

$$\mathcal{M}_H(W_S^+ W_S^- \to Z_S Z_S) = -\frac{s}{v^2}\left(\frac{s}{s - m_H^2}\right), \tag{3.216}$$

which for finite Higgs mass cancels the growing s-dependence of (3.215) so that

$$\mathcal{M}(W_S^+ W_S^- \to Z_S Z_S) = \mathcal{M}_{\text{gauge}} + \mathcal{M}_H = \frac{s}{v^2} - \frac{s}{v^2}\left(\frac{s}{s - m_H^2}\right) \to -\frac{m_H^2}{v^2} \tag{3.217}$$

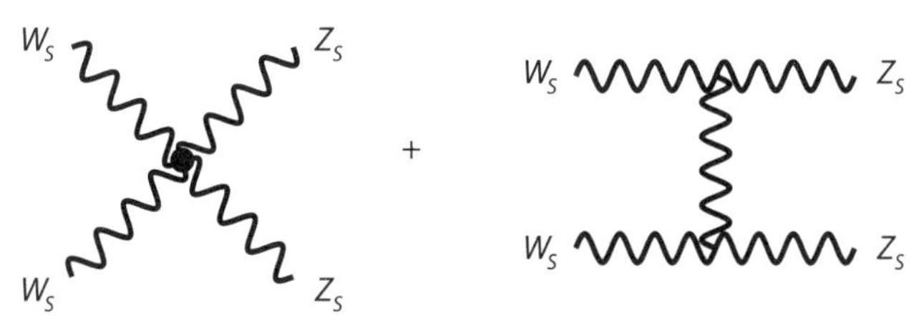

FIGURE 3.8. Scattering diagrams involving longitudinally (S) polarized W and Z bosons for the process $W^+W^- \to ZZ$.

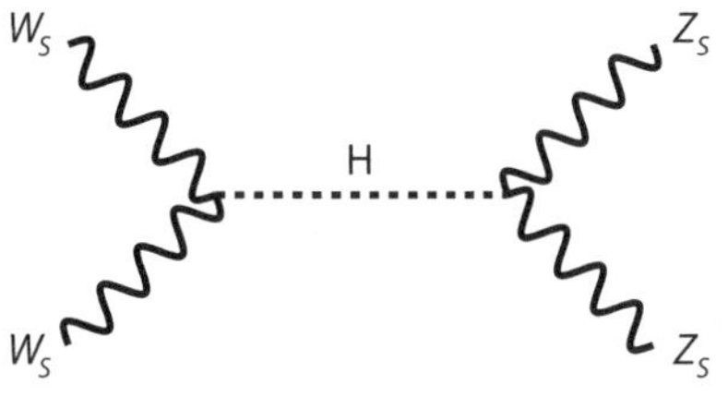

FIGURE 3.9. An additional scattering diagram involving the Higgs boson H and longitudinally (S) polarized W and Z bosons in the $W^+W^- \to ZZ$ process.

for $s \gg m_H^2$. In the absence of the $\mathcal{M}_H$ amplitude, the equivalent upper bound on the growth of the $\mathcal{M}_{\text{gauge}}$ amplitude gives $\sqrt{s} = 4\sqrt{\pi}\,v \approx 1.7$ TeV/c^2. Therefore, in the absence of a Higgs boson below the TeV scale, the longitudinal-longitudinal scattering of massive weak bosons saturates the perturbative S-matrix expansion and the theory becomes strongly coupled. Therefore, in the absence of a sub-TeV mass Higgs boson, electroweak interactions will take on a completely different character at small length scale and introduce new experimentally measureable phenomena such as weak diboson bound states and resonances.

3.16 Anomaly Cancellation

In this final section, we take a brief look at one of the pillars of theoretical work that underscores why the predictions of the Standard Model are finite beyond the tree-level calculations described so far. Anomaly cancellation as will be shown in examples below places constraints on the collection of elementary particles charged under the $SU(3)_C \times SU(2)_L \times U(1)_Y$ gauge interactions.

An *anomaly* is a situation in which an invariance derived from the equations of motion (or equivalently from the Lorentz-invariant Lagrangian density via Noether's theorem) is formally no longer valid after the application of quantum corrections. The divergence of quantum corrections, or "loops," in the theory can involve multiple gauge interactions and, in particular, chiral gauge interactions. If elementary particles are arranged in groups whose summed contributions to the divergent loops cancel, then the anomaly will be canceled. In the Standard Model, anomaly cancellation provides a connection between what would be otherwise independent gauge degrees of freedom of $SU(3)_C$ and $SU(2)_L \times U(1)_Y$. The requirement that a theory be anomaly free is not guaranteed by a gauge theory, and yet the Standard Model is anomaly free. It is believed in this respect that anomaly cancellation is one of the defining principles of elementary particle interactions.

Quoting a result, for a chiral gauge theory, the *axial-vector current anomaly* arising from the chiral transformation of the Dirac Lagrangian

$$\psi(x) \to \exp(i\alpha\gamma^5)\psi(x) \tag{3.218}$$

destroys the Noether current of the theory, and consequently generates divergent gauge-boson mass terms. We consider here the fermions that couple to the gauge bosons X^a through the currents

$$J^a_\mu = \bar{\Psi}_R \gamma_\mu T^a_R \Psi_R + \bar{\Psi}_L \gamma_\mu T^a_L \Psi_L \tag{3.219}$$

where the right-handed and left-handed currents can be different in chiral gauge theories. Recall that $SU(2)_L$ has different representations for left- and right-handed fermions. Similarly, $U(1)_Y$ also has different charges depending on chirality. In particular, triangle diagrams arising from the axial-vector current involving three gauge bosons can be divergent. When considering crossed and uncrossed diagrams, as shown in the figures below, we have the cancellation condition that

$$\sum_{\text{all left-handed fermions}} (T^a T^b + T^b T^a) T^c - \sum_{\text{all right-handed fermions}} (T^a T^b + T^b T^a) T^c = 0 \qquad (3.220)$$

where $T = T_3$ for $SU(2)_L$, $T = Y$ for $U(1)_Y$, and $T = \lambda_3/2$ for $SU(3)_C$. For example, the diagrams in figure 3.10 summed over all fermions must cancel, which according to equation (3.220) gives

$$\sum_{\text{left}} (T_3)^2 Y - \sum_{\text{right}} (T_3)^2 Y = \sum_{\text{left}} (T_3)^2 Y = 0 \qquad (3.221)$$

or, equivalently, when summing over all left-handed fermions $\boldsymbol{L}$ and $\boldsymbol{Q}$

$$\sum_{\text{all left-handed fermions}} Y = 2(-1) + 3 \cdot 2 \cdot \frac{1}{3} = 0 \qquad (3.222)$$

with the factor of 3 for the number of color states of the quarks.

A similar set of diagrams for $U(1)_Y$ acting at all vertices, shown in figure 3.11, gives

$$\sum_{\text{left}} Y^3 - \sum_{\text{right}} Y^3 = \left[2(-1)^3 + 3 \cdot 2\left(\frac{1}{3}\right)^3\right] - \left[(-2)^3 + 3\left(\frac{4}{3}\right)^3 + 3\left(-\frac{2}{3}\right)^3\right] = 0. \qquad (3.223)$$

If we consider a diagram involving two $SU(3)_C$ gauge bosons and one $U(1)_Y$ boson, then we get the following condition:

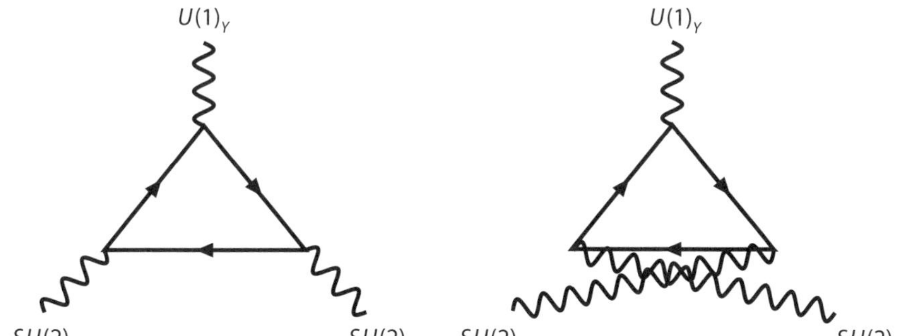

FIGURE 3.10. Triangle diagrams with chiral $U(1)_Y$ and $SU(2)_L$ interactions.

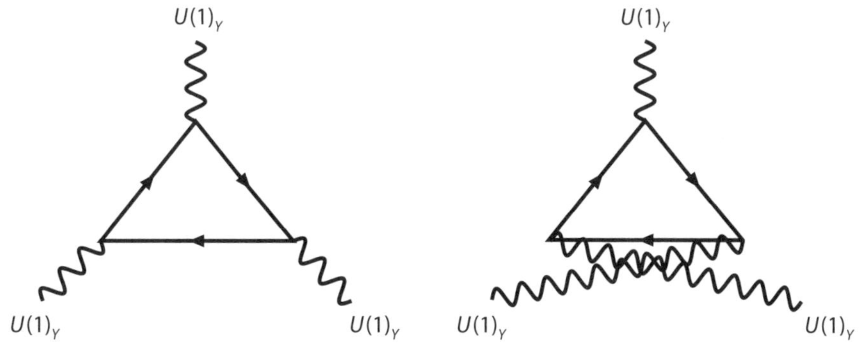

FIGURE 3.11. Triangle diagrams with chiral $U(1)_Y$ interactions.

$$\sum_{\text{left}} (\lambda_3/2)^2 Y - \sum_{\text{right}} (\lambda_3/2)^2 Y = 0 \tag{3.224}$$

or, equivalently, when summing over the quarks

$$\sum_{\text{left-handed quarks}} Y - \sum_{\text{right-handed quarks}} Y = 3 \cdot \left[2\left(\frac{1}{3}\right) - \frac{4}{3} - \left(-\frac{2}{3}\right)\right] = 0. \tag{3.225}$$

Similarly, the crossed and uncrossed diagrams with two $SU(3)_C$ gauge bosons and one $SU(2)_L$ boson yield the condition

$$\sum_{\text{left}} (\lambda_3/2)^2 T_3 - \sum_{\text{right}} (\lambda_3/2)^2 T_3 = 0 \tag{3.226}$$

or, equivalently, when summing over the left-handed quarks

$$\sum_{\text{left-handed quarks}} T_3 = 3 \cdot \left[\frac{1}{2} - \frac{1}{2}\right] = 0. \tag{3.227}$$

The Standard Model of $SU(3)_C \times SU(2)_L \times U(1)_Y$ is a chiral gauge theory free of axial anomalies. The cancellation of anomalies requires that leptons and quarks appear in complete generations

$$(u_R, Q, d_R, L, e_R) \tag{3.228}$$

where all but the gauge singlet right-handed neutrinos ν_R are active in loop diagrams. It was this principle, the cancellation of anomalies and the need for completed generations of matter, that guided precision electroweak measurements to constrain the top quark mass in advance of the direct production of top quarks at high-energy hadron colliders.

3.17 Exercises

1. Follow the procedure from equations (3.28–3.32) and show that the correct gauge transformation for an $SU(2)$ interaction is given by

   $$\mathbf{W}'_\mu(x) = \mathbf{W}_\mu(x) + \partial_\mu \boldsymbol{\alpha}(x) + g\mathbf{W}_\mu(x) \times \boldsymbol{\alpha}(x)$$

   to first order in g.

2. Compute the commutator of two-covariant derivatives $[\mathcal{D}^\mu, \mathcal{D}^\nu]$ with $\mathcal{D}^\mu = \partial^\mu - ig\mathbf{W}^\mu \cdot \boldsymbol{\tau}/2$. It is always assumed that the commutator is an operator acting on a Dirac wave function.

3. The $SU(2)$ field tensor is defined as

   $$\mathbf{F}^{\mu\nu} = \partial^\mu \mathbf{W}^\nu - \partial^\nu \mathbf{W}^\mu + g\mathbf{W}^\mu \times \mathbf{W}^\nu.$$

   Show that the kinetic energy term

   $$\mathcal{L}_{\text{kinetic}} = -\frac{1}{4}\mathbf{F}^{\mu\nu}\mathbf{F}_{\mu\nu}$$

is invariant under the local $SU(2)$ gauge transformation given in problem 1. Evaluate whether the mass term is gauge invariant

$$\mathcal{L}_{\text{mass}} = \frac{1}{2} M^2 \mathbf{W}_\mu \mathbf{W}^\mu .$$

4. For an $SU(2)$ triplet, the covariant derivative has the form

$$(\mathcal{D}^\mu)_{lm} = \partial^\mu \delta_{lm} + g\epsilon_{klm} W_k^\mu .$$

What are the three column vectors corresponding to the isospin components $T_3 = -1$, 0, +1 for the triplet in the adjoint representation?

5. Show that there is one remaining gauge degree of freedom $U(1)_{\text{EM}}$ in the Higgs vacuum $\mathbf{\Phi}_{\text{min}}$.

Guide: Given the gauge transformations for the matter fields under $SU(2)_L \times U(1)_Y$

$$\begin{aligned} \text{under } SU(2)_L: \quad & \mathbf{\Phi}' = \exp(ig\boldsymbol{\alpha}(x)\cdot\boldsymbol{\tau}/2)\mathbf{\Phi} \\ \text{under } U(1)_Y: \quad & \mathbf{\Phi}' = \exp(ig'\chi(x)Y_\Phi/2)\mathbf{\Phi} \end{aligned}$$

find the unitary transformation U such that the complex doublet Higgs field $\mathbf{\Phi}$ with four real fields $(\sigma_1(x), \sigma_2(x), \sigma_3(x), H(x))$ is rotated to $\mathbf{\Phi}'$ as follows:

$$\mathbf{\Phi}'(x) = \frac{1}{\sqrt{2}}\begin{pmatrix} 0 \\ v + H(x) \end{pmatrix} = U\mathbf{\Phi}(x) = U\frac{1}{\sqrt{2}}\begin{pmatrix} \sigma_1(x) + i\sigma_2(x) \\ v + H(x) + i\sigma_3(x) \end{pmatrix}$$

by making the three massless real scalar fields $\sigma_1(x)$, $\sigma_2(x)$, $\sigma_3(x)$ functions of $\alpha_1(x)$, $\alpha_2(x)$, $\alpha_3(x)$, and $\chi(x)$. Determine the $U(1)$ unitary transform U' such that

$$\mathbf{\Phi}'(x) = \frac{1}{\sqrt{2}}\begin{pmatrix} 0 \\ v + H(x) \end{pmatrix} = U'\frac{1}{\sqrt{2}}\begin{pmatrix} 0 \\ v + H(x) \end{pmatrix}$$

where U' is a local gauge transformation with a space-time dependent $U(1)$ phase $\chi'(x)$. Writing U' as an $SU(2)_L \times U(1)_Y$ gauge transformation, what are the values of $\boldsymbol{\alpha}(x)$ and $\chi(x)$ in terms of $\chi'(x)$? Using these relationships for $\boldsymbol{\alpha}(x)$ and $\chi(x)$ in terms of $\chi'(x)$, apply the corresponding gauge transformations on the physical fields recalling that the gauge transformations of the W_i^μ and B^μ fields are

$$\begin{aligned} \mathbf{W}'_\mu(x) &= \mathbf{W}_\mu(x) + \partial_\mu \boldsymbol{\alpha}(x) + g\mathbf{W}_\mu(x) \times \boldsymbol{\alpha}(x) \\ B'_\mu(x) &= B_\mu(x) + \partial_\mu \chi(x) \end{aligned} \tag{3.229}$$

and the definition of the physical fields $W^\pm$, Z^μ, and A^μ in terms of W_i^μ and B^μ. Show that the resulting gauge degree of freedom in the physical fields corresponds to the electromagnetic interaction $U(1)_{\text{EM}}$.

6. Number of light neutrino families. The measured width of the Z boson is $\Gamma_Z = 2.495 \pm 0.002$ GeV. Assume that the observed total width corresponds to exactly three neutrino families $(\nu_e, \nu_\mu, \nu_\tau)$, three charged lepton families (e, μ, τ), and five quark flavors (u, d, c, s, b). What would the value of Γ_Z have been if instead there had been four flavors of light neutrinos (and no other new light quarks or charged leptons)? Assume that $\sin^2\theta_W = 0.232$, where θ_W is the Weinberg angle. Recall that the electroweak neutral current is given by

$$j_Z^\mu = \frac{g}{\cos\theta_W}\bar{\Psi}\gamma^\mu\left[C_L\frac{(1-\gamma^5)}{2} + C_R\frac{(1+\gamma^5)}{2}\right]\Psi = \frac{g}{2\cos\theta_W}\bar{\Psi}\gamma^\mu(g_V - g_A\gamma^5)\Psi \qquad (3.230)$$

with

$$C_L = T^3 - Q\sin^2\theta_W, \quad C_R = -Q\sin^2\theta_W, \qquad (3.231)$$

or, alternatively, with

$$g_V = C_L + C_R = T^3 - 2Q\sin^2\theta_W, \quad g_A = C_L - C_R = T^3. \qquad (3.232)$$

7. **The Meissner effect.** The propagation of photons in a superconductor is inhibited due to the nonzero electric charge of scalar Cooper pairs (e^-e^-) that have condensed in the "vacuum" of the superconductor. The Higgs mechanism acting on QED is a model for describing this phenomenon. The Lagrangian describing the kinetic energy and self-interaction of the Cooper pairs ϕ and the kinetic energy term of photons in QED is given by

$$\mathcal{L} = (D_\mu\phi^\dagger)(D^\mu\phi) + \frac{\mu^2}{2}\phi^\dagger\phi - \frac{\lambda}{4}(\phi^\dagger\phi)^2 - \frac{1}{4}F^{\mu\nu}F_{\mu\nu} \qquad (3.233)$$

where the minimal gauge interaction of ϕ and the electomagnetic field tensor are given by

$$D_\mu\phi = (\partial_\mu - i2eA_\mu)\phi \quad \text{and} \quad F^{\mu\nu} = \partial^\mu A^\nu - \partial^\nu A^\mu, \qquad (3.234)$$

respectively.

(a) Assuming a zero vacuum expectation for ϕ, compute Maxwell's equations within the volume of the superconductor for a static A^μ and a constant potential $A^0 = \text{constant}$. The current that couples to A^μ is given by the Noether current

$$2ej_\nu = i2e\left[(D_\mu\phi^\dagger)\phi - \phi^\dagger(D_\mu\phi)\right]. \qquad (3.235)$$

(b) In the presence of spontaneous symmetry breaking, show that the current $J^\nu = 2ej^\nu$ satisfies the London equation

$$J = (2e)^2 v^2 \mathbf{A} \qquad (3.236)$$

where v is the vacuum expectation value for the classical minimum of the ϕ potential.

(c) Derive the Meissner effect

$$\nabla^2\mathbf{B} = (2e)^2 v^2 \mathbf{B} \qquad (3.237)$$

and show that an external magnetic field entering the volume of the superconductor will exponentially decay with a length scale given by the photon mass. Show the photon mass has the form of a coupling times a vacuum expectation value as is typical of the Higgs mechanism.

8. **Triplet Higgs.** As opposed to the usual Standard Model, in which the Higgs field is in a doublet representation of $SU(2)_L$ with hypercharge $Y = 1$, assume the Higgs boson is in a *triplet* representation of $SU(2)_L$ with hypercharge $Y = 2$. Aside from that, the content of the model is the same as for the Standard Model. The generators of the doublet representation of $SU(2)_L$ are

$$T^1 = \frac{1}{2}\begin{pmatrix} 0 & 1 \\ 1 & 0 \end{pmatrix}, \; T^2 = \frac{1}{2}\begin{pmatrix} 0 & -i \\ i & 0 \end{pmatrix}, \; T^3 = \frac{1}{2}\begin{pmatrix} 1 & 0 \\ 0 & -1 \end{pmatrix}, \tag{3.238}$$

whereas for the triplet representation they are

$$T^1 = \begin{pmatrix} 0 & 0 & 0 \\ 0 & 0 & -i \\ 0 & i & 0 \end{pmatrix}, \; T^2 = \begin{pmatrix} 0 & 0 & i \\ 0 & 0 & 0 \\ -i & 0 & 0 \end{pmatrix}, \; T^3 = \begin{pmatrix} 0 & -i & 0 \\ i & 0 & 0 \\ 0 & 0 & 0 \end{pmatrix}. \tag{3.239}$$

(a) Construct a tree-level potential which permits spontaneous breaking of $SU(2)_L \times U(1)_Y \to U(1)_{EM}$ gauge symmetry. Derive the masses of the W and Z gauge bosons. Perform your analysis at *tree level*.

(b) What is the low-energy spectrum of scalars and gauge bosons? Give the fields and their masses in terms of the gauge couplings and any parameters you have defined in your potential.

In the usual Standard Model, the ratio of the masses of the Z and W bosons is $M_Z/M_W = \cos^{-1}\theta_W$, where θ_W is the weak mixing angle, i.e., $\tan\theta_W = g'/g$. In our alternative model with a Higgs triplet, is there a value of the Higgs expectation value such that the ratio of masses satisfies this relation? State two measurements which can distinguish this spectrum from the conventional Standard Model ones.

(c) Construct all possible Yukawa interactions in this alternative model. Which fermions, if any, acquire a mass?

9. **Light top quark.** The measured width of the Z boson is $\Gamma_Z = 2.495 \pm 0.002$ GeV. Assume that the observed total width corresponds to exactly three neutrino families $(\nu_e, \nu_\mu, \nu_\tau)$, three charged lepton families (e, μ, τ), and five quark flavors (u, d, c, s, b). What would be the value of Γ_Z if the top quark was light and therefore the decay of a Z boson to a top and an antitop quark were kinematically allowed, adding to the five other flavors of hadronic Z decays? Assume that the fermion masses (including the top quark) are negligible compared to the Z boson mass, giving an identical phase space for all fermion-antifermion decays. Use the parameter $\sin^2\theta_W = 0.232$, where θ_W is the Weinberg angle. Refer to equations (3.230), (3.231), and (3.232) from a similar exercise above.

10. **Pair annihilation at high energy.**

(a) Draw the two tree-level Feynman diagrams for pair annihilation into two photons, $e^+e^- \to \gamma\gamma$.

(b) Compute the total differential cross section for the process $e^+e^- \to \gamma\gamma$ in the center-of-mass system in the limit that the electron mass is negligible compared to its energy. Assume unpolarized electron-positron beams and sum over all possible photon polarizations in the final state.

(c) Replace one of the photons in the final state with the massive neutral Z boson to form the pair annihilation process $e^+e^- \to \gamma + Z$. Compute the total differential cross section in the center-of-mass system, neglecting the mass of the electron.

(d) What is the relative magnitude of $e^+e^- \to \gamma\gamma$ to $e^+e^- \to \gamma + Z$ for a scattering angle of $\theta = \pi/2$ in the lab frame for a center-of-mass energy of $\sqrt{s} = 2M_Z$. Use $M_Z =$ 91.2 GeV/c^2.

3.18 References and Further Reading

A selection of graduate-level texts in particle physics can be found here: [5, 6, 7, 8, 9, 10, 11, 12, 13, 14, 15].

More details about Lie groups can be found here: [16, 17].

A selection of reference texts and articles about supersymmetry and related topics can be found here: [18, 19, 20, 21, 22, 23, 24, 25].

[1] Akira Tonomura, Nobuyuki Osakabe, Tsuyoshi Matsuda, Takeshi Kawasaki, Junji Endo, Shin-ichiro Yano, and Hiroji Yamada. *Phys. Rev. Lett.* **56**(1986):792–795.

[2] S. L. Glashow, *Nucl. Phys.* **22** (1961) 579; S. Weinberg, *Phys. Rev. Lett.* **19** (1967) 1264; A. Salam, "Weak and Electromagnetic Interactions," in *Elementary Particle Theory*, Ed. N. Svartholm. Almquist and Wiksell, 1968, 367, LCCN 68055064; H. D. Politzer, *Phys. Rev. Lett.* **30** (1973) 1346; D. J. Gross and F. Wilczek, *Phys. Rev. Lett.* **30** (1973) 1343.

[3] K. Nakamura et al. (Particle Data~Group). *J. Phys. G* **37** (2010). http://pdg.lbl.gov.

[4] P. W. Higgs, *Phys. Lett.* **12** (1964) 132, *Phys. Rev. Lett.* **13** (1964) 508, and *Phys. Rev.* **145** (1966) 1156; F. Englert and R. Brout, *Phys. Rev. Lett.* **13** (1964) 321.

[5] I.J.R. Aitchison and A.J.G. Hey. *Gauge Theories in Particle Physics.* IOP Publishing, 2003. Vol. 1, ISBN 0-7503-0864-8; vol. 2, ISBN 0-7503-0950-4.

[6] W. N. Cottingham and D. A. Greenwood. *An Introduction to the Standard Model of Particle Physics.* Cambridge University Press, 1998. ISBN 0-521-58832-4.

[7] D. Bailin and A. Love. *Introduction to Gauge Field Theory*. IOP Publishing, 1993. ISBN 0-7503-0281-X.

[8] P. Becher, M. Böhm, and H. Joos. *Gauge Theories of Strong and Electroweak Interactions.* Wiley, 1984. ISBN 0-471-10429-9.

[9] Ta-Pei Cheng and Ling-Fong Li. *Gauge Theory of Elementary Particle Physics.* Oxford University Press, 1984. ISBN 0-19-851961-3.

[10] Ta-Pei Cheng and Ling-Fong Li. *Gauge Theory of Elementary Particle Physics: Problems and Solutions.* Oxford University Press, 2000. ISBN 0-19-850621-X.

[11] Francis Halzen and Alan D. Martin. *Quarks and Leptons.* Wiley, 1984. ISBN 0-471-88741-2.

[12] Howard Georgi. *Weak Interactions and Modern Particle Theory.* Benjamin/Cummings Publishing, 1984. ISBN 0486469042.

[13] E. D. Commins and P. H. Bucksbaum. *Weak Interactions of Leptons and Quarks.* Cambridge University Press, 1983. ISBN 0-521-27370-6.

[14] Chris Quigg. *Gauge Theories of the Strong, Weak, and Electromagnetic Interactions.* Westview Press, 1983. ISBN 0-201-32832-1.

[15] L. B. Okun. *Leptons and Quarks.* Elsevier Science Publishers, 1982. ISBN 0-444-86924-7.

[16] Howard Georgi. *Lie Algebras in Particle Physics.* Benjamin/Cummings Publishing, 1982. ISBN 0-8053-3153-0.

[17] Robert Hermann. *Lie Groups for Physicists.* Benjamin/Cummings Publishing, 1966. ISBN 0-8053-3951-5.

[18] Gordon L. Kane, editor. *Perspectives on Supersymmetry II.* World Scientific Publishing, 2010. ISBN 978-981-4307-49-9.

[19] Gordon L. Kane, editor. *Perspectives on Supersymmetry.* World Scientific Publishing, 1998. ISBN 981-02-3553-4.

[20] Pierre Binétruy. *Supersymmetry.* Oxford University Press, 2006. ISBN 978-0-19-850954-7.

[21] Manuel Drees, Rohini M. Godbole, and Probir Roy. *Theory and Phenomenology of Sparticles.* World Scientific Publishing, 2004. ISBN 981-256-531-0.

[22] Rabindra Mohapatra. *Unification and Supersymmetry.* Springer-Verlag, 2003. ISBN 0-387-95534-8.

[23] D. Bailin and A. Love. *Supersymmetric Gauge Field Theory and String Theory.* IOP Publishing, 1994. ISBN 0-7503-0267-4.

[24] Julius Wess and Jonathan Bagger. *Supersymmetry and Supergravity.* Princeton University Press, 1983. ISBN 0-691-08326-6.

[25] Y. A. Golfand and E. P. Likhtman, *Sov. Phys. JETP* **13** (1971) 323; D. V. Volkhov and V. P. Akulov, *Phys. Lett.* **B 46** (1973) 109; J. Wess and B. Zumino, *Nucl. Phys.* **B 70** (1974) 39; P. Fayet and S. Ferrara, *Phys. Rep.* **C 32** (1977) 249; A. Salam and J. Strathdee, *Fortschr. Phys.* **26** (1978) 57.

4 Hadrons

The quarks and gluons of QCD are charged under color. A bare color charge and the null mass of the gluons suggests the possibility of a long-range strong interaction. However, long-range color forces are not observed in Nature. In going from the $U(1)_{EM}$ gauge symmetry of electromagnetism to the $SU(3)_C$ symmetry of QCD, a dramatic change occurs in the properties of the interaction. The most important of these properties is the antiscreening of the color charge and the corresponding confinement of quarks and gluons into color-neutral bound states, known as *hadrons*.

4.1 Color Antiscreening and Quark Confinement

If an individual quark were placed in the vacuum, as with electric charge the color charge would be screened by pulling quark-antiquark pairs out of the vacuum. However, unlike electromagnetism, the gluons also carry color charges and the vacuum polarization effects of the eight gluons are in the opposite sense from fermion-antifermion pairs. Namely, the gluons strengthen the color charge at long distances by an amount that grows with distance. The gluon contribution outweighs the quark-antiquark screening due to the finite number of quark flavors, resulting in an overall antiscreening of the color charge.

The color charge cannot grow indefinitely at long distances by energy conservation, and so the physically stable configurations of matter are dynamical bound states that are in color-neutral configurations in order to truncate the otherwise growing color fields at long distances. The length scale for this truncation is given by the typical hadron size, known as Λ_{QCD}.

The effective strength of the QCD coupling $\alpha_s = g^2/4\pi$ at finite distances is given to leading order by the expression

$$\alpha_s(Q^2) = \frac{12\pi}{(33 - 2n_f)\ln(Q^2/\Lambda^2_{\text{QCD}})} \tag{4.1}$$

where Q^2 is the magnitude of the four-momentum squared that characterizes the length scale of the charge being probed. The observed value of Λ_{QCD} is approximately 0.1–0.3 GeV. To quote a value of α_s below $Q^2 = \Lambda^2_{\text{QCD}}$ makes no sense, as this is effectively the smallest momentum scale probed by the strong interaction due to the dynamical truncation. Expression (4.1) gives a value of $\alpha_s(M_Z^2) \approx 0.12$ at the electroweak scale, where M_Z is the mass of the Z boson. Note that the value of n_f in equation (4.1) is the number of active quarks (fermions) participating in the charge screening; this includes all quarks at very high Q^2 and for lower scales does not include the heaviest quarks. A length-scale dependence of the coupling strength is also experimentally observed in QED, but is substantially smaller and runs in the opposite direction, i.e., α_{QED} increases slowly with Q^2 while α_s decreases rapidly with Q^2. At the electroweak scale, the QED coupling strength is $\alpha_{\mathrm{QED}}^{-1}(M_Z^2) \approx 129$ as compared with the $Q^2 = 0$ value of $\alpha_{\mathrm{QED}}^{-1}(Q^2 = 0) \approx 137$. The *running* of gauge coupling factors (g, g', g_s) versus Q^2 is a prediction of quantum field theories and originates from the higher-order loop and vertex corrections to the fundamental tree-level scattering processes.

4.2 Light Mesons and Baryons

In section 3.5, $SU(3)$ is shown to have two generators whose matrix representations are diagonal, and therefore states in $SU(3)$ can be labeled by the eigenvalues of the two diagonal matrices, $\lambda_3/2$ and $\lambda_8/2$. A quark is in the fundamental representation of the $SU(3)_C$ color symmetry group, denoted as **3**. An antiquark is in the $\bar{\mathbf{3}}$ representation. These representations are plotted in terms of $(\lambda_3/2, \lambda_8/2)$ in figure 3.3.

The direct product of quark and antiquark representations result in the color octet **8**, represented by the corners of a hexagon in the $(\lambda_3/2, \lambda_8/2)$-plane plus two entries at the origin, and color singlet **1** representations

$$\mathbf{3} \otimes \bar{\mathbf{3}} = \mathbf{8} \oplus \mathbf{1}. \tag{4.2}$$

The color singlet for the $q\bar{q}$ combination is given by

$$\psi_C(q\bar{q}) = \frac{1}{\sqrt{3}}(r\bar{r} + g\bar{g} + b\bar{b}). \tag{4.3}$$

The *color factor*, as introduced in the chapter on QCD, for the $q\bar{q}$ interaction in the color singlet is computed from the single-gluon exchange diagrams with vertex coupling factors from equation (3.110) where there is a sign reversal of the vertex color factor for anticolor relative to color. The contribution to the color factor from the $c\bar{c} \rightarrow c\bar{c}$ diagram comes from G_3^μ and G_8^μ and gives

$$\begin{aligned}
&\text{Computed for } r\bar{r} \text{ or } g\bar{g}\text{:}\\
&\quad (\text{from } G_3^\mu\text{:})\left(\frac{1}{2}\right)\left(-\frac{1}{2}\right) + (\text{from } G_8^\mu\text{:})\left(\frac{1}{2\sqrt{3}}\right)\left(-\frac{1}{2\sqrt{3}}\right) = -\frac{1}{3}\\
&\text{Computed for } b\bar{b}\text{:}\\
&\quad (\text{from } G_8^\mu\text{:})\left(-\frac{1}{\sqrt{3}}\right)\left(\frac{1}{\sqrt{3}}\right) = -\frac{1}{3}
\end{aligned} \tag{4.4}$$

independent of the choice of $c\bar{c}$, as expected. There are also contributions to the color wave function (4.3) from $c\bar{c} \rightarrow c'\bar{c}'$ from $G_{c\bar{c}'}^\mu$ where there are two equally contributing possibilities for the primed color c':

$$2 \cdot \left(\frac{1}{\sqrt{2}}\right)\left(-\frac{1}{\sqrt{2}}\right) = -1. \tag{4.5}$$

Therefore, the overall color factor for $q\bar{q}$ in a singlet configuration from summing (4.4) and (4.5) is negative:

$$C_F^{q\bar{q}}(1) = -\frac{4}{3} \tag{4.6}$$

where the color wave function normalization accounts for the contribution from the three terms in (4.3). The negative sign indicates an attractive potential and supports the observation that color-neutral bound states of quarks and antiquarks, mesons, are found in Nature. A similar analysis for the color octet combination of $q\bar{q}$ shows that the octet configuration is repulsive, and therefore is not a candidate pairwise interaction for a bound-state configuration.

The combination of two quarks gives

$$\mathbf{3} \otimes \mathbf{3} = \mathbf{6} \oplus \bar{\mathbf{3}} \tag{4.7}$$

where the $\mathbf{6}$ is symmetric and the $\bar{\mathbf{3}}$ is antisymmetric under the exchange of quarks. The fact that the direct product of two $\mathbf{3}$ representations admits a $\bar{\mathbf{3}}$ representation makes the qq pairwise interaction a potential candidate for bound-state configurations with a third quark. The inverting of the $\mathbf{3}$ triangle can be seen by adding the corresponding $(\lambda_3/2, \lambda_8/2)$ coordinates of $rb = (\lambda_3^r/2 + \lambda_3^b/2, \lambda_8^r/2 + \lambda_8^b/2)$, bg, and gr. The addition of a third quark yields one multiplet, which is a color singlet

$$\mathbf{3} \otimes \mathbf{3} \otimes \mathbf{3} = (\mathbf{6} \otimes \mathbf{3}) \oplus (\bar{\mathbf{3}} \otimes \mathbf{3}) = \mathbf{10} \oplus \mathbf{8} \oplus \mathbf{8} \oplus \mathbf{1}. \tag{4.8}$$

The decuplet $\mathbf{10}$, represented by a triangle in the $(\lambda_3/2, \lambda_8/2)$-plane, is symmetric and the color singlet is antisymmetric. The two octets have mixed symmetry. The three-quark color singlet is given by

$$\psi_C(qqq) = \frac{1}{\sqrt{6}}(rgb - rbg + gbr - grb + brg - bgr), \tag{4.9}$$

whose form can be constructed by simply circularly permuting the three colors and for each permutation to reverse the order of the last two colors and flip the sign. Expression (4.9) can be rewritten to show the explicit combination of the $\bar{\mathbf{3}}$ representation of the diquark qq multiplet from (4.7) and the third quark:

$$\psi_C(q[qq]) = \frac{1}{\sqrt{6}}(r(gb - bg) + g(br - rb) + b(rg - gr)). \tag{4.10}$$

The fact that the diquark is in an anticolor $\bar{\mathbf{3}}$ representation shows the strong similarity of the baryon and meson color singlet configurations (4.3). The color factor for the qq pairwise interaction in the $\bar{\mathbf{3}}$ representation is computed from the diagram for $cc' \to cc'$ and subtracting the contribution from the exchange diagram $cc' \to c'c$, owing to the antisymmetry of the $\bar{\mathbf{3}}$. We therefore reverse the sign of the terms (4.4) and (4.5) for cc' scattering and subtract the two to account for the antisymmetry under exchange. This gives a color factor for the qq pairwise interaction in the qqq color singlet configuration that is negative:

$$C_F^{qq}(\bar{3}) = -\frac{2}{3} \tag{4.11}$$

where again the color wave function normalization accounts for the six contributing diquark terms in (4.10). The color factor for the sextet **6** combination is repulsive.

The known bound states of the strong interaction are the quark-antiquark states called *mesons* and the three quark configurations known as *baryons*. This observation is supported by the sign of the color factors evaluated for the pairwise interactions of the $q\bar{q}$ color singlet and the qq color configurations in the qqq color singlet. From the analysis above, one can consider constructing color singlet configurations with attractive pairwise interactions in bound states involving more than three quarks or antiquarks. The $\bar{q}\bar{q}$ and qq in the $\bar{q}\bar{q}qq$ bound state can be placed in **3** and $\bar{\mathbf{3}}$ configurations, respectively. Similarly, the two $\bar{q}\bar{q}$ diquarks of $q\bar{q}\bar{q}\bar{q}\bar{q}$ can be placed in **3** configurations, as can the three diquarks in the $\bar{q}\bar{q}\bar{q}\bar{q}\bar{q}\bar{q}$ configuration. However, no experimental evidence for bound states or resonances involving more than three quarks/antiquarks has stood the test of time. These states are either too weakly bound or too wide to have a clear experimental signature relative to incoherent meson and baryon production with similar quark/antiquark compositions.

4.3 Flavor Symmetry

Flavor symmetry was one of the first symmetries identified in hadron spectroscopy. Flavor symmetry is used to explain the observed hadron multiplets and properties of hadron-hadron interactions. Through the use of Fermi statistics, the flavor wave functions of baryons were used to infer the existence of an $SU(3)_C$ group before QCD was formulated as a gauge theory. Indeed, flavor symmetry produces a rich spectroscopy of hadrons, many of which are known only to the particle data book [1], but the lightest of which, such as the pions and kaons, are commonly produced in high-energy particle collisions.

4.3.1 Nuclear Isospin

One of the earliest predictions of the existence of mesons comes from attempts to describe the nuclear force as the exchange of particles between nucleons. The pion-nucleon interaction Lagrangian can generally be written as the sum of several possible Yukawa interaction terms

$$\mathcal{L}_{\text{int}} = g_{pn}p^{\dagger}n\pi^{+} + g_{np}n^{\dagger}p\pi^{-} + g_{pp}p^{\dagger}p\pi^{0} + g_{nn}n^{\dagger}n\pi^{0} \tag{4.12}$$

where each term describes a three-point vertex with two nucleons and a pion. The nucleons come in two flavors, the proton and neutron (p,n), and the pion has three charge states $(\pi^{+},\pi^{0},\pi^{-})$. The short range of the nucleon-nucleon interaction led Yukawa in 1935 to predict an exponentially decaying force whose length scale was described by the pion mass

$$V(r) = -\alpha_s \frac{\exp(-m_\pi r)}{r} \tag{4.13}$$

as would be predicted for a spin-0 propagator $1/(p^2 - m_\pi^2)$ exchanged between nucleons. This prediction gave an accurate estimate for the pion mass $m_\pi \sim \mathcal{O}(0.1)$ GeV and led to the eventual discovery of the pion in 1947.

The relationship between the coupling constants in the interaction Lagrangian (4.12) is predicted by an $SU(2)_I$ symmetry of nuclear isospin. In this symmetry, the first known flavor symmetry, we have a doublet of indistinguishable nucleons and a triplet of indistinguishable pions, where here we treat the electromagnetic interaction as small relative to the nuclear force at the scale of the pion mass,

$$N = \begin{pmatrix} p \\ n \end{pmatrix} \quad \text{and} \quad \pi = \begin{pmatrix} \pi_1 \\ \pi_2 \\ \pi_3 \end{pmatrix}. \tag{4.14}$$

An $SU(2)_I$-invariant Lagrangian can be written as

$$\mathcal{L}_{\text{int}} = g(N^\dagger \boldsymbol{\tau} N) \cdot \boldsymbol{\pi} \tag{4.15}$$

where

$$\boldsymbol{\tau} \cdot \boldsymbol{\pi} = \tau_1 \pi_1 + \tau_2 \pi_3 + \tau_3 \pi_3 = \begin{pmatrix} \pi^0 & -\sqrt{2}\pi^+ \\ -\sqrt{2}\pi^- & -\pi^0 \end{pmatrix} \tag{4.16}$$

using $\pi^\pm = (-\pi_1 \pm i\pi_2)/\sqrt{2}$ and $\pi^0 = \pi_3$. The $SU(2)_I$-invariant Lagrangian (4.15) correctly predicts that the pion couplings are related according to the proportionality

$$g_{pn} : g_{np} : g_{pp} : g_{nn} \propto \sqrt{2} : \sqrt{2} : -1 : 1. \tag{4.17}$$

Ultimately, the nuclear isospin symmetry is a reflection of the underlying $SU(2)$ flavor symmetry of up- and down-quarks in the limit that we neglect the electromagnetic and electroweak couplings as small compared to QCD interactions and we neglect the mass differences of the up- and down-quarks as small compared to Λ_{QCD}. These approximations are quite accurate and therefore the up- and down-quarks can be treated as indistinguishable quarks in one $SU(2)_I$ flavor doublet. The flavor composition hadrons, and fundamentally the concept of the quark, was originally proposed in 1964 by Gell-Mann and independently by Zweig.

4.3.2 Meson Wave Functions

The pion with a mass of $m_\pi \sim 140$ MeV is the lightest of a class of meson states called the *pseudoscalars*. A pseudoscalar is a spin-0 particle with negative parity and zero angular momentum. The flavor wave function of the pion is an $I = 1$ flavor isospin vector, hence the triplet of states (π^+, π^0, π^-). The explicit construction of the total pion wave function from the possible quark-antiquark states is a product of spatial, spin, flavor, and color wave functions

$$\psi(\text{total}) = \psi(\text{space}) \times \psi(\text{spin}) \times \psi(\text{flavor}) \times \psi(\text{color}) \tag{4.18}$$

where the universal color wave function for mesons is given in equation (4.3). The spatial wave function is a symmetric S wave and the negative parity of the wave function comes from the Dirac equation. The parity of a Dirac particle is negative relative to the antiparticle, as determined from the γ^0 parity operator. The spin-0 wave function is the antisymmetric sum of two spin-1/2 particles

$$\psi^{\pi}(\text{spin}) = \frac{1}{\sqrt{2}}(\uparrow\downarrow - \downarrow\uparrow). \tag{4.19}$$

The flavor wave function is composed of up- and down-quarks and therefore transforms under the $SU(2)_I$ isospin symmetry group. However, the pion is the flavor sum of a quark and an antiquark, and therefore the direct product is given by

$$\mathbf{2} \otimes \bar{\mathbf{2}} = \mathbf{3} \oplus \mathbf{1} \tag{4.20}$$

where the antiquark representation is given by $\bar{\mathbf{2}}$.

The antiflavor $SU(2)_I$ doublet of $\bar{d}$ and $\bar{u}$ looks much the same as the particle doublet with some exceptions. First, there is a reversal of additive quantum numbers and therefore the anti-down-quark $\bar{d}$ has $I_3 = \frac{1}{2}$ and the anti-up-quark $\bar{u}$ has $I_3 = -\frac{1}{2}$. The antiparticle doublet must transform under the $SU(2)$ unitary transformations. The finite transformation of a particle doublet around the I_2 axis is given by

$$q' = \begin{pmatrix} u' \\ d' \end{pmatrix} = \exp(-i\theta\tau_2/2)\begin{pmatrix} u \\ d \end{pmatrix} = \begin{pmatrix} \cos\theta/2 & -\sin\theta/2 \\ \sin\theta/2 & \cos\theta/2 \end{pmatrix}\begin{pmatrix} u \\ d \end{pmatrix}. \tag{4.21}$$

If we multiply out the transformation (4.21), replace u and d with $\bar{u}$ and $\bar{d}$, respectively, and multiply the $\bar{d}'$ transformation by minus one, then we get

$$\begin{aligned} -\bar{d}' &= -\,\bar{u}\sin\theta/2 - \bar{d}\cos\theta/2, \\ \bar{u}' &= \bar{u}\cos\theta/2 - \bar{d}\sin\theta/2, \end{aligned} \tag{4.22}$$

where here we have written the $\bar{d}'$ transformation on top and the $\bar{u}'$ on bottom in preparation for the next step. In matrix form, equations (4.22) can be written

$$\bar{q}' = \begin{pmatrix} -\bar{d}' \\ \bar{u}' \end{pmatrix} = \begin{pmatrix} \cos\theta/2 & -\sin\theta/2 \\ \sin\theta/2 & \cos\theta/2 \end{pmatrix}\begin{pmatrix} -\bar{d} \\ \bar{u} \end{pmatrix} = \exp(-i\theta\tau_2/2)\begin{pmatrix} -\bar{d} \\ \bar{u} \end{pmatrix}, \tag{4.23}$$

giving precisely the same $SU(2)$ transformation for the antiparticle doublet $\bar{\mathbf{2}}$:

$$\bar{q} = \begin{pmatrix} -\bar{d} \\ \bar{u} \end{pmatrix}. \tag{4.24}$$

The antiparticle $\bar{\mathbf{2}}$ representation of $SU(2)_I$ therefore transforms identically to the particle $\mathbf{2}$ representation. This property does not hold for $SU(N)$ for $N > 2$, as clearly shown in figure 3.3 for $SU(3)$.

We can now return to the direct product (4.20) and write the **3** representation in terms of the quark-antiquark doublet. For reference, the $SU(2)$ spin-symmetric **3** combination is given by

$$\begin{aligned} |1,1\rangle &= \uparrow\uparrow, \\ |1,0\rangle &= \frac{1}{\sqrt{2}}(\uparrow\downarrow + \downarrow\uparrow), \\ |1,-1\rangle &= \downarrow\downarrow, \end{aligned} \tag{4.25}$$

and using this as a reference we can replace the first spin with up- and down-flavor and the second spin with the antiflavor doublet. This gives

$$\begin{aligned} \psi^{\pi^+}(\text{flavor}) &= -u\bar{d}, \\ \psi^{\pi^0}(\text{flavor}) &= \frac{1}{\sqrt{2}}(u\bar{u} - d\bar{d}), \\ \psi^{\pi^-}(\text{flavor}) &= d\bar{u}, \end{aligned} \tag{4.26}$$

for the pion flavor wave functions.

If one starts with the spatial, flavor, and color wave functions of the pion and inserts instead the spin-1 wave function of $SU(2)$, then the resulting total wave function describes the vector meson called the $\rho = (\rho^+, \rho^0, \rho^-)$. The ρ is the lightest of the vector mesons with a mass of $m_\rho \sim 770$ MeV. The extension to a third quark flavor, the strange quark, changes the flavor symmetry group from $SU(2)_I$ isospin to $SU(3)$ flavor. Note that $SU(2)_I$ is a subgroup of $SU(3)_{\text{flavor}}$. The $SU(3)$ flavor group is the largest group for which the multiplet structure of the symmetry group is not greatly modified by large quark mass differences. In fact, the strange quark mass, $m_s \sim 0.3$ GeV, is already beyond the Λ_{QCD} scale, and hence $SU(3)_{\text{flavor}}$ is only a partially respected symmetry.

Introducing $SU(3)_{\text{flavor}}$ expands the number of pseudoscalar mesons from the triplet of pions to an octet of pions $(\pi^\pm, \pi^0)$, kaons $(K^\pm, K^0, \bar{K}^0)$, and the η meson. The masses are $m_K \sim 500$ MeV and $m_\eta \sim 550$ MeV. The octet is typically drawn using the two diagonal matrices of $SU(3)$ to indicate the coordinates in the plane of isospin $I_3 = \lambda_3/2$ and hypercharge $Y = \lambda_8/\sqrt{3}$. Another representation of the pseudoscalar meson octet M^{0^-} is the adjoint representation (3.110) as used for the gluons

$$M^{0^-} = \begin{pmatrix} \frac{\pi^0}{\sqrt{2}} + \frac{\eta}{\sqrt{6}} & \pi^+ & K^+ \\ \pi^- & -\frac{\pi^0}{\sqrt{2}} + \frac{\eta}{\sqrt{6}} & K^0 \\ K^- & \bar{K}^0 & -\frac{2\eta}{\sqrt{6}} \end{pmatrix}. \tag{4.27}$$

This form shows more clearly that the η meson is an admixture of up-, down-, and strange quark-antiquark pairs. The corresponding $SU(3)_{\text{flavor}}$ singlet combination is believed to be the η' meson $m_{\eta'} \sim 958$ MeV. The corresponding $SU(2)$ octet and singlet of vector mesons corresponds to the ρ $(\rho^\pm, \rho^0)$, $K^*(K^{*\pm}, K^{*0}, \bar{K}^{*0})$, and ω and ϕ neutral states. The masses are $m_{K^*} \sim 892$ MeV, $m_\omega \sim 782$ MeV, and $m_\phi \sim 1020$ MeV. The ω and ϕ are believed to be mixed states of $(u\bar{u} + d\bar{d})/\sqrt{2}$ and $s\bar{s}$ flavor wave functions with the ϕ having a larger fraction of $s\bar{s}$.

A list of significant decay modes for commonly known light mesons, kaons, and other strange mesons, as can be found in the particle data book, is given in table 4.1.

Table 4.1 Decay modes, branching fractions, and properties of commonly known light mesons, kaons, and other strange mesons [1].

Meson	Decay modes	Br (%)	$J^{P(C)}$	Mass (MeV)	Width/Lifetime
π^+/π^-	$\to \mu^+\nu_\mu$	~100	0^-	139.6	26.0 ns
π^0	$\to \gamma\gamma$	98.8	0^{-+}	135.0	8.4×10^{-5} ps
$K^+(u\bar{s})/K^-(\bar{u}s)$	$\to \mu^+\nu_\mu$	63.4	0^-	493.7	12.4 ns
	$\to \pi^+\pi^0$	20.9			
	$\to \pi^+\pi^+\pi^-$	5.6			
	$\to \pi^0 e^+ \nu_e$	5.0			
	$\to \pi^0\mu^+\nu_\mu$	3.3			
	$\to \pi^+\pi^0\pi^0$	1.8			
K_S^0	$\to \pi^+\pi^-$	69.2	0^-	497.6	89.6 ps
	$\to \pi^0\pi^0$	30.7			
K_L^0	$\to \pi^+e^-\bar{\nu}_e$	40.5	0^-	497.6	51.1 ns
	$\to \pi^+\mu^-\bar{\nu}_\mu$	27.0			
	$\to \pi^0\pi^0\pi^0$	19.6			
	$\to \pi^+\pi^-\pi^0$	12.6			
	$\to \pi^+\pi^-$	0.2			
	$\to \pi^0\pi^0$	0.1			
η	$\to \gamma\gamma$	39.4	0^{-+}	547.5	1.3 keV
	$\to \pi^0\pi^0\pi^0$	32.5			
$\rho(770)$	$\to \pi^\pm\pi^0(\pi^+\pi^-)$	~100	1^{--}	~770	146.4 MeV
$\omega(782)$	$\to \pi^+\pi^-\pi^0$	89.1	1^{--}	~782	8.5 MeV
	$\to \pi^0\gamma$	8.9			
$K^*(892)$	$\to K\pi$	~100	1^-	~892	50 MeV
$\eta'(958)$	$\to \pi^+\pi^-\eta$	44.5	0^{-+}	~958	0.2 MeV
	$\to \rho^0\gamma\,(\pi^+\pi^-\gamma)$	29.4			
	$\to \pi^0\pi^0\eta$	20.8			
$\phi(1020)(s\bar{s})$	$\to K^+K^-$	49.2	1^{--}	~1020	4.3 MeV
	$\to K_L^0K_S^0$	34.0			
	$\to \rho^\pi\pi^\mp\,(\pi^+\pi^-\pi^0)$	15.3			

4.3.3 *Baryon Wave Functions*

An important new constraint comes in the construction of the three-quark qqq bound states. Baryons are fermions and the total wave function must be antisymmetric under the interchange of quarks. Out of the possible 27 $SU(3)$ flavor combinations of $\mathbf{3} \otimes \mathbf{3} \otimes \mathbf{3}$, all but 18 are excluded by the Pauli exclusion principle. As we will see, the antisymmetry of the baryon wave function brings together an admixture of multiple spin and flavor configurations to form physical states, most notably the lightest baryons, the proton and neutron.

As with the mesons, the total baryon wave functions consist of spatial, spin, flavor, and color wave functions. The universal color wave function for baryons is given by equation (4.9). The low-lying states will have zero angular momentum $L=0$ and symmetric spatial wave functions. The possible $SU(2)$ spin configurations are given by the direct product

$$\mathbf{2} \otimes \mathbf{2} \otimes \mathbf{2} = \mathbf{4} \oplus \mathbf{2}^{\mathrm{MA}} \oplus \mathbf{2}^{\mathrm{MS}} \tag{4.28}$$

where the MA and MS labels indicate mixed asymmetric and mixed symmetric, respectively. The spin-3/2 combination, or **4**, is totally symmetric under the exchange of any two spins. The explicit forms of the MA and MS wave functions are given by

$$\begin{aligned} \left|\frac{1}{2}^{\mathrm{MA}}, \frac{1}{2}\right\rangle &= \frac{1}{\sqrt{2}}(\uparrow\downarrow\uparrow - \downarrow\uparrow\uparrow), \\ \left|\frac{1}{2}^{\mathrm{MA}}, -\frac{1}{2}\right\rangle &= -\frac{1}{\sqrt{2}}(\downarrow\uparrow\downarrow - \uparrow\downarrow\downarrow), \\ \left|\frac{1}{2}^{\mathrm{MS}}, \frac{1}{2}\right\rangle &= \frac{1}{\sqrt{6}}(\uparrow\downarrow\uparrow + \downarrow\uparrow\uparrow - 2\uparrow\uparrow\downarrow), \\ \left|\frac{1}{2}^{\mathrm{MS}}, -\frac{1}{2}\right\rangle &= -\frac{1}{\sqrt{6}}(\downarrow\uparrow\downarrow + \uparrow\downarrow\downarrow - 2\downarrow\downarrow\uparrow), \end{aligned} \tag{4.29}$$

showing that the (anti)symmetry is present only under the exchange of the first two spins. A similar situation of mixed symmetry occurs in the $SU(3)_{\text{flavor}}$ octets formed from the direct product

$$\mathbf{3} \otimes \mathbf{3} \otimes \mathbf{3} = \mathbf{10} \oplus \mathbf{8}^{\mathrm{MA}} \oplus \mathbf{8}^{\mathrm{MS}} \oplus \mathbf{1} \tag{4.30}$$

where the mixed symmetry is in the first two flavors. One can therefore form a totally symmetric flavor octet of spin-1/2 baryons with the combination

$$\psi(\text{flavor}) \times \psi(\text{spin}) = \frac{1}{\sqrt{2}}(\mathbf{8}^{\mathrm{MA}} \times \mathbf{2}^{\mathrm{MA}} + \mathbf{8}^{\mathrm{MS}} \times \mathbf{2}^{\mathrm{MS}}). \tag{4.31}$$

With a symmetric spatial wave function, the overall antisymmetry of the baryon is provided by the color wave function (4.9). The elements of the baryon octet can be written in the adjoint representation as

$$B^{\frac{1}{2}^+} = \begin{pmatrix} \frac{\Sigma^0}{\sqrt{2}} + \frac{\Lambda}{\sqrt{6}} & \Sigma^+ & p \\ \Sigma^- & -\frac{\Sigma^0}{\sqrt{2}} + \frac{\Lambda}{\sqrt{6}} & n \\ \Xi^- & \Xi^0 & -\frac{2\Lambda}{\sqrt{6}} \end{pmatrix}. \tag{4.32}$$

The baryon octet has a total angular momentum J and parity P given by $J^P = \frac{1}{2}^+$. The masses are $m_p \sim m_n \sim 939$ MeV, $m_\Sigma \sim 1193$ MeV, $m_\Lambda \sim 1116$ MeV, and $m_\Xi \sim 1318$ MeV. Note that the members of the baryons octet with strange quark content, all but the proton and neutron, are commonly referred to as *hyperons*.

The baryon flavor decuplet with $J^P = \frac{3}{2}^+$ is the totally symmetric combination of flavors where the extreme multiplet components correspond to $\Delta^{++}(uuu)$, $\Delta^-(ddd)$, and $\Omega^-(sss)$. The presence of the Δ^{++} with the extreme spin component $\uparrow\uparrow\uparrow$ directly led to the conjecture of the totally antisymmetric color wave function to satisfy Fermi-Dirac statistics, predating

Table 4.2 Decay modes, branching fractions, and properties of commonly known light baryons and Hyperons [1].

Baryon	Decay modes	Br (%)	$J^{P(C)}$	Mass (MeV)	Width/Lifetime
$p(uud)$	stable		$\frac{1}{2}^+$	938.3	$> 6 \times 10^{33}$ years
$n(uud)$	$\to pe^-\bar{\nu}_e$	100	$\frac{1}{2}^+$	939.6	885.7 s
$\Delta(1232)$	$\to N\pi$	~100	$\frac{3}{2}^+$	~1232	~118 MeV
$\Lambda(uds)$	$\to p\pi^-$	63.9	$\frac{1}{2}^+$	1115.7	263 ps
	$\to n\pi^0$	35.8			
$\Sigma^+(uus)$	$\to p\pi^0$	51.6	$\frac{1}{2}^+$	1189.4	80 ps
	$\to n\pi^+$	48.3			
$\Sigma^0(uds)$	$\to \Lambda\gamma$	~100	$\frac{1}{2}^+$	1192.6	7.4×10^{-8} ps
$\Sigma^-(dds)$	$\to n\pi^-$	99.8	$\frac{1}{2}^+$	1197.4	148 ps
$\Xi^0(uss)$	$\to \Lambda\pi^0$	99.5	$\frac{1}{2}^+$	1314.8	290 ps
$\Xi^-(dss)$	$\to \Lambda\pi^-$	99.9	$\frac{1}{2}^+$	1321.3	164 ps
$\Sigma^*(1385)$	$\to \Lambda\pi$	87.0	$\frac{3}{2}^+$	~1385	36 MeV
	$\to \Sigma\pi$	11.7			
$\Xi^*(1530)$	$\to \Xi\pi$	~100	$\frac{3}{2}^+$	~1530	9 MeV
$\Omega^-(sss)$	$\to \Lambda K^-$	67.8	$\frac{3}{2}^+$	1672.5	82 ps
	$\to \Xi^0\pi^-$	23.6			
	$\to \Xi^-\pi^0$	8.6			

the development of QCD. The elements of the baryon decuplet consist of the Δ (Δ^{++}, $\Delta^+, \Delta^0, \Delta^-$), Σ^* ($\Sigma^{*\pm}, \Sigma^{*0}$), Ξ^*(Ξ^{*-}, Ξ^{*0}), and Ω^-. The masses are $m_\Delta \sim 1232$ MeV, $m_{\Sigma^*} \sim 1318$ MeV, $m_{\Xi^*} \sim 1384$ MeV, and $\Omega^- \sim 1672$ MeV. The baryon octet (4.31) and the decuplet with a flavor $\times$ spin wave function of $\mathbf{10} \times \mathbf{4}$ are the only symmetric representations of $SU(3)_{\text{flavor}} \times SU(2)_{\text{spin}}$ and therefore are the only representations allowed for ground-state baryons.

A list of significant decay modes for commonly known light baryons and hyperons, as can be found in the particle data book, is given in table 4.2.

4.4 Heavy Flavors, Quarkonia, and Meson Factories

The $SU(3)$ flavor symmetry of the (u, d, s) quarks is an approximate symmetry whose interval of validity extends up to Λ_{QCD}, where the quark masses satisfy $m_{u,d} \ll \Lambda_{\text{QCD}}$ and $m_s \lesssim \Lambda_{\text{QCD}}$. The heavy quark flavors consist of the charm, bottom, and top quarks. The top quark has a decay width that is an order of magnitude larger than Λ_{QCD}, and therefore decays on a time scale shorter than the QCD hadronization time. The charm and bottom quarks, however, have unique properties due to their mass and width, $m_{c,b} \gg \Lambda_{\text{QCD}}$ and $\Gamma_{c,b} \ll \Lambda_{\text{QCD}}$. The charm and bottom quarks form heavy mesons and baryons whose properties

Table 4.3 Properties of commonly known heavy-flavor mesons [1]. The *slow pion* decay of the D^* meson is a unique signature of charm production.

Meson	Decay modes	Br (%)	$J^{P(C)}$	Mass (MeV)	Width/Lifetime
$D^+(c\bar{d})/D^-(\bar{c}d)$	$\to \mu^+\nu_\mu$	0.044	0^-	1869.6	1.04 ps
	$\to (\bar{K}^0/\bar{K}^{*0})e^+\nu_e$	14.1			
	$\to (\bar{K}^0/\bar{K}^{*0})\mu^+\nu_\mu$	14.7			
	$\to K_S^0\pi^+(\pi^0)$	8.3			
	$\to K_L^0\pi^+$	1.5			
	$\to K^-\pi^+\pi^+(\pi^0)$	15.2			
	$\to K_S^0\pi^+\pi^+\pi^-$	3.0			
$D^0(c\bar{u})/\bar{D}^0(\bar{c}u)$	$\to (K^-/K^{*-})e^+\nu_e$	5.7	0^-	1864.8	0.41 ps
	$\to (K^-/K^{*-})\mu^+\nu_\mu$	5.3			
	$\to K^-\pi^+(\pi^0)$	17.8			
	$\to K_S^0\pi^0(\pi^0)$	2.0			
	$\to K_L^0\pi^0$	1.0			
	$\to K^-\pi^+\pi^+\pi^-(\pi^0)$	12.4			
	$\to K_S^0\pi^+\pi^-(\pi^0)$	8.3			
D^{*+}/D^{*-}	$\to D^0\pi^+$ (slow pion)	67.7	1^-	2010.3	96 keV
	$\to D^+\pi^0$	30.7			
$D^{*0}/\bar{D}^{*0}$	$\to D^0\pi^0$	61.9	1^-	2007.0	< 2.1 MeV
	$\to D^0\gamma$	38.1			
$D_s^+(c\bar{s})/D_s^-(\bar{c}s)$	$\to \tau^+\nu_\tau$	6.6	0^-	1968.5	0.50 ps
	$\to K^+\bar{K}^0$	4.4			
	$\to K^+K^-\pi^+(\pi^0)$	11.1			
D_s^{*+}/D_s^{*-}	$\to D_s^+\gamma$	94.2	1^-	2112.3	< 1.9 MeV
$J/\Psi(1S)\,(c\bar{c})$	$\to e^+e^-/\mu^+\mu^-$	5.9/5.9	1^{--}	3096.9	93 keV
$B^+(u\bar{b})/B^-(\bar{u}b)$	$\to (e^+\nu_e/\mu^+\nu_\mu)+X$	10.9/10.9	0^-	5279.2	1.6 ps
	$\to \bar{D}^0+X$	~79			
$B^0(d\bar{b})/\bar{B}^0(\bar{d}b)$	$\to (e^+\nu_e/\mu^+\nu_\mu)+X$	10.4/10.4	0^-	5279.5	1.5 ps
	$\to D^-+X$	36.9			
	$\to \bar{D}^0+X$	47.4			
B^*	$\to B\gamma$	~100	1^-	5325.0	< 6 MeV
$B_s^0(s\bar{b})/\bar{B}_s^0(\bar{s}b)$	$\to D_s^-+X$	~94	0^-	5366.3	1.5 ps
$B_s^{*0}/\bar{B}_s^{*0}$	$\to B_s\gamma$	~100	1^-	5415.4	< 6 MeV
$B_c^+(c\bar{b})/B_c^-(\bar{c}b)$	$\to J/\Psi(1S)\ell^+\nu_\ell+X$	0.005	0^-	~6277	0.45 ps
$\Upsilon(1S)\,(b\bar{b})$	$\to e^+e^-/\mu^+\mu^-/\tau^+\tau^-$	2.4/2.5/2.7	1^{--}	9460.3	54 keV
$\Upsilon(2S)\,(b\bar{b})$	$\to e^+e^-/\mu^+\mu^-/\tau^+\tau^-$	1.9/1.9/2.0	1^{--}	10023.3	32 keV
$\Upsilon(3S)\,(b\bar{b})$	$\to \mu^+\mu^-/\tau^+\tau^-$	2.2/2.3	1^{--}	10355.2	20 keV
$\Upsilon(4S)\,(b\bar{b})$	$\to B^+B^-$	51.6	1^{--}	10579.4	20.5 MeV
	$\to B^0\bar{B}^0$	48.4			
$\Upsilon(10860)\,(b\bar{b})$	$\to B\bar{B}+X$	~59	1^{--}	~10865	110 MeV
	$\to B_s^{(*)}\bar{B}_s^{(*)}+X$	~19.3			

Table 4.4 Properties of commonly known heavy-flavor baryons [1].

Baryon	Decay modes	Br (%)	J^P	Mass (MeV)	Width/Lifetime
$\Lambda_c^+(udc)$	$\to \Lambda(e^+\nu_e/\mu^+\nu_\mu)$	2.1/2.0	$\frac{1}{2}^+$	2286.5	0.2 ps
	$\to pK^-\pi^+$	5.0			
$\Sigma_c^{++}(uuc)$	$\to \Lambda_c^+\pi^+$	~100	$\frac{1}{2}^+$	~2454	~2.2 MeV
$\Sigma_c^+(udc)$	$\to \Lambda_c^+\pi^0$	~100	$\frac{1}{2}^+$	~2454	~2.2 MeV
$\Sigma_c^0(ddc)$	$\to \Lambda_c^+\pi^-$	~100	$\frac{1}{2}^+$	~2454	~2.2 MeV
$\Xi_c^+(usc)$	$\to \Xi^-\pi^+\pi^+$	(obs.)	$\frac{1}{2}^+$	2467.8	0.44 ps
$\Xi_c^0(dsc)$	$\to \Xi^-\pi^+$	(obs.)	$\frac{1}{2}^+$	2470.9	0.11 ps
$\Omega_c^0(ssc)$	$\to \Omega^-\pi^+$	(obs.)	$\frac{1}{2}^+$	2695.2	0.069 ps
$\Lambda_b^0(udb)$	$\to \Lambda_c^+(e^-\bar{\nu}_e/\mu^-\bar{\nu}_\mu)+X$	10.6/10.6	$\frac{1}{2}^+$	5620.2	1.4 ps
	$\to \Lambda_c^+\pi^-$	0.88			
$\Sigma_b^+(uub)/\Sigma_b^-(ddb)$	$\to \Lambda_b^0\pi$	~100	$\frac{1}{2}^+$	5807.8	—
$\Sigma_b^{*+}/\Sigma_b^{*-}$	$\to \Lambda_b^0\pi$	~100	$\frac{3}{2}^+$	5829.0	—
$\Xi_b^0(usb)$	$\to \Xi^-\ell^-\bar{\nu}_\ell+X$	(obs.)	$\frac{1}{2}^+$	5792.4	1.4 ps
$\Xi_b^-(dsb)$	$\to \Xi^-J/\Psi(1S)$	(obs.)	$\frac{1}{2}^+$	5792.4	1.4 ps
$\Omega_b^-(ssb)$	$\to \Omega^-J/\Psi(1S)$	(obs.)	$\frac{1}{2}^+$	~6165	—

are largely determined by the heavy quark(s) in the bound state. The *b*-quark containing mesons are known as *B* mesons and the *c*-quark containing mesons are denoted *D* mesons. A list of significant decay modes for commonly known heavy-flavor mesons, as can be found in the particle data book, is given in table 4.3. Baryons carry a subscript of *b* or *c* to indicate their heavy quark content, such as the Λ_c^+ and the Λ_b^0 baryons. A partial list of heavy-flavor baryons is given in table 4.4.

Quark flavor is conserved under the strong, electromagnetic, and neutral-current interactions. Therefore, e^+e^- and hadron colliders will predominantly produce heavy quarks in $Q\bar{Q}$ pairs. Near threshold, the $Q\bar{Q}$ pair will form bound states known as *quarkonia*. A chart of the charmonium and bottom quarkonium states is given in table 4.5 and plotted in figure 4.1 for bottom quarkonia. The $J^{PC}=1^{--}$ states have the same quantum numbers as the photon and hence can be thought of as "heavy photons."

Some decay properties are unique to vector mesons and in some cases serve to suppress decays and therefore narrow the decay width. For example, the decay of a massive vector meson into a pair of massless spin-1 particles (photons or gluons) is forbidden by a theorem due to C. N. Yang based on the angular momentum addition of massless particles. This can be seen as follows. The $J_3=0$ states of the two-photon system are shown in figure 4.2. The $J=1$ state cannot have $J_3=\pm 2$ projections. If we consider a π-rotation about the x-axis, then the transformation properties of the two-photon wave function are

$$P_J(\cos\theta) \xrightarrow{R_x(\pi)} P_J(-\cos\theta) = (-1)^J P_J(\cos\theta). \tag{4.33}$$

Therefore, $J=1$ is forbidden since the two-photon state must be even under $R_x(\pi)$ exchange, as required for a pair of identical bosons in the final state.

Table 4.5 Quarkonium bound states of $c\bar{c}$ and $b\bar{b}$.

L	S	Charmonium	Bottom quarkonium
0	1	$J/\psi(n=1), \psi(nS)\,(n>1)$	$\Upsilon(nS)$
	0	$\eta_c(nS)$	$\eta_b(nS)$
1	1	$\chi_{cJ}(nP)$	$\chi_{bJ}(nP)$
	0	$h_c(nP)$	$h_b(nP)$

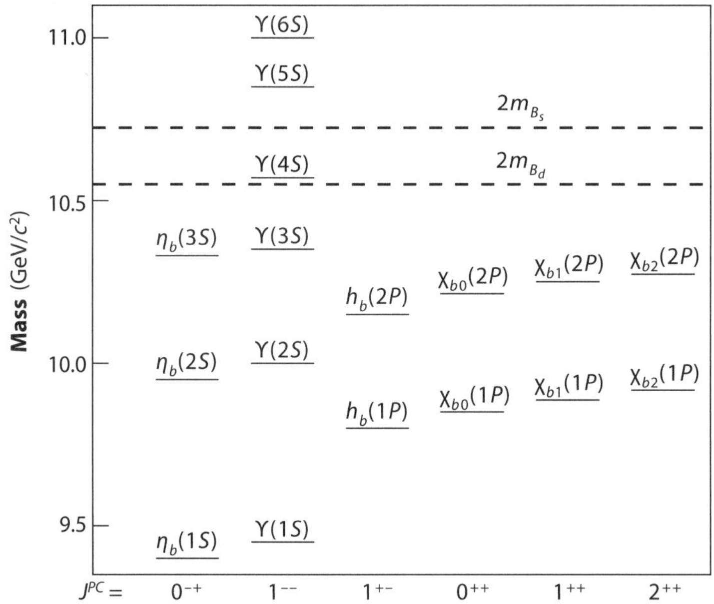

FIGURE 4.1. Bottom quarkonium states. The masses of the states are shown for different J^{PC} and L quantum numbers. The $\Upsilon(4S)$ is above the threshold to decay into a pair of B_d mesons. The $\Upsilon(5S)$ and $\Upsilon(6S)$ states are above the thresholds to decay to pairs of B_s or B_d mesons.

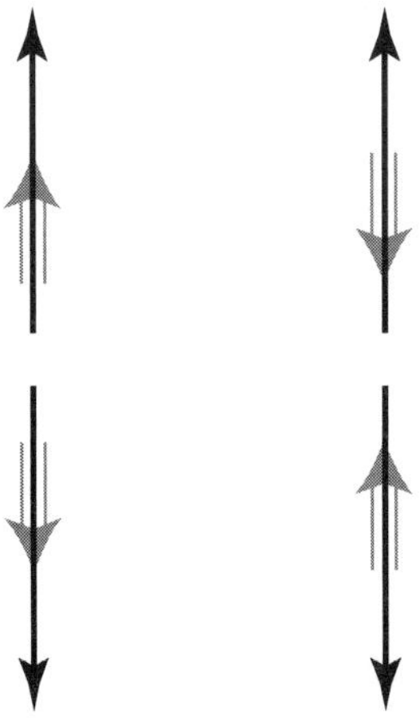

FIGURE 4.2. The $J_3 = 0$ states of the two-photon system from vector meson decay, showing the direction of the outgrowing 3-momenta (solid arrows) and J_3 spin projections (open arrows) of the photons along the axis defined by the corresponding three-momenta.

A vector meson resonance can be narrow due to a large binding energy, reducing the mass of the state and therefore kinematically excluding decays to heavier final states. For example, the low-lying 1^{--} decays can be subthreshold for the decay into two pseudoscalar massive mesons with the corresponding leading-quark flavor of the vector meson. This forces the quark-antiquark state in the vector meson to annihilate to preserve flavor in strong decays. Vector meson decays of the diagrammatic topology that have no fermion lines that run directly from the initial to final states are suppressed, an effect identified by Okubo, Zweig, and Iizuka (OZI). OZI suppression can also be used to explain above-threshold decay preferences, and explains why the $\phi(s\bar{s})$ state with $J^{PC} = 1^{--}$ prefers to decay to a pair of kaons rather than three pions, despite the substantially smaller phase space to decay to a pair of kaons. Even the heavy-flavor vector mesons have low-lying states that are predicted to be narrow and, indeed, the decay width of the $\Upsilon(1S\text{–}3S)$ ranges from 50 to 20 keV while the $\Upsilon(4S)$ is 20 MeV wide. The $\Upsilon(4S)$ is sufficiently massive to decay strongly to a (non-OZI suppressed) pair of B_d mesons and similarly the $\Upsilon(5S)$ is 110 MeV wide decaying to both B_d and B_s mesons. A similar situation exists for charmonium where the $J/\Psi(1S)$ has a width of 93 keV, the $\Psi(2S)$ has width 337 keV, and the $\Psi(3770)$, above the threshold for D meson pair production, has a width of 23 MeV.

The direct coupling to *s*-channel virtual photons in e^+e^- collisions allows the heavy photon spectrum to be scanned with center-of-mass energy, as shown in figure 4.3 from data taken by the CLEO experiment at CESR. The 1 nb cross section of the $\Upsilon(4S)$ is the primary operating point for the asymmetric B meson factories where hundreds of fb^{-1} of integrated luminosity have been accumulated, corresponding to hundreds of millions of $B\bar{B}$ pairs. Production of the $B_s^0\text{–}\bar{B}_s^0$ states, however, requires running on the $\Upsilon(5S)$ resulting in a substantially lower yield, and hence the B_s^0 meson is studied primarily at hadron colliders.

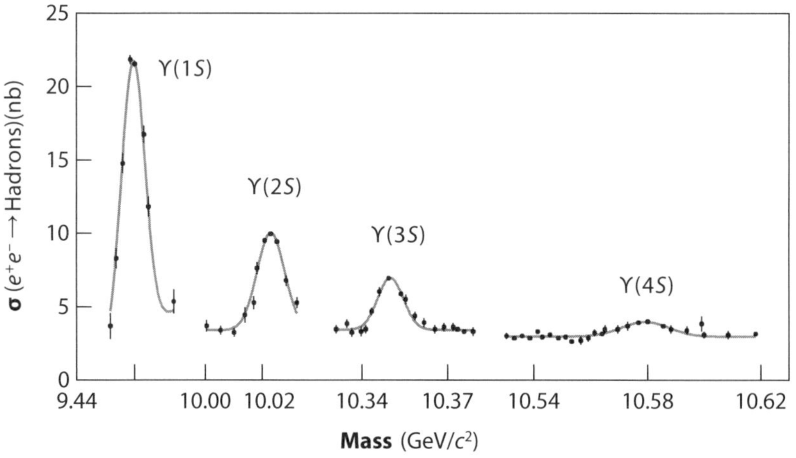

FIGURE 4.3. The cross section for low-lying $J^{PC} = 1^{--}$ $b\bar{b}$-quarkonium states decaying into hadrons in e^+e^- collisions (Credit: CLEO).

4.5 Exercises

1. The magnetic moment operator is given in equation (2.32), where the evaluation can be made along the σ_3 direction. Compute the ratio of the neutron to proton magnetic moment in the approximation $m_u = m_d$.

2. Draw the coordinates of the following $SU(3)_{\text{flavor}}$ multiplets in the $(\lambda_3/2, \lambda_8/2)$ plane: **1**, **3**, $\bar{\mathbf{3}}$, **6**, **8**, **10**.

3. **Scalar superbaryons.** In supersymmetry, there is postulated to be a spin-0 quark, called a *squark*, corresponding to each of the Standard Model spin-1/2 quark flavors. Imagine that the scalar quarks, denoted $\tilde{q}$, are stable. Construct a three-squark $(\tilde{q}\tilde{q}\tilde{q})$ color-singlet state in a spin-0, s-state $(L=0)$ with integer electric charge, the supersymmetric version of a baryon. Write one example of a flavor $\times$ color wave function for a superbaryon.

4. **Pentaquarks.**

   (a) Diquarks are combinations of two quarks (or two antiquarks). What are the possible spin, flavor, and color configurations of ground-state (orbital angular momentum $L=0$) diquarks composed of the three light quarks (u, d, and s)? Present your results for the flavor wave functions in the form of states in $I_3 - Y$ space, labeled by their quark content, where I_3 is the third component of the isospin and $Y = B + S$ is the hypercharge = baryon number + strangeness.

   Can diquarks exist as free particles?

   (b) Experimental data (of erratic quality) suggests that there is a (baryon) resonance in K^+n scattering at 1540 MeV center-of-mass energy. This cannot correspond to a particle in any plausible three-quark model of baryons (as no appropriate companion states have been observed), which has led to speculations as to the existence of pentaquark baryon states consisting of two diquarks plus an antiquark: $(qq)(qq)\bar{q}$.

   One pentaquark model, due to Robert Jaffe and Frank Wilczek (*Phys. Rev. Lett.* **91**, 232003 (2003)), supposes that the constituent diquark states are of the type considered in part (a), with the additional restriction that the diquark wave functions are antisymmetric separately with respect to quark interchange in each of the three degrees of freedom: color, flavor, and spin. Furthermore, the overall flavor and spin wave functions of the pentaquarks are symmetric with respect to diquark interchange.

   What flavor multiplets of pentaquarks are permitted in this model? Present your results in the form of states in $I_3 - Y$ space, labeled by their quark content. Which pentaquark state could correspond to the K^+n resonance, which is sometimes called the $\theta^+(1540)$?

5. **Fully leptonic decays of pseudoscalar mesons.** A charged pion $\pi^-(d\bar{u})$ is a spin-0 particle with negative intrinsic parity (pseudoscalar meson). The d- and anti-u-quarks can annihilate through the charged current interaction to produce a lepton and lepton antineutrino in the final state. As the incoming d and anti-u quarks originate

from a bound state, the free-quark current $j_q^\mu = \bar{\nu}\gamma^\mu(1-\gamma^5)u$ is replaced by an effective hadronic current j_π^μ constructed out of the only available four-vector describing the initial state, the pion four-momentum q^μ. The pion four-momentum is related by energy-momentum conservation to the final-state lepton p_ℓ^μ and antineutrino p_ν^μ four-momenta, $q^\mu = p_\ell^\mu + p_\nu^\mu$. Therefore, we parameterize the pion hadronic current

$$j_\pi^\mu = if_\pi V_{ud} q^\mu = if_\pi V_{ud}(p_\ell^\mu + p_\nu^\mu) \tag{4.34}$$

where $f_\pi \approx 90$ MeV is called the *pion decay constant* and V_{ud} is the CKM matrix element for a $d\bar{u}$-charged current interaction.

(a) Compute the ratio of the decay rate of $\pi^- \to \mu^- \bar{\nu}_\mu$ to $\pi^- \to e^- \bar{\nu}_e$. Use $m_\pi = 139.6$ MeV/c^2, $m_\mu = 105.6$ MeV/c^2, $m_e = 0.511$ MeV/c^2, $m_\nu \approx 0$.

(b) Repeat the same calculation as in part (a), but apply this to the charm meson system. Compute the ratio of the decay rate of $D^- \to \tau^- \bar{\nu}_\tau$ to $D^- \to \mu^- \bar{\nu}_\mu$. Use $m_{D^-} = 1.869$ GeV/c^2 and $m_\tau = 1.777$ GeV/c^2.

(c) The charged kaon $K^-(s\bar{u})$ is also a pseudoscalar meson. If we approximate the kaon decay constant f_K to be equal to the pion decay constant $f_K \approx f_\pi$ (using $SU(3)$ flavor symmetry), then we can directly compare the partial decay widths of charged pions and charged kaons. Compute the ratio of the partial decay widths of $K^- \to \mu^- \bar{\nu}_\mu$ and $\pi^- \to \mu^- \bar{\nu}_\mu$. Use $m_K = 494$ MeV/c^2, $V_{ud} \approx \cos\theta_C = 0.975$, $V_{us} \approx \sin\theta_C = 0.221$.

(d) Do you expect the charged kaon lifetime to be longer or shorter than the charged pion lifetime (what factors influence the kaon lifetime relative to the pion)?

6. **Pionium decay.** Here, the first and most important thing to remember, this problem treats the charged pion as a fundamental spin-0 charged particle interacting in QED. No quarks, please!

In this problem, you are asked to estimate the lifetime of the ground state of pionium, a bound state of $\pi^+\pi^-$. The mass of π^+ is 139.56 MeV.

(a) There are three diagrams at tree-level for $\pi^+\pi^- \to \gamma\gamma$. Two of these can be ignored in the rest frame of the pions. Why?

(b) The lifetime τ of a bound state is related to the cross section by $1/\tau = n\sigma v$, with $n = |\Psi(0)|^2$ where $\Psi(0)$ is the value of the wave function at the origin. (*Hint:* Approximate the two-body wave function as a (properly normalized) simple decaying exponential. Use the wave function to estimate the velocity v.)

(c) Estimate the lifetime of pionium.

4.6 References and Further Reading

A selection of reference texts on bound states of the strong interaction can be found here: [2, 3, 4, 5, 6, 7, 8].

[1] K. Nakamura et al. (Particle Data Group). *J. Phys. G* **37**, (2010). http://pdg.lbl.gov.

[2] Bogdan Povh, Klaus Rith, Christoph Scholz, and Frank Zetsche. *Particles and Nuclei*. Springer, 2006. ISBN 978-3-540-36683-6.

[3] Stephen Gasiorowicz. *Quantum Physics*. Wiley, 2003. ISBN 0-471-05700-2.

[4] John Donoghue, Eugene Golowich, and Barry Holstein. *Dynamics of the Standard Model*. Cambridge University Press, 1994. ISBN 0-521-47652-6.

[5] Sheldon Stone, editor. *B Decays*. World Scientific Publishing, 1994. ISBN 981-02-1897-4.

[6] T. D. Lee. *Particle Physics and Introduction to Field Theory*. Harwood Academic Publishers, 1988. ISBN 3-7186-0033-1.

[7] Kurt Gottfried and Victor F. Weisskopf. *Concepts of Particle Physics*. Oxford University Press, 1986. ISBN 0-19-504373-1.

[8] F. E. Close. *An Introduction to Quarks and Partons*. Academic Press, 1979. ISBN 0-12-175152-X.

5 Detectors and Measurements

Contrary to how we have discussed elementary particle physics up to this point, the basic experimental observables in a detector are not four-vectors. It is also rare for one single type of detector to provide enough information to identify the particles coming from a high-energy interaction and measure their energies and momenta. High-energy experiments are, therefore, collections of detectors that each provide some level of particle identification (PID) and kinematic measurement. To get a global view of what particles have been produced subsequent to a given high-energy interaction requires piecing together the information from a half dozen different detector subsystems. Rather than go over different types of detectors and their use as a starting point, the discussion below is organized from the perspective of the type of particle or object to be measured, followed by a review of the different experimental handles that exist for these measurements. Discussions of major detector topics are invoked as needed to explain the experimental methods that are involved.

5.1 Photons and Electromagnetic Calorimeters

Photons are primarily detected by their interaction with an *electromagnetic calorimeter*. To understand the function of a calorimeter, treat each particle incident on a volume of matter as a fixed-target experiment. The behavior of a particle interacting with the material will be governed by the largest cross-section processes. For electromagnetically interacting particles, the largest cross sections occur at low-momentum transfer ($1/q^2$ in the photon propagator) in regions of the material where there is a strong Coulomb source from nonzero net nuclear charge:

$$\frac{1}{r_{\text{atom}}} < |q| < \frac{1}{r_{\text{nucleus}}}. \tag{5.1}$$

A two-vertex electromagnetic interaction at low transverse momentum has a cross section set by the classical electron radius, $r_e = \alpha_{\text{QED}}/m_e$, in the Thomson formula $8\pi r_e^2/3 \approx 0.665$

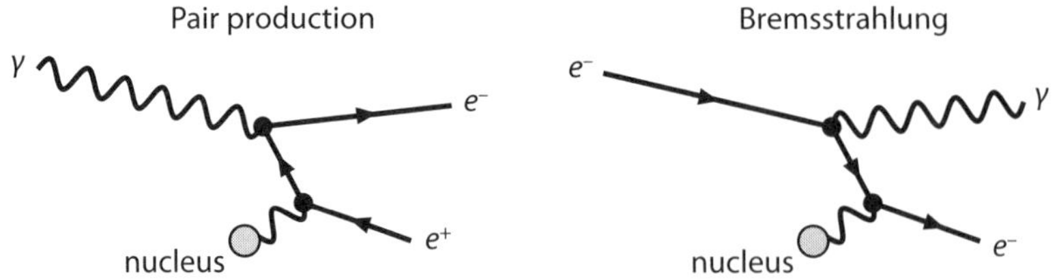

FIGURE 5.1. On the left, pair production of an electron-positron pair from a photon interacting with the charge of a nucleus. On the right, Bremsstrahlung, or "braking," radiation of a photon off of an electron or positron interacting with the charge of a nucleus.

barn. A radiative process, i.e., one with an extra photon external leg or external Coulomb source, will have an additional power of α_{QED}/π and a photon propagator integral over the length scales where there is a Coulomb source from unscreened nonzero nuclear charge in the volume of the material, as shown in figure 5.1.

The electron-positron pair production cross section, as for *Bremsstrahlung*, meaning "braking radiation," can be approximated by applying one power of α_{QED} per interaction vertex and a photon propagator momentum integral over the appropriate inverse length scales

$$\sigma_{\text{radiative}} \approx \pi r_e^2 \left[\left(\frac{Z^2\alpha}{\pi} \right) \int_{1/r^2_{\text{atom}}}^{1/r^2_{\text{nucleus}}} \frac{dq^2}{q^2} \right] = 2\alpha \left(\frac{\alpha}{m_e} \right)^2 Z^2 \ln\left(\frac{r_{\text{atom}}}{r_{\text{nucleus}}} \right) \tag{5.2}$$

where the pair production cross section is a factor of 7/9 smaller than that for Bremsstrahlung due to differing terms in the invariant amplitude. For a material with N atoms per cm^3, there will be on average n collisions in a path length L related by

$$n = N\sigma_{\text{radiative}} L. \tag{5.3}$$

Setting $n = 1$ and writing $L = X_0$, the inverse *radiation length*, X_0^{-1}, is given by

$$\frac{1}{X_0} = N\sigma_{\text{radiative}} \approx 4N\alpha \left(\frac{\alpha}{m_e} \right)^2 Z^2 \ln\left(\frac{183}{Z^{1/3}} \right), \tag{5.4}$$

where the additional factor of 2 comes from the inverse dependence of the atomic and nuclear radii on the nuclear charge in the logarithm, i.e., $r_{\text{atom}} \propto Z^{-1/3}$ while $r_{\text{nucleus}} \propto Z^{1/3}$. The factor of 183 in the logarithm is a numerical approximation for the square root of the ratio of atomic and nuclear length scales. Therefore, for energetic photons, well above the mass threshold for pair production, the interaction with matter takes on the form of an *electromagnetic cascade* or *shower*, whereby at a characteristic distance of X_0 in a material, all photons in the shower pair produce and all electrons and positrons undergo Bremsstrahlung. Thus, the average energy of a particle in the shower drops to $E_0/2^n$, corresponding to the total multiplicity of particles in the electromagnetic shower at that depth, where E_0 is the incident photon energy and n is the number of X_0 into the shower. Once the energy drops down to on order 10 MeV, a growing fraction of $1 \to 1$ processes occur, such as *Compton scattering, photoelectric interactions,* and *ionization.* Only about 25% of the total energy

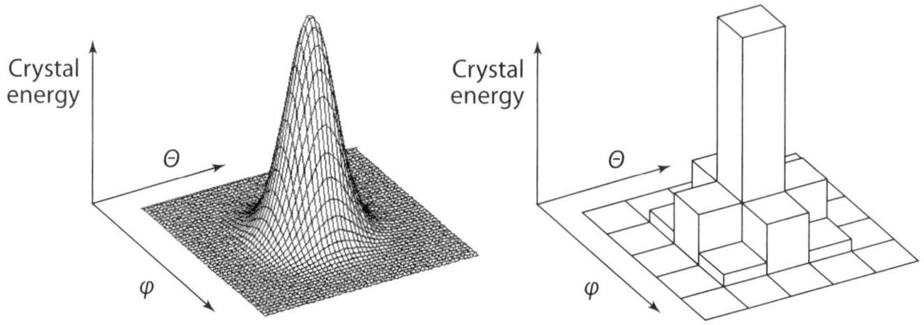

FIGURE 5.2. Transverse shower shape of a photon in an electromagnetic calorimeter composed of inorganic scintillating crystals, shown on the left. The two axes are the cylindrical coordinates of a barrel geometry calorimeter. The right-hand side shows the typical granularity of a detector where the cell size is comparable to the Molière radius of the shower in the material. Typically nine crystals contain 99% of the lateral shower energy.

deposited by the shower in the material is from positron interactions. There are two orders of magnitude more liberated atomic electrons in the shower than positrons. The large multiplicity of particles in the shower allows one to use the transverse and longitudinal shape of the shower to identify the incident particle as electromagnetic, i.e., an electron, positron, or photon. Figure 5.2 shows the transverse shower profile of a photon and a typical transverse segmentation of an actual detector, where the segmentation is set comparable to a characteristic transverse size known as the Molière radius. The thickness of the electromagnetic calorimeter is typically 20–25 radiation lengths, ~25 cm for high-density crystal calorimeters, in order to keep the longitudinal containment fluctuations below ~0.5%.

For a calorimeter based on scintillating crystals, the statistics of the light yield and photodetection factor into the stochastic term S in the energy measurement. An equivalent term arises in ionization-based cryogenic liquid calorimeters where the linear dimension, d_{active} [mm] of the liquid gap, typically 2–3 mm, and the *sampling fraction* f_{samp}, typically 1–2%, of the energy loss in the liquid compared to the passive absorbers give a stochastic term $S \approx 2.7\% \sqrt{d_{\text{active}}[\text{mm}]/f_{\text{samp}}}$. Other contributions to the energy resolution, added in quadrature, are from electronic noise N and a percentage error on the energy measurement, known in the fractional energy resolution as a *constant term* C. The fractional energy resolution is parameterized by

$$\frac{\sigma_E}{E} = \frac{N}{E} \oplus \frac{S}{\sqrt{E}} \oplus C, \tag{5.5}$$

where σ_E is the Gaussian standard deviation of the energy measurement relative to the mean response with E in units of GeV. The symbol $\oplus$ means addition in *quadature*:

$$\frac{N}{E} \oplus \frac{S}{\sqrt{E}} \oplus C \equiv \sqrt{\left(\frac{N}{E}\right)^2 + \frac{S^2}{E} + C^2}. \tag{5.6}$$

Ultimately, the constant term is the limiting factor on the resolution of high-energy measurements. The constant term comes from intrinsic nonuniformities in the calorimeter response, including from cell-to-cell energy calibration inaccuracies. An electromagnetic calorimeter will typically have a stochastic term varying from $S = 3$–15%, with ~ 3–5% typical for crystal calorimeters and ~ 10–15% typical for sampling calorimeters, and a constant term $C = 0.5$–2% with an electronic noise in the range $N = 0.1$–0.5 GeV.

5.2 Electrons, Tracking, *dE/dx*, and Transition Radiation Detectors

An electron is distinguished from a photon primarily by the presence of a *charged track* pointing directly at the center of the electromagnetic shower. A *charged track* is a term used for a measurement of the trajectory of a charged particle using detectors known as trackers.

When a charged particle passes in proximity to an atomic electron, there is a finite probability that the atomic electron is excited into the conduction band of the material. In a gaseous medium, the atomic electron is liberated in this process, thereby leaving an ionized molecule, or ion. If an ambient electric field is applied across the tracker medium, then the ionized charges in a gas, or electron-hole pairs in a semiconductor, will separate. Under normal material densities and electric field strengths, the electrons due to a finite mean-free path length will drift at a constant velocity, known as the *drift velocity*, and similarly for the ions, but at a substantially lower velocity. The charge is drifted to anodes and cathodes and collected. If the drift velocity is sufficiently low, then the drift time, the time that elapses before the charge is incident on the collection electrode, is used as a means to compute the spatial coordinate of where the initial ionization occurred within the tracking volume. In a silicon strip tracker, a thin, typically 300 μm thick silicon wafer with metallization for charge collection at a fine spacing, such as 50 μm, will detect ionization charge across several closely spaced strips forming a *charge cluster*. Correcting for possible shifts from an effect known as the *Lorentz angle* due to drift in ambient electric and magnetic fields, the center of the charge cluster is a spatial measurement of the location where the charged particle passed through the silicon detector.

A tracker is, therefore, a device that measures spatial coordinates along the trajectory of a charged particle. The spatial measurement is made directly with charge clusters or indirectly with drift time measurements. The spatial coordinates along a charged particle trajectory need to be associated with the same track in order to accurately measure the properties of the trajectory, such as the proximity of the track to the collision point, the impact point of the track on the calorimeter, and any bending or curvature that the track has due to motion in an ambient magnetic field. A tracker, unlike a calorimeter, competes with the necessity to make measurements without interfering with the properties of the charged particle.

An ideal helicoidal track requires five parameters, where one choice of parameters in the bending plane is the curvature, the azimuthal angle of the momentum at the position of closest approach to a reference point and the distance of closest approach to the reference point. For a circular trajectory in the bending plane, the transverse momentum p_{T} (in GeV/c) is related to the radius of curvature R (in m) in a magnetic field B (in T) via

$$p_{\mathrm{T}} = 0.2998BR \tag{5.7}$$

where the significant digits come from the speed of light $c = 2.99793 \times 10^8$ m/s. The *sagitta* $s \approx 0.3BL^2/(8p_{\mathrm{T}})$ is the maximum deviation of the track from a straight line, which for a tracker lever arm, or chord, $L = 1$ m, a magnetic field $B = 4$ T, and $p_{\mathrm{T}} = 1000$ GeV/c gives a sagitta of 150 μm. In the nonbending plane for an arbitrary helicoidal trajectory, an offset and slope are typical parameters.

The most important degrading effect from material on the performance of the tracking system is called *multiple scattering*. If a charged particle bending in a magnetic field passes through a length of tracker material, then it will be randomly deflected with a characteristic opening angle depending on the density and thickness of the material. Fortunately, the larger the magnetic field, the smaller the percentage deflection of the charged particle relative to the deflection expected from the motion in the field. Therefore, multiple scattering is reduced for high magnetic fields down to a typical level of 0.5–2% of the track momentum transverse to the direction of the magnetic field. This momentum resolution limit is flat over a finite range of track momenta. Below typically 50–100 MeV/c in p_{T}, the track curls in the magnetic field and eventually stops in the material. For large transverse momenta, the track will have little deflection in the magnetic field and the finite curvature cannot be distinguished from a straight trajectory, leading to the possibility of a mismeasurement of the sign of the curvature, known as *charge confusion*. One can parameterize the transverse momentum resolution of a tracker as

$$\frac{\sigma_{p_{\mathrm{T}}}}{p_{\mathrm{T}}} = c_0 \oplus c_1 \cdot p_{\mathrm{T}} \tag{5.8}$$

added in quadrature according to equation (5.6) where $\sigma_{p_{\mathrm{T}}}$ is the Gaussian standard deviation, or uncertainty, of the measurement relative to the mean p_{T}, the component of the momentum orthogonal to the ambient magnetic field. The multiple-scattering term dominates for most of the low-p_{T} momentum range with a constant percentage error $c_0 = 0.5$–2%. The second term is a limitation of the momentum resolution from the measurement of the finite curvature of the track where for a value of $c_1 = 10^{-3}$–10^{-4} (GeV/c)$^{-1}$ will begin to dominate for p_{T} greater than 100 GeV/c. Typically, LHC experiments design their tracking systems to maintain 10% momentum resolution or better at a momentum of 1 TeV/c. The term c_0 grows as the square root of the thickness of the tracker in units of radiation length, but depends on the inverse of the magnetic field strength and therefore leads to high-field magnets to compensate for tracker material. The term c_1 is inversely proportional to $BL^2/\sigma_{\mathrm{hit}}$, where L is the tracker radius or "lever arm" of the curvature measurement and is typically $L = 1$–2 m. The single hit resolution σ_{hit} for a silicon tracker can be as low as 10–50 μm, or even smaller for the centroid of a charge cluster, while a gas drift chamber is typically 100–150 μm. For electron measurements, it is useful to know when the momentum resolution of the tracker, which is given by the multiple-scattering term at low p_{T}, is superseded by the energy measurement resolution of the electromagnetic calorimeter. It's not uncommon for two measurement resolutions to be equal in the vicinity of 30–40 GeV depending on the detector parameters. Typically, the tracker elements are first aligned with straight, zero-ambient magnetic-field charged tracks. Then, the tracker is used to cross-calibrate the calorimeter up to the energy scale where the resolutions are comparable.

When the integrated tracker thickness is comparable to a radiation length, then electrons will undergo the Bremsstrahlung process within the tracker volume and liberate photons along its trajectory. Similarly, photons will have a finite probability to "convert" into closely spaced electron-positron tracks through the pair-production process. Therefore, for the measurement of electron trajectories there is a competing trade-off between increasing the number of spatial measurements along the trajectory and reducing the probability for radiative processes that degrade the ability to match the charged track to a well-shaped electromagnetic shower in the calorimeter. Tracker-based methods for identifying electrons in advance of the electromagnetic calorimeter are therefore also important. One method relies on measuring the magnitude of the ionization by a charged particle in the tracker layers, a technique known as dE/dx. The magnitude of dE/dx for muons is given by the Bethe-Bloch formula [1] plotted in figure 5.3. The rate of dE/dx energy loss has a shallow minimum starting at $\beta\gamma \approx 1$ extending to $\beta\gamma = 100$–1000. The lowest value of dE/dx is at $\beta\gamma = 3$–3.5, known as *minimum ionizing*, with a typical value of $(1.5/\rho)$ MeV cm^2/g normalized to the denisty of the material. Copper, for example, has a density of $\rho = 9$ g/cm^3 giving $dE/dx = 13.5$ MeV/cm. Most long-lived charged particles produced in collider experiments, typically at momenta of a few to tens of GeV/c, such as muons, pions, and kaons, are *minimum ionizing particles (MIPs)* over a wide range of momentum as set by $\beta\gamma$. However, the electron, which has additional scattering processes with atomic electrons and significant radiative losses due to its small mass, is typically 1.3–$1.6 \times$ minimum ionizing over the same range of momenta. The dE/dx measurement in combination with the momentum measurement can therefore be used for particle

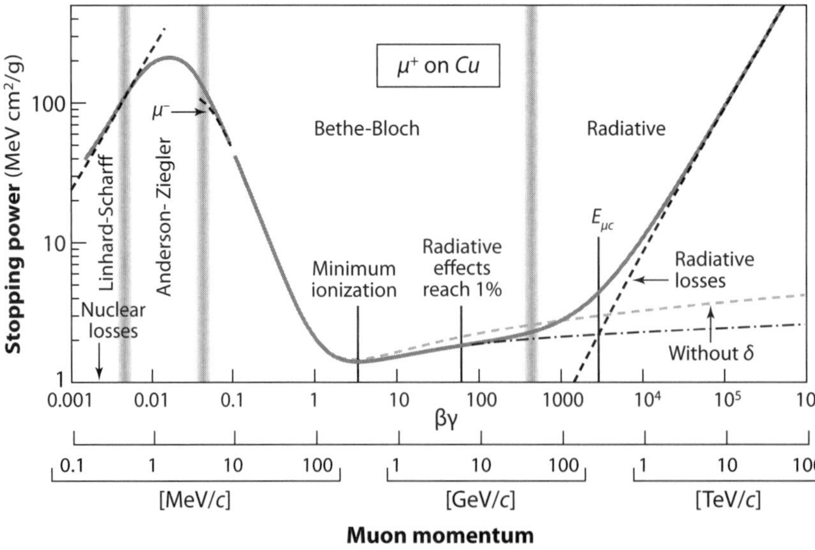

FIGURE 5.3. Rate of energy loss for a μ^+ traveling through copper normalized to the density of copper (Credit: PDG) [1].

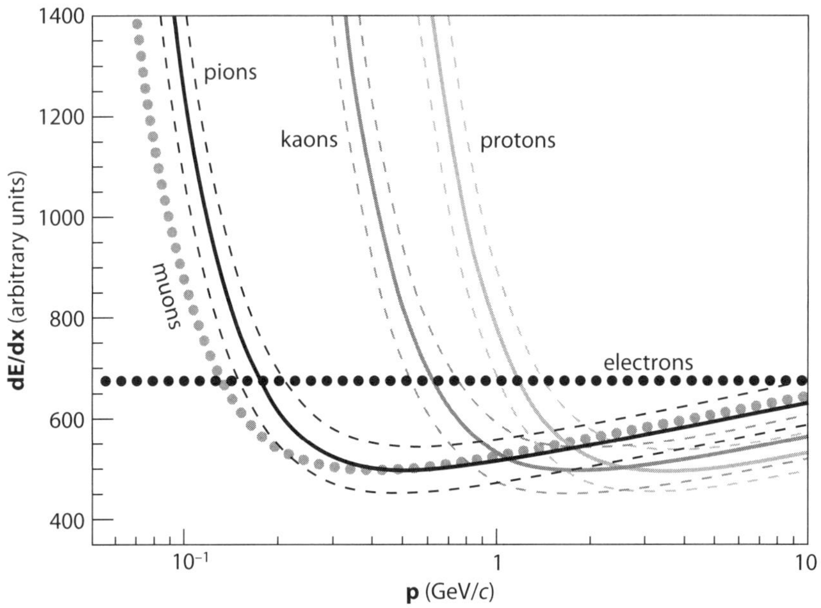

FIGURE 5.4. The measurement of dE/dx for different particle types as a function of momentum (Credit: BABAR). The dashed bands indicate the typical measurement resolutions.

identification at low momentum, as shown in figure 5.4. A second class of particle identification methods relies on the large Lorentz boost of the electron in the lab frame. An effect discovered in 1939, known as *transition radiation* [1, 2], causes relativitistic electrons to emit X-rays as they pass from a dielectric into a gas-filled vacuum. The flux of emitted X-rays when absorbed by a high-Z gas layer effectively results in the enhancement of the dE/dx signal from electrons, thereby improving the electron identification.

5.3 Single Hadrons, Time-of-Flight, Cherenkov Radiation, and Hadron Calorimeters

The total interaction cross section of a hadron is an order of magnitude smaller, roughly 40 mb, than for an *electromagnetic particle,* i.e., an electron, positron, or photon, interacting with the same calorimeter material. The large spatial extent of hadronic showers forces hadronic calorimeters to be thick, massive detectors, using dense materials such as copper, iron, lead, uranium, and tungsten to provide shower containment. The particular choice of absorber material can depend on the mechanical strength requirement, such as the need to maintain a precise gap spacing (i.e., no sagging) over a large distance, the need for nonmagnetic materials, or the use of self-calibrating fission signals. Unlike an electromagnetic shower, the particle multiplicity per interaction in a hadronic shower is not constant. The first interaction will have the highest particle multiplicity, on order 10 secondary particles for an incident pion of 300 GeV, with the multiplicity growing logarithmically with incident pion energy. The second set of interactions will drop down, for the given example, to the level of 6 secondary particles per interaction on average, and the third set to $1 \rightarrow 3$ processes. By the fourth interaction length, $1 \rightarrow 1$ processes begin to dominate and the number of showering particles begins to drop off. The average shape of a hadron shower thus has a much sharper rise of energy deposition in the first nuclear interaction length than an electromagnetic shower. And, unlike a radiation length, the nuclear interaction length is less universal. Typically one refers to the proton nuclear interaction length when describing calorimeters, while, in fact, the pion nuclear interaction length is 1.5 times longer and can lead to a small probability for charged pions to fully penetrate the hadron calorimeter, an occurrence known as *punch-through*. Typically and for practical reasons, hadron calorimeters are limited to ~10 (proton) nuclear interactions lengths in thickness, i.e., 1.1–1.7 m.

Hadronic showers are challenging to measure due to the irregularity of their shape and fluctuations in particle content and multiplicity. A large number of neutrons are liberated in the shower from nuclear break-up and only a fraction of the neutron energy is collected within the integration time of the calorimeter energy measurement. There are fluctuations in the particle multiplicity per interaction, and an important effect comes from the asymmetry in charged pion and neutral pion energy depositions. Nuclear interactions will allow nuclear isospin quantities to "tumble" freely in the shower. However, if a charged pion tumbles into a π^0 or a π^0 is produced, the π^0 decays promptly via $\pi^0 \rightarrow \gamma\gamma$ and, therefore, immediately initiates an electromagnetic shower. Thus, for every nuclear interaction length into the hadron calorimeter, there is a finite probability for an electromagnetic shower to occur within the hadronic shower. The π^0 fraction f_{π^0}, the so-called electromagnetic component of hadronic showers, grows nonlinearly with incident hadron energy and imparts a nonlinear energy response in calorimeters due to unequal efficiencies for measuring electromagnetic (e) shower energies as compared to hadronic (h) energy depositions and ionization, known as the e/h-value. A calorimeter with $h/e = 1$ is known as *compensating*. Noncompensating calorimeters typically have larger stochastic and constant terms than the ideal, compensating calorimeters. In fact, out of geometric necessity, the electromagnetic calorimeter typically acts as the first interaction length of the hadron calorimeter measurement and this can further contribute to a nonuniform response if the EM calorimeter has a different e/h-value or there are significant amounts of dead material between the EM and hadronic compartments. One nuclear interaction length will typically not be sufficient to contain a hadronic shower (exceptions are early electromagnetic showering from *charge-exchange processes* where a charge pion converts directly to a π^0). Therefore, hadrons can be separated from electrons by requiring some fraction of the total calorimeter energy to be deposited in the hadronic calorimeter, the so-called H/E-ratio of hadronic to electromagnetic calorimeter energy. The energy measurement resolution of a hadron calorimeter can be expressed in the same form as electromagnetic calorimeters (5.5). However, in contrast to an EM calorimeter, the parameters for a hadron calorimeter are typically an order of magnitude larger with the stochastic term in the range $S = 45$–125%, the constant term $C = 4$–8%, and the noise term 1–5 GeV.

Particle identification for hadrons can be achieved with detectors that sit in front of the calorimeters and make measurements related to the finite velocity of the hadron. In recent years, the level of precision of spatial measurements, now down to the sub-10 μm level, has been complemented with time-of-flight systems that can achieve sub-10 ps timing resolution. These systems use thin devices with low transit-time readout electronics, such as fast radiators coupled to silicon photomultiplier detectors or chambers that look for ionization-induced discharges across a thin high-voltage gap. Two different hadrons originating at the same vertex with the same momentum p, such as a charged pion and a charged kaon, will have a time-of-flight difference at a distance L from a common vertex given by

$$\Delta t \approx \frac{Lc}{2p^2}\left(m_K^2 - m_\pi^2\right), \tag{5.9}$$

which shows that for flight distances of $L = 1$ m, a high-precision time-of-flight detector separates kaons from pions for momenta up to a few GeV/c.

The fast radiator is a substance such as quartz with a high index of refraction. A form of radiation known as *Cherenkov radiation* is emitted when a particle travels in a medium at a velocity that exceeds the maximum propagation velocity of electromagnetic radiation within that material. The flux of Cherenkov light, whose frequency spectrum starts in the blue and goes into the ultraviolet, has a conical opening angle that depends on the quantity

$$\cos\theta_c = \frac{1}{\beta n(\lambda)} \tag{5.10}$$

where $\beta = v/c$ is the velocity of the particle normalized to the speed of light in vacuum and $n(\lambda)$ the wavelength-dependent index of refraction of the medium. The opening angle of the radiation, θ_c, is used for a type of particle identification system known as a Ring-Imaging CHerenkov (RICH) detector, where the rings correspond to the time expansion of the Cherenkov radiation and subsequent measurement with a dense spatial array of UV-sensitive photodetectors. In its most compact form, ring-imaging information can be preserved and transported using a highly polished quartz panel exhibiting internally reflected Cherenkov light. The momentum-dependent particle identification system using the Detection of Internally Reflected Cherenkov (DIRC) radiation is shown in figure 5.5.

Hadrons that decay promptly or within the tracker volume can be identified using resonance reconstruction, as in the case of $K_S^0 \to \pi^+\pi^-$ and $\pi^0 \to \gamma\gamma$ shown in figures 5.6 and 5.7, respectively.

5.4 Muons and Muon Spectrometers

The muon is unique in that it is the only long-lived charged particle with minimal cross section to interact with matter for momenta up to nearly 1 TeV/c. A muon will therefore pass through the central tracker, the electromagnetic and hadronic calorimeters, and the

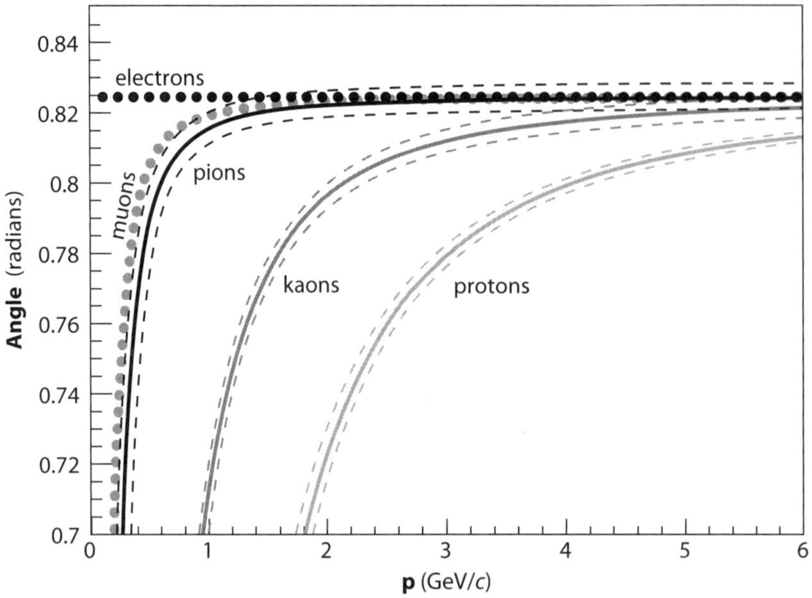

FIGURE 5.5. The opening angle of Cherenkov radiation emitted from different particle types as a function of momentum (Credit: BABAR). The Cherenkov light is internally reflected in a DIRC detector before being measured. The dashed bands indicate the typical measurement resolutions.

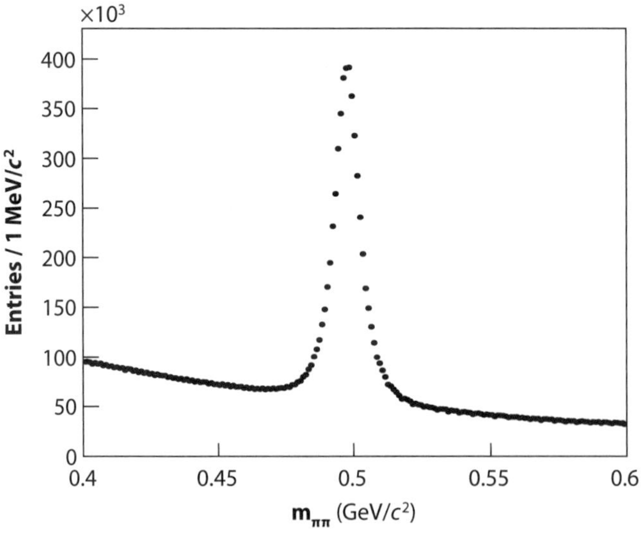

FIGURE 5.6. Invariant mass distribution of a $\pi^+\pi^-$ secondary vertex showing a resonance peak at the K_S^0 mass in $\sqrt{s} = 7$ TeV data collected by the CMS experiment (Credit: CMS) [3].

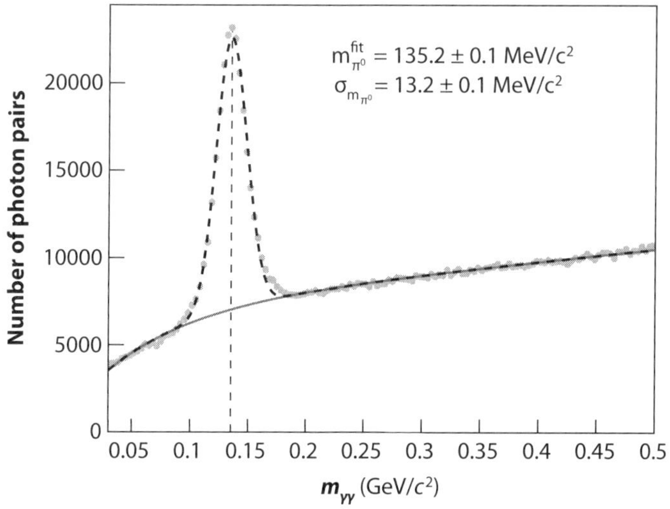

FIGURE 5.7. The diphoton reconstructed invariant mass from π^0 decays collected by the CMS experiment using $\sqrt{s} = 7$ TeV data (Credit: CMS) [3]. The baseline under the peak comes mainly from the combinatorical background of diphoton selection.

magnet coil with only *minimum ionizing* energy loss, typically losing 2–3 GeV of total energy in the calorimeters. An example trajectory of a muon in the CMS experiment and a summary of several of the particle measurement techniques is shown in figure 5.8 for a slice of the many detector subsystems comprising the experiment. The bending of the muon reverses sign outside of the solenoid due to the return flux of the magnetic field concentrated within the layers of iron interspersed between tracking chambers, known as the *muon spectrometer* or instrumented return flux. The additional lever arm of the muon spectrometer complements the combined tracking momentum resolution by reducing the effective c_1 term in equation (5.8) relative to using only the inner tracker and thereby improving momentum measurements for charged particles of several hundred GeV/c to beyond 1 TeV/c. At low momenta, the multiple scattering term, c_0, is substantially larger for the stand-alone muon spectrometer and, consequently, the primary function of the muon chambers is to identify the muon as a minimum ionizing charged particle escaping the calorimeters. The calorimeters are typically at least 10 nuclear interaction lengths thick to suppress the probability that a charged hadron can *punch through* and leave a track in the muon chambers.

5.5 Jets and Jet Algorithms

Quarks and gluons that are produced from an elementary interaction will propagate for on order 10^{-23}s before the energy uncertainty of the strongly interacting system begins to

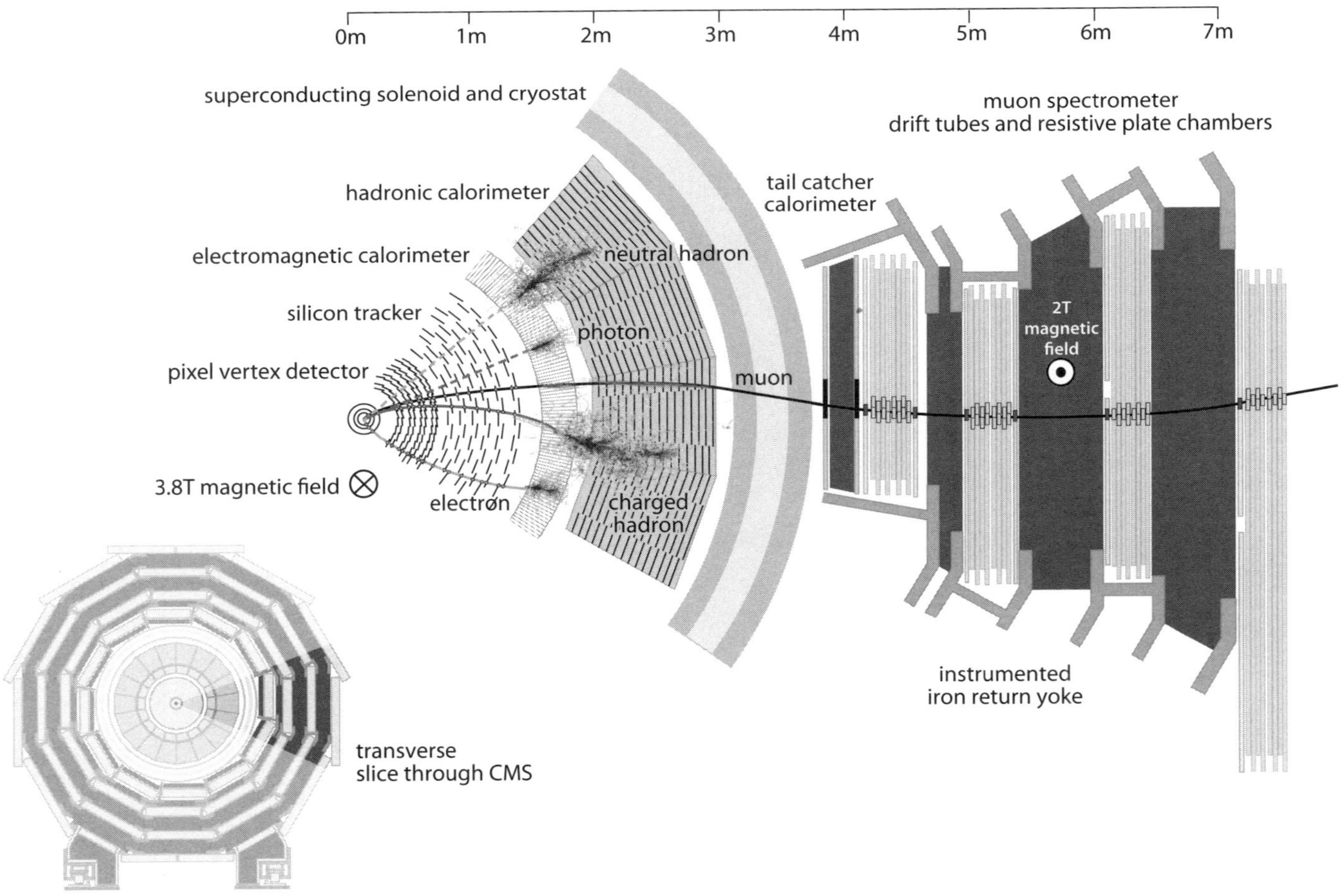

Figure 5.8. A schematic view of a slice of the CMS experiment and the corresponding interactions of muons, electrons, photons, and charged and neutral hadrons in elements of the detector (Credit: CMS/D. Barney) [3].

be comparable to the natural widths of color-neutral hadronic bound states that can be directly formed from quark-antiquark pair production in the color potential. Within this time scale, the initial quark or gluon transfers its energy and momentum into a collection of hadrons, some of which promptly decay to other hadrons while others will propagate for much longer time scales. Long-lived hadrons are those that decay through the weak or electromagnetic interactions, or are stable, as in the case of the proton and antiproton. The phenomenon of hadron production initiated by a strongly interacting quark or gluon is known as *QCD jet production.* The bulk of the particles in jets are pions, with an equal number (nuclear isospin independent) on average of π^+, π^- and promptly decaying $\pi^0 \to \gamma\gamma$. The heavier mesons and vector mesons are produced at an order of magnitude smaller fractions, on average. Typically, only 10% of the particles produced are baryons and antibaryons. As the overall energy and momentum of the collection of hadrons is set by the initial strongly interacting particle, known as a *parton,* the jet properties are an indirect measure of the initial parton four-momentum. Moreover, the *virtuality* of the initial parton and the effects of λ_{QCD}-scale interactions contribute to a finite jet mass, as computed from the four-momenta of the jet of hadrons. One particular parton, however, the top quark will not directly initiate a QCD jet. Top quark physics will be discussed separately in a later chapter. The weak charged-current decay of the top quark is on a shorter time scale than the strong interaction and, therefore, the top quark does not hadronize before decaying to a *b*-quark and a W boson.

The kinematical quantities corresponding to a QCD jet involve three distinct levels of development. QCD matrix elements assign a four-momentum to the initial parton forming the jet and associated *parton showering* processes such as gluon radiation and quark-antiquark pair production above a perturbative momentum scale k_{T}. The *perturbative QCD* stage of jet development creates a shower of partons, known as parton-level jets. After the jet momentum sharing among particles in the low-p_{T} development of the jet, known as *fragmentation,* and the associated *hadronization* into color-neutral hadrons has occurred, the jet is referred to as a particle-level jet. The particle-level jet consists of all stable particles, optionally excluding prompt noninteracting particles such as neutrinos from semileptonic heavy quark decay, and corresponds to what an ideal calorimeter (and muon system) is expected to measure. Detector-level jets formed from a combination of calorimeter and tracking measurements for charged particles, as in *particle-flow* techniques, are treated similarly. The calorimeter-level or detector-level jet is composed of the measured energy depositions from the electromagnetic and hadronic showering in the calorimeter and muons measured with tracking detectors and muon spectrometers. These stages of jet development are shown in figure 5.9. Many factors, measurement-related and algorithmic, introduce nonideal jet energy response. The jet energy scale corrections and scale uncertainties for measured jets are typically factorized into calorimeter-to-particle level and particle-to-parton level. The particle-to-parton-level correction depends on the hypothesis for the parton-type initiating the jet and includes energy radiated out of the cone of the reconstructed jet. The jet energy scale corrections depend on every step of jet reconstruction from thresholds in the calorimeter, to jet algorithm and seed thresholds, to detector response and shower containment, to parton-type hypothesis.

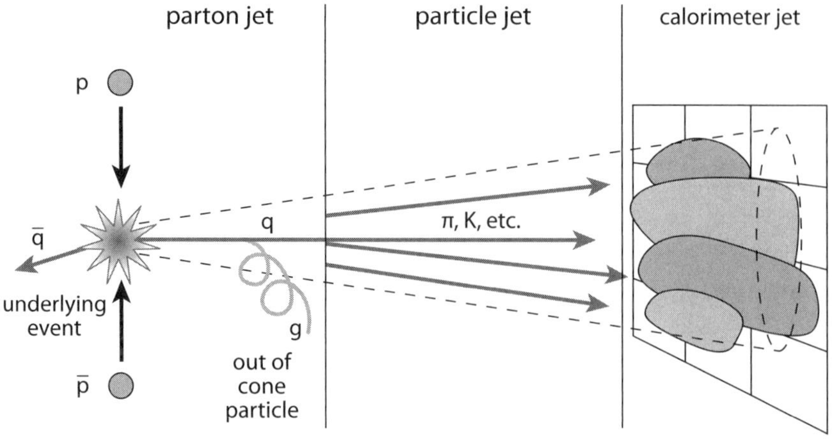

FIGURE 5.9. The three levels of jet development corresponding to QCD matrix element (parton-level), fragmented and hadronized stable particles (particle-level) and detector measurements (calorimeter-jets) (Credit: CDF).

The first jet algorithms for hadron physics, proton-proton and proton-antiproton colliders, were simple cones with a cone radius defined in coordinates that are invariant under Lorentz boosts along the momentum direction of the incoming proton. The *rapidity* y is given by

$$y = \frac{1}{2}\ln\left(\frac{E + p_z}{E - p_z}\right) \tag{5.11}$$

and is explicitly invariant under longitudinal boosts as will be described later in the section on hadron collider physics. In the limit of massless particles

$$\begin{aligned} E \pm p_z &= E(1 \pm p_z/E) = E(1 \pm \beta\cos\theta), \\ y \to \eta &= \frac{1}{2}\ln\left(\frac{1+\beta\cos\theta}{1-\beta\cos\theta}\right) \overset{\beta\to 1}{=} -\ln\tan\frac{\theta}{2}, \end{aligned} \tag{5.12}$$

where η is called the *pseudorapidity*. The R-measure, or cone radius for simple cone jet algorithms, is in terms of η and ϕ, the azimuthal angular separation in the plane transverse to the incoming proton direction, and is given by

$$R = \sqrt{(\eta_1 - \eta_2)^2 + (\phi_1 - \phi_2)^2} \tag{5.13}$$

where the 2π periodicity of ϕ must be explicitly removed by taking ϕ differences in the range $[-\pi, \pi]$.

Over the last two decades, cone and clustering techniques have advanced in sophistication and variety. A number of jet reconstruction algorithms are in common use, including, for example, the iterative cone and variants thereof, midpoint-cone, SIScone (Seedless Infrared Safe Cone), the inclusive k_T, the anti-k_T, and the Cambridge-Aachen jet algorithms. The jet algorithms may be used with different *recombination schemes* for adding the constituents. In the *energy scheme*, constituents are added as four-vectors, where the final four-vector describing the jet has a finite mass.

To give an explicit example, in the iterative cone algorithm an E_T-ordered list of input objects (particle tracks or calorimeter towers) is created, where E_T is shorthand for the transverse momentum of a massless calorimeter object. A cone of size R in (η, φ)-space is cast around the input object having the largest transverse energy above a specified *seed threshold*. The objects inside the cone are used to calculate a *proto-jet* direction and energy. The computed direction is used to seed a new proto-jet. The procedure is repeated until the energy of the proto-jet changes by less than 1% between iterations and the direction of the proto-jet changes by $\Delta R < 0.01$. When a stable proto-jet is found, all objects in the proto-jet are removed from the list of input objects and the stable proto-jet is added to the list of jets. The whole procedure is repeated until the list contains no more objects with an E_T above the seed threshold. The cone size and the seed threshold are parameters of the algorithm. When the algorithm is terminated, a recombination scheme, such as the energy scheme, is applied to jet constituents to define the jet kinematic properties.

Cone algorithms, such as the midpoint cone, can be designed to facilitate the splitting and merging of jets for a pair of jets closely spaced in R. In the midpoint cone, contrary to the iterative cone algorithm described above, no object is removed from the input list. This can result in overlapping proto-jets (a single input object may belong to several proto-jets). To enhance the *collinear* and *infrared safety* of the algorithm, a second iteration of the list of stable jets is done. For every pair of proto-jets that are closer than the cone diameter, a *midpoint* is calculated as the direction of the combined momentum. These midpoints are then used as additional seeds to find more proto-jets. When all proto-jets are found, the following splitting and merging procedure is applied, starting with the highest E_T proto-jet. If the proto-jet does not share objects with other proto-jets, it is defined as a jet and removed from the proto-jet list. Otherwise, the transverse energy shared with the highest E_T neighbor proto-jet is compared to the total transverse energy of this neighbor proto-jet. If the fraction is greater than f (typically 50%), the proto-jets are merged; otherwise the shared objects are individually assigned to the proto-jet that is closest in (η, φ)-space. The procedure is repeated, again always starting with the highest E_T proto-jet, until no proto-jets are left. This algorithm implements the energy scheme to calculate the proto-jet properties but a different recombination scheme may be used for the final jet. The parameters of the algorithm include a seed threshold, a cone radius, a threshold f on the shared energy fraction for jet merging, and also a maximum number of proto-jets that are used to calculate midpoints.

The inclusive k_T, anti-k_T, and Cambridge-Aachen jet algorithms are cluster-based jet algorithms [4]. The cluster procedure starts with a list of input objects, stable particle tracks, or calorimeter cells. For each object i and each pair (i,j) the following distances are calculated:

$$\begin{aligned} d_{iB} &= (k_{T,i})^{2p}, \\ d_{ij} &= \min\{k_{T,i}^{2p}, k_{T,j}^{2p}\} R_{ij}^2 / R^2 \quad \text{with} \quad R_{ij}^2 = (\eta_i - \eta_j)^2 + (\varphi_i - \varphi_j)^2, \end{aligned} \tag{5.14}$$

where $k_{T,i}$ is the transverse momentum of the ith particle with respect to the beam axis. The parameter R is typically set in the range $R = 0.1$ for extremely narrow jet substructure to

$R = 1.0$ to encompass the bulk of the total particle flow from all particles produced from a jet generated by a quark or gluon. The parameter $p = 1, 0, -1$ is set to a value that corresponds to the inclusive k_T, Cambridge-Aachen, and anti-k_T jet algorithms, respectively. Here $p = 1$ begins clustering with the lowest k_T objects while $p = -1$ begins with the highest k_T objects. The quantity d_{iB} is called the particle-beam distance as it sets a reference for object i for the transverse momentum relative to the beam direction. The clustering algorithm searches for the smallest d_{iB} or d_{ij}. If a value of type d_{ij} is the smallest, the corresponding objects i and j are removed from the list of input objects, merged using one of the recombination schemes and filled into the list as one new object. If a distance of type d_{iB} is smaller or equal to d_{ij}, then the corresponding object i is removed from the list of input objects and filled into the list of final jets. The procedure is repeated until all objects are included in jets.

Once a jet list is formed, it is natural to evaluate the energy resolution of the resulting objects as they relate to the expected four-vector summation for the particles forming the jet, and also at a deeper level to the quark or gluon that initiated the jet. The jet energy resolution at hadron colliders is typically fitted with the following functional form:

$$\frac{\sigma_{E_T}}{E_T} = \frac{N}{E_T} \oplus \frac{S}{\sqrt{E_T}} \oplus C, \tag{5.15}$$

added in quadrature according to equation (5.6) where the first term is due to fixed energy fluctuations in the cone from electronics noise and also, as described in chapter 8, from minbias pile-up, and *underlying-event* energy; the second term comes from the stochastic fluctuations of the hadronic calorimeter measurements; and the last term is the constant term from residual nonuniformities and nonlinearities in the detector response.

The average particle p_T in a jet predicted for LHC 7 TeV-on-7 TeV proton-proton collisions is plotted in figure 5.10 versus jet p_T for $R = 1.0$ iterative cone jets. Similarly, the particle multiplicity in jets for $R = 0.5$ and $R = 1.0$ iterative cone jets is plotted in figure

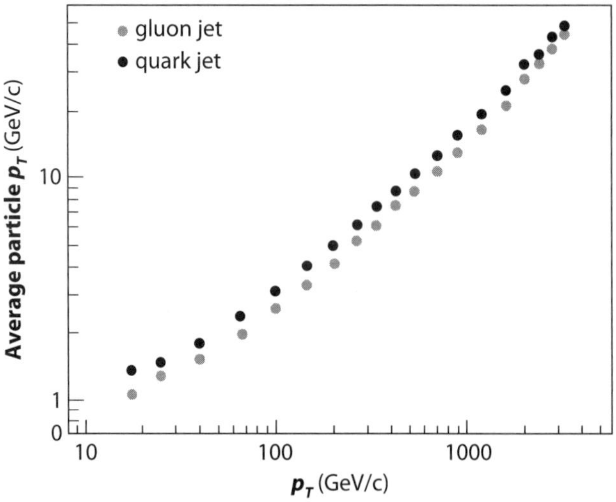

FIGURE 5.10. Average particle p_T in a gluon or light quark jet as a function of jet p_T, as estimated from simulation.

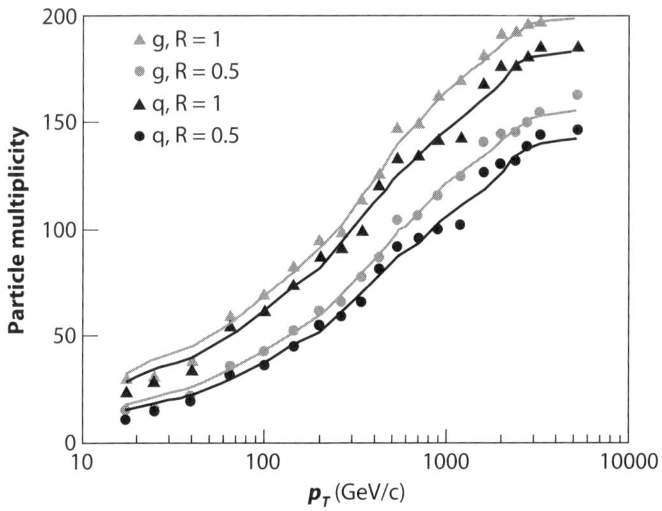

FIGURE 5.11. Average particle multiplicity in gluon (g) and light quark (q) jets within a cone radius of $R = 0.5$ and $R = 1.0$ as a function of jet p_T, as estimated from simulation.

5.11 where the particle content of the jets, in terms of particles impacting the calorimeter, is dominated by charged pions and photons from π^0 decay. Excluding jets with leptons from semileptonic decay, one sees typically ~75% of the visible energy flow in hadrons, about 80% of which are charged hadrons. The remaining ~25% of the visible energy flow is in photons, where there are large jet-to-jet fluctuations on the fraction of π^0 versus $\pi^{\pm}$.

5.6 *b*-Jets and Vertex Detectors

The finite lifetime of *B*-hadrons introduces a unique ability to identify jets originating from *b*-quarks, so-called *heavy flavor* jets. A typical beam pipe radius in a collider experiment is at least 4 cm. A *B*-hadron produced with an energy of 35 GeV after fragmentation will have a decay length of ~2–3 mm, well within the beam pipe. Therefore, the tracking detectors must accurately extrapolate charged particle trajectories back into the vacuum region of the beam pipe to determine their point of origin. To accomplish this, a vertex detector is positioned directly outside the beam pipe, where the beam pipe is typically made out of beryllium or other low multiple-scattering materials. A vertex detector must handle a high flux of closely spaced particles coming from jets, multiple interactions, and highly boosted short-lived particles as well as processes such as photon conversion in the beam pipe.

The strategy of using a vertex detector is to provide unambiguous three-dimensional spatial points unique to each charged-particle trajectory and to provide at least three radial layers of measurements to enable a minimal level of track-finding capability. This is achieved with the use of a silicon pixel detector. Pixel sensors are typically 200–300 μm thick with individual pixel dimensions of 50–100 μm in width in the bending plane and 150–400 μm in length along the beam pipe axis. The total number of individual channels in an LHC pixel detector is nearly 100 megapixels.

Pixel detector coordinates simultaneously improve the accuracy of the track extrapolation back into the collision region and the momentum resolution of the combined tracker, despite the relatively minor increase in measurement lever arm for the track curvature. Several parameters are needed to determine the helicoidal trajectory of a charged particle. A measurement improvement on the distance of closest approach, as determined from a pixel detector, can therefore translate into an improvement in the momentum resolution when the track parameters are heavily correlated. To take into account finite scattering in the tracking material, a *Kalman filter* or *Gaussian-sum filter* is used to propagate nonideal charged-particle trajectories.

One of the most important measurements of the tracking system is the *primary vertex* location. The primary vertex is the spatial location of where two particles in the colliding beams interacted. With a pixel detector, the primary vertex can be determined to better than 10–50 μm in all three spatial coordinates depending on the track momenta and multiplicity. In the case of heavy flavor production there are multiple vertices in the event. The first is a collection of charged tracks produced promptly at the primary vertex from strong-interaction decays. The second set are *secondary vertices* that originate from the B-hadron decay, one from each jet, and there is potentially a third set of *tertiary vertices* that come from the finite lifetime of D mesons and kaons produced in B-hadron decay. Figure 5.12 shows an example of a b-tagging in a heavy flavor jet. A typical measure of the sensitivity for detecting secondary vertices is the resolution of the miss distance or *impact parameter* of a track to the primary vertex. The impact parameter resolution typically varies from 10 μm to 100 μm corresponding to 100 GeV/c and 1 GeV/c track transverse momenta p_T, respectively. A collection of tracks with significant impact parameters and therefore low

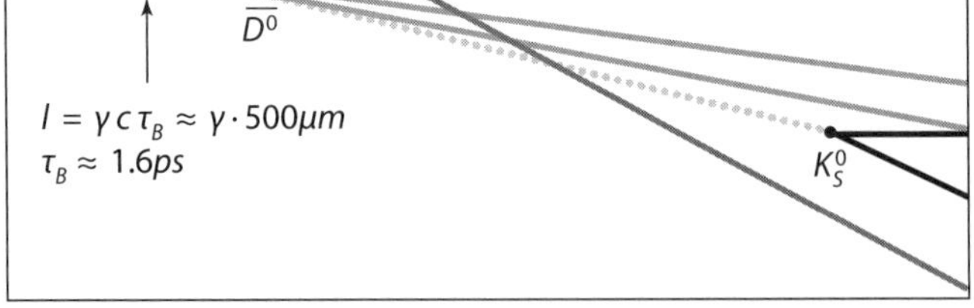

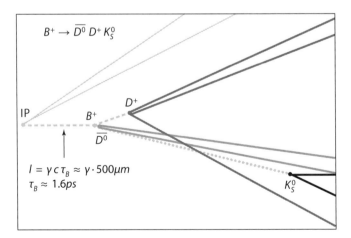

FIGURE 5.12. The figure shows the reconstructed primary vertex, also known as the interaction point (IP), to the left. In the center is the secondary vertex from the B-hadron decay. Additional tertiary vertices are found for the D-meson and neutral kaon decays (Credit: ALEPH).

probabilities to come from the primary vertex are used to compute the probability that a jet contains a heavy flavor hadron. A jet containing a secondary vertex has a high probability to be a jet originating from an energetic b-quark parton or, with lower sensitivity, a c-quark. There is, however, a finite probability that heavy flavor hadrons are formed in the hadronization stage of jet production, such as in the process of *gluon splitting* to a heavy quark-antiquark pair.

Impact parameters will have a *lifetime sign* indicating whether the decay length from the track is on the side of the jet relative to the primary vertex or whether it is in the *negative lifetime* region. The negative lifetime distribution is a valuable measure of the finite resolution for secondary vertex tagging and the expected mistag rate for measuring a false positive-lifetime secondary vertex. For a 60% lifetime b-tagging efficiency for heavy flavor jets, the typical mistag rate is 1% or less for light quark (excluding charm) and gluon jets. The mistag rate performance curve of the lifetime b-tagging algorithm typically increases exponentially with b-tagging efficiency where it is common to use mistag rates between 0.1% and a few percent depending on the efficiency needed. An alternative method for identifying b-jets is to search for the semileptonic decay of a B-hadron producing a charged lepton, electron or muon, with a large momentum and relatively large transverse momentum with respect to the jet axis owing to the large mass of the B-hadron. The *soft lepton* tagging method, which has 10–15% efficiency for a 1% mistag rate, is often used to estimate the performance of the lifetime tagging algorithm.

5.7 τ-Leptons

The τ-lepton is unique among the leptons in that it can decay into hadrons. Table 5.1 lists the charged-lepton properties and several of the τ-lepton decay modes. A hadronically decaying τ-lepton has properties similar to QCD jets, but with tight restrictions on the invariant mass, angular width, and charge multiplicity. Typically, the decay products are well-contained within a $\Delta R = 0.3$ cone jet, and the charge multiplicity is dominantly one-prong and three-prong with an average visible energy of 80% of the initial τ-lepton energy. The primary hadronic-τ categories are one-prong with no EM depositions in the calorimeter, one-prong with at least one EM shower, and inclusive three-prong final states. Typically, hadronic τ-leptons can be identified with an overall efficiency of 40% with a QCD jet fake rate of 0.1%.

The dominant one-prong and three-prong decays of τ-leptons and the expected visible mass distribution are shown in figures 5.13 and 5.14, respectively, for selected τ-lepton candidates in Tevatron data. The identification of hadronic-τ decays relies on the narrowness and isolation of the τ-jet. Figures 5.15 and 5.16 show the definition and performance, respectively, of a "shrinking cone" method for separating τ-jets from QCD jets. Due to the finite mass of the τ-lepton, a higher p_{T} boost will result in a narrower jet of particles from its decay.

The leptonic decays appear much the same as single-electron or single-muon production, providing a wide range of visible momenta in the final-state charged lepton.

Table 5.1 Decay modes, branching fractions, and properties of the charged leptons. A summary of τ-lepton branching fractions in 1-, 3- and 5-pronged charged-track final states is listed.

Lepton	Decay modes	Br (%)	Mass (MeV)	Decay Lifetime
e^-	stable		0.511	$> 4 \times 10^{26}$ years
μ^-	$\to e^-\bar{\nu}_e\nu_\mu$	~100	105.7	2.197 μs
τ^-	$\to e^-\bar{\nu}_e\nu_\tau$	17.8	1777	0.29 ps
	$\to \mu^-\bar{\nu}_\mu\nu_\tau$	17.4		
	$\to \pi^-\nu_\tau$	10.9		
	$\to \pi^-\pi^0\nu_\tau$ (including ρ^-)	25.5		
	$\to \pi^-\pi^0\pi^0\nu_\tau$	9.3		
	$\to \pi^-\pi^0\pi^0\pi^0\nu_\tau$	1.0		
	$\to \pi^-\pi^+\pi^-\nu_\tau$	9.0		
	$\to \pi^-\pi^+\pi^-\pi^0\nu_\tau$	2.7		
	$\to K^-\nu_\tau$	0.7		
	$\to \bar{K}^0\pi^-\nu_\tau$	0.9		
	other	4.8		
τ^-	all 1-prong (non-leptonic)	50.1		
	all 3-prong	14.6		
	all 5-prong	0.1		

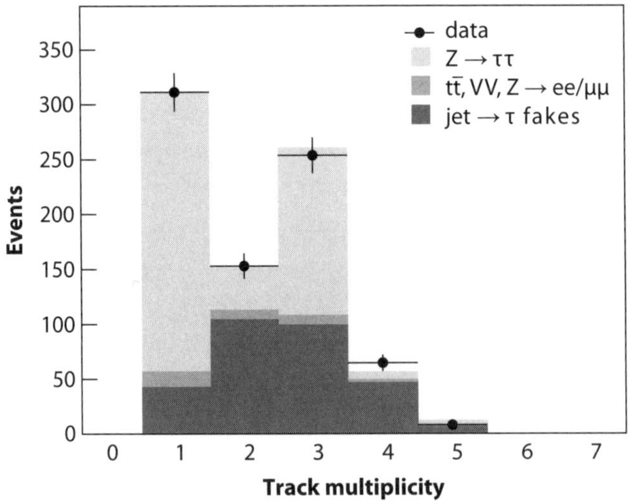

FIGURE 5.13. Track multiplicity in selected τ-lepton candidates before requiring an oppositely charged pair of τ-leptons in the event. Even numbers of tracks indicate track finding inefficiencies in the measurement (Credit: CDF/J. Conway).

However, decays to $\tau^+\tau^-$ or $\tau^+\nu_\tau$ can provide constraints for identifying and even partially recovering the missing momentum carried by the τ-neutrinos. The finite lifetime of the τ-lepton provides the possibility that heavily boosted τ-leptons, including those that decay leptonically, can be identified by the finite impact parameters of the charged particles in their decays.

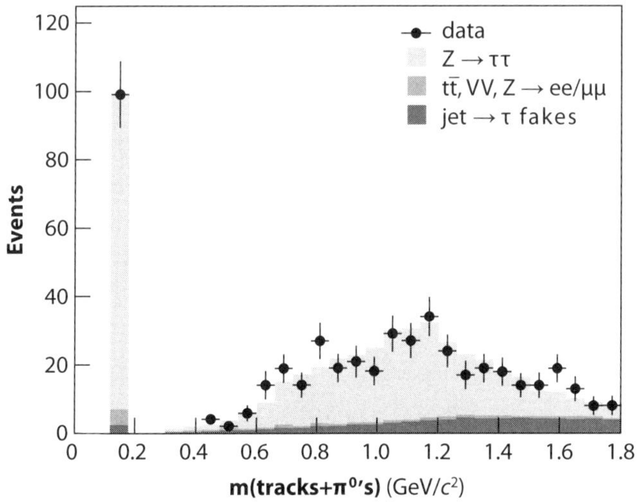

FIGURE 5.14. Reconstructed visible mass of selected τ-lepton candidates. The spike at the pion mass corresponds to one-prong pion decays (Credit: CDF/J. Conway).

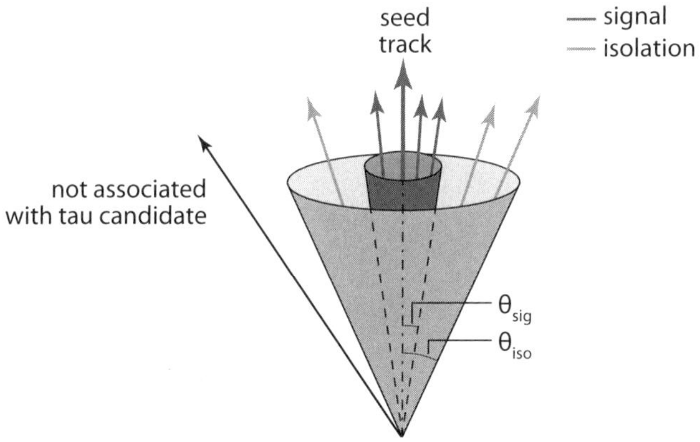

FIGURE 5.15. Cone definitions used for τ-lepton identification with a central signal cone containing the τ-lepton decay products and the outer isolation cone used to veto against QCD jet fakes (Credit: CDF/J. Conway).

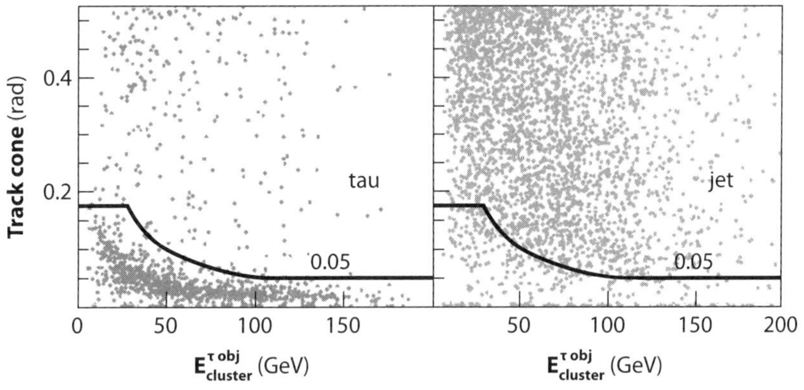

FIGURE 5.16. Jet/τ-lepton separation showing that a dynamically shrinking signal cone as a function of the visible τ-lepton energy is a powerful discriminator against QCD jet fakes (Credit: CDF/J. Conway).

5.8 Missing Transverse Energy

Beginning with the UA1 experiment at CERN (1981–1993), all major detectors at hadron colliders have been designed to cover as much solid angle as practically possible with calorimetry. The primary motivation of this is to provide as complete a picture as possible of the event, including the presence of one or more energetic neutrinos or other weakly interacting stable particles through apparent missing energy. Energetic particles produced in the direction of the beam pipe make it impossible to directly measure missing energy longitudinal to the beam direction; however, the transverse energy balance can be measured with an accuracy good enough to help establish a physics signature involving one or more noninteracting particles. The W boson was discovered and its mass determined to 3% with just six events due to the ability of UA1 to infer the presence of 40 GeV neutrinos with a resolution of a few GeV. Since the time of the W discovery, measurement of missing transverse energy has been a major tool in the search for new phenomena at hadron colliders.

The missing transverse energy vector is calculated by summing individual calorimeter towers having energy E_n, polar angle θ_n, and azimuthal angle ϕ_n:

$$\vec{E}_{\mathrm{T}}^{\mathrm{miss}} = -\Sigma(E_n \sin\theta_n \cos\phi_n \hat{\mathbf{i}} + E_n \sin\theta_n \sin\phi_n \hat{\mathbf{j}}) = E_x^{\mathrm{miss}}\hat{\mathbf{i}} + E_y^{\mathrm{miss}}\hat{\mathbf{j}} \tag{5.16}$$

or, alternatively, using pseudorapidity

$$\vec{E}_{\mathrm{T}}^{\mathrm{miss}} = -\Sigma\left(\frac{E_n \cos\phi_n}{\cosh\eta_n}\hat{\mathbf{i}} + \frac{E_n \sin\phi_n}{\cosh\eta_n}\hat{\mathbf{j}}\right) = E_x^{\mathrm{miss}}\hat{\mathbf{i}} + E_y^{\mathrm{miss}}\hat{\mathbf{j}}. \tag{5.17}$$

Reconstructed muons are taken into account by subtracting the expected minimum ionizing calorimeter deposit (typically 2–3 GeV) and vectorally adding the muon track transverse momentum as measured by the muon spectrometer. When jet energy scale calibrations are applied to the event, the corresponding (type-1) correction to the $E_{\mathrm{T}}^{\mathrm{miss}}$ is given by

$$E_{\mathrm{T}x(y)}^{\mathrm{miss}} = -\left[E_{\mathrm{T}x(y)}^{\mathrm{raw}} + \sum_{\mathrm{jets}}\left(p_{\mathrm{T}x(y)}^{\mathrm{corr.\,jet}} - p_{\mathrm{T}x(y)}^{\mathrm{raw\,jet}}\right)\right]. \tag{5.18}$$

An alternative to applying type-1 corrections is to calibrate the $E_{\mathrm{T}}^{\mathrm{miss}}$ object inputs before running the algorithms, and thus avoid having a jet p_{T} threshold when it comes to applying energy scale corrections. An advanced method for the calibration of input objects is known as *particle flow*. The particle-flow technique uses a wide range of detector information from the tracking system and calorimeters to identify and extract the energy and momentum of the individual stable particles, hadrons, photons, and leptons, in the event. The collection of particle-flow objects provides a complete picture of the particle types, energies and momenta and, thus, form the basis for all measurements in the high-energy event.

The $E_{\mathrm{T}}^{\mathrm{miss}}$ resolution is dependent on the overall activity of the event, characterized by the scalar sum of transverse energy in all calorimeter cells, denoted ΣE_{T}. The resolution of $E_{\mathrm{T}}^{\mathrm{miss}}$ for an intrinsically balanced event can be measured by the width along any fixed spatial direction, i.e., the x- or y-component (E_x^{miss} or E_y^{miss}) of missing transverse energy, as a function of $\sum E_{\mathrm{T}}$. The resolution for balanced *QCD dijet events*, as arise from $2 \to 2$

scattering, is observed to be described by the function $\sigma_{E_x^{\rm miss}} = S'\sqrt{\Sigma E_{\rm T}}$, $E_{\rm T}$ in units of GeV, owing to the stochastic effects of calorimeter showers where S' is a parameter related to the calorimeter energy resolution, typically $S' = 40$–100% for calorimeter-based inputs. For particle-flow inputs, on the other hand, the effective stochastic term can be substantially reduced due to the nonstochastic nature of tracking system measurements. In the high-scalar-sum $E_{\rm T}$ region, eventually the constant term, typically 1–2% added in quadrature according to equation (5.6), will dominate the $E_{\rm T}^{\rm miss}$ resolution. For nonzero $E_{\rm T}^{\rm miss}$, as in the case of neutrinos from W-boson decay, Z bosons decaying to neutrinos, and the possibility of heavy weakly interacting particles produced in collisions, it is common to estimate the significance of a nonzero $E_{\rm T}^{\rm miss}$ with the $E_{\rm T}^{\rm miss}$ *significance* given by

$$\mathrm{Sig}(E_{\rm T}^{\rm miss}) \propto \frac{|E_{\rm T}^{\rm miss}|}{\sqrt{\Sigma E_{\rm T}}} \tag{5.19}$$

where a large significance is more likely to originate from a weakly interacting particle than from mismeasurement in an intrinsically $p_{\rm T}$-balanced process. A more detailed evaluation of the $E_{\rm T}^{\rm miss}$ significance involves the angles of the $E_{\rm T}^{\rm miss}$ and the measured objects, where the angular measurements are generally more precise than the energy measurements and, therefore, the significance is enhanced if the measured $E_{\rm T}^{\rm miss}$ points in a direction inconsistent with the energy measurement uncertainties alone.

It is also common practice to form event variables that describe the total scalar sum $p_{\rm T}$ and the missing transverse momentum $p_{\rm T}^{\rm miss}$ that are constructed directly from high-$p_{\rm T}$ objects, such as jets, leptons, and photons. The variable, $H_{\rm T}$, is the scalar sum of the jet $p_{\rm T}$ for all jets with a jet $p_{\rm T} > p_{\rm T,min}$, where $p_{\rm T,min}$ is the minimum jet $p_{\rm T}$ threshold, typically 10–30 GeV. As jets have a finite jet mass, it is important to consider the $p_{\rm T}$ of the jet, rather than simply the $E_{\rm T}$ as is used for calorimeter cells. Simiarly, one can define $MH_{\rm T}$ as the missing transverse momentum vector computed from the negative of the vector sum of the jet $p_{\rm T}$ vectors, for jets above a minimum jet $p_{\rm T}$ threshold. The properties of these quantities are similar to $\sum E_{\rm T}$ and $E_{\rm T}^{\rm miss}$ but are less sensitive to electronic noise in the calorimeters and low-$p_{\rm T}$ isolated hadrons.

5.9 Neutrinos and Water Cherenkov Detectors

The water Cherenkov detector is a novel detector geometry consisting of a large volume of water or mineral oil viewed from all directions with UV-sensitive photomultiplier tubes (PMTs). The first large-scale experiments utilizing this concept, the IMB and Kamioka experiments, were originally designed to detect the $SU(5)$ GUT prediction of proton decay via lepto-quark interactions in the mode $p \to e^+\pi^0$. The energetic EM shower from the positron emitted in the hypothetical proton decay would generate Cherenkov radiation with a conical opening angle along the three-momentum direction of the positron. A similar set of radiation cones are generated from the photons of the π^0 decay. One of the primary backgrounds in the proton decay experiments comes from cosmic muons and *spallation* products of nearby interactions. A relativistic muon passing through the water volume will produce a similar Cherenkov cone, better defined in angle due to the reduced angular scattering of muons passing through water relative to that of electrons, with a high brightness due to the large path length and with a distribution of momentum directions characteristic of the muon source, where cosmic muons are dominantly downward going. The patterns of Cherenkov light from different sources are detected with a large array of PMTs pointing inward to the center of the water volume from all directions. Example patterns are shown in figure 5.17.

The large volume and powerful particle identification capabilities of water Cherenkov detectors are particularly well-suited for neutrino experiments, and, indeed, the IMB and Kamioka experiments were the first to detect a neutrino burst from a supernova. Large-volume water Cherenkov detectors, such as the 30,000 tons of water in the Super-Kamiokande detector, are capable of measuring signals from interactions induced by neutrinos originating from a wide range of sources including weak processes in the Sun,

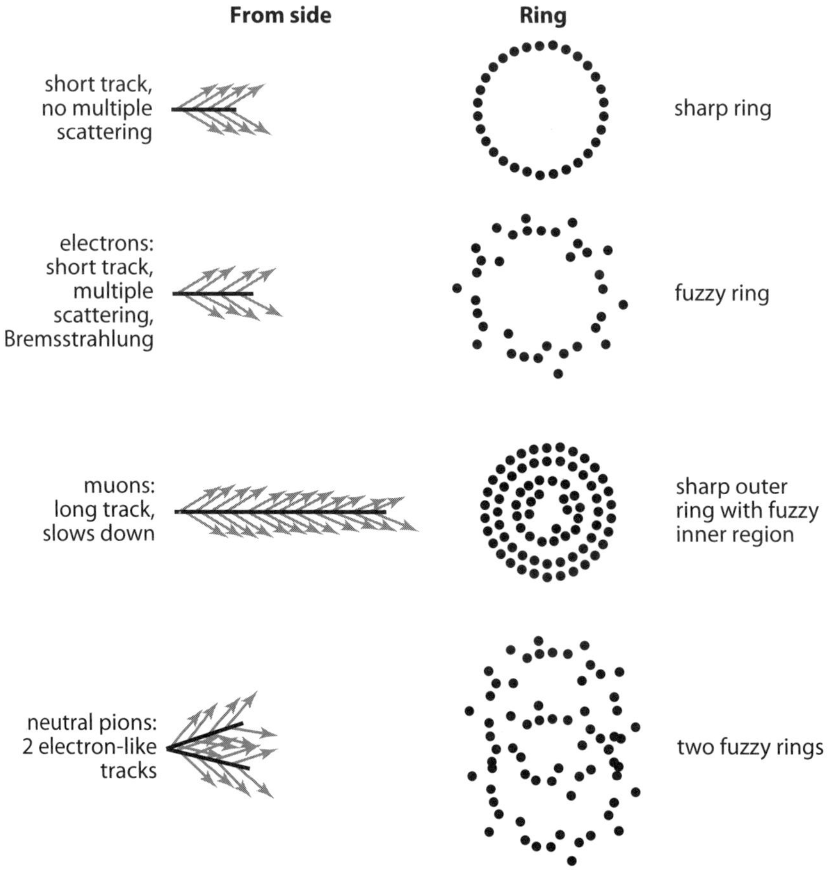

FIGURE 5.17. Cherenkov radiation patterns from different particles traveling through water or mineral oil at relativistic velocities. The patterns are measured by spatial arrays of photomultiplier tubes that point inward toward the center of the volume of liquid from all directions (Credit: BOONE/J. Conrad)

neutrinos produced from cosmic rays in Earth's atmosphere, and accelerator-based neutrino sources.

5.10 Dark Matter Detectors

Not all the matter in the universe is accounted for in the Standard Model of particle physics. Nearly one-quarter of the mass in the universe is attributed to nonbaryonic cold *dark matter*. The low level of self-interaction of the cold dark matter is consist with a particle physics model for direct detection known as Weakly Interacting Massive Particles (WIMPs). WIMP particles are predicted to fill all of space in the form of an extremely low density, electrically neutral gas having particle number densities of approximately $10^3/\mathrm{m}^3$ with a particle mass in the range 40–1000 GeV/c^2.

One technique for WIMP detection is to look for anomalous nuclear recoils in a low-background Earth-based detector. The kinetic energy of the WIMPs is set by the velocity at which Earth travels through the *galactic halo*, yielding kinetic energies of order 10 keV for WIMPs in the rest frame of an Earth-based detector and introducing a directionality in the scattering process. Low-temperature, typically 20 mK, solid-state germanium and silicon detectors such as those used in the CDMS and Edelweiss experiments can measure keV-scale energy depositions to high precision. The low thermal capacity of solids at low temperature enables these detectors to be sensitive to *phonons* from nuclear recoil, quanta of heat dissipation that propagate as sound waves in the detector. Cryogenic solid-state detectors can isolate the very low-ionization nuclear recoils expected in WIMP interactions from background processes with larger ionization signals typical of β-decay and gamma-ray sources from radiative contaminants and cosmic-muon induced spallation backgrounds. Neutron and spallation backgrounds are reduced substantially by locating these detectors deep underground. Neutrons can be further separated from WIMPs by their tendency to scatter multiple times in the detector volume.

Cryogenic noble liquids are also effective dark matter detectors. Although the energy deposition from ionization has lower calorimetric energy resolution than solid-state detectors, the ionization electrons can be drifted out of the noble liquid into a gaseous region where single-electron gas-gain amplification can be used to enhance the detection sensitivity to ionization signals. The lower density of noble gas liquids relative to germanium can be compensated by larger detector volumes employing geometries similar to the water Cherenkov detectors described above. The location of the nuclear recoil within the active detector volume can be determined using the coordinate of the drifted ionization and the drift time relative to an associated light scintillation signal produced in the noble liquid and detected with a dense spatial array of photodetectors facing the active liquid volume. The light scintillation signals generated in the noble liquid from nuclear recoil act in much the same sense as the phonon signals in solid-state detectors. The ratio of the ionization to scintillation is much lower for nuclear recoils than for β-decay or gamma-ray sources. The pulse shape and timing of the scintillation relative to the ionization is also an effective discriminate to identify the time-delayed nuclear recoil scintillation signal.

The push for large-volume, low-background dark matter detectors located deep underground is a rapidly expanding field of particle physics.

5.11 Exercises

1. A photon of energy 10 GeV interacts with the material of the beam pipe to pair produce into an electron and positron, each with $p_T = 5$ GeV/c. Compare the expected electromagnetic calorimeter energy resolution for a 10 GeV photon with $N = 0.1$ GeV, $S = 5\%$, and $C = 0.05\%$ with a tracker measurement of the total p_T of the electron-positron tracks from the converted photon. For the tracking system, use $c_0 = 1.5\%$ and $c_1 = 2 \times 10^{-4}$ (GeV/c)$^{-1}$. For what photon energy is the energy resolution of the calorimeter equal to that of the converted photon using the tracking system to measure the p_T?

2. For some measurements, the electron energy measurement is made with the electromagnetic calorimeter, and the tracking system is primarily used to determine the sign of the electron, i.e., whether it is an electron or positron. For high-p_T tracks there will be a finite probability for *charge confusion*, namely, the mismeasurement of the charge q. This is most clearly seen plotting the q/p_T distribution which is proportional to the sagitta of the tracks. If $\sigma_{p_T}/p_T = 100\%$ for a given track p_T, what is the average charge confusion at this value of p_T? Recall that 68% of a Gaussian distribution is within one standard deviation of the mean. The rate of charge confusion is typically lower for a muon spectrometer at high p_T relative to an inner tracking system. Describe an algorithm for using the muon chambers with a finite but lower charge confusion rate to determine the charge confusion rate of the inner tracker for a given sample of muons having a range of measured p_T values.

3. Compute the ratio of the charged pion-to-electron response, π/e, for a pion shower with a π^0 fraction, $f_{\pi^0} = 0.7$, and a noncompensating calorimeter with an e/h value of $e/h = 1.4$.

4. Compute the number of Gaussian sigma separation between the reconstructed mass peaks of a W boson and a Z boson decaying at rest to a pair of jets. Assume the calorimeter has a jet energy resolution with $N = 1$ GeV, $S = 30\%$, $C = 4\%$, and a negligible angular resolution. Repeat the same calculation for $S = 50\%$.

5. A particle with a mass of 115 GeV decays to a pair of b-jets. If the particle is produced with $p_T = 350$ GeV and decays into a pair of jets, what is the ΔR separation of the jets for equal energy jets in the lab frame? Suppose the jet reconstruction is run with a large jet cone size of $R = 1.5$. How could one reconstruct a mass peak of the 115 GeV particle using large jets?

6. Assume the Z boson is produced at rest and decays to a pair of τ-leptons. If the τ^- decays to $\rho^- \nu_\tau$ with the ρ^- carrying 80% of the initial τ-lepton energy, estimate the

separation of the π^- and π^0 from the ρ decay in the lab frame assuming the π^- and π^0 share equally the initial ρ^- energy. Repeat the analysis for the photon separation in the lab frame for the subsequent $\pi^0 \to \gamma\gamma$ decay assuming equal energy photons.

7. A 1 TeV/c muon is produced and measured with an inner tracking system with parameters $c_0 = 1.5\%$ and $c_1 = 2 \times 10^{-4}$ (GeV/c)$^{-1}$. The muon spectrometer also measures the same muon with a combined tracking system p_T resolution of $c_0 = 3.0\%$ and $c_1 = 10^{-4}$ (GeV/c)$^{-1}$. What is the muon momentum resolution for a 1 TeV/c muon for the inner tracking system alone and for the combined inner plus muon spectrometer tracking system?

8. A J/Ψ resonance with a mass of approximately 3.1 GeV/c^2 is produced with an energy of 30 GeV in the lab frame and decays to a pair of electrons. For equal electron energies and an electromagnetic calorimeter with a radius of 1.2 m and a Molière radius of 2 cm, will the two electromagnetic showers be distinguished? Consider the two tracks from the J/Ψ at a radius of 4 cm at the location of the first layer of pixel detectors. What is the maximum spatial separation of the charged tracks, neglecting magnetic field, and how does it compare with a pixel spacing of 50 μm?

9. **Soft-pion tagging of charm production.** The mass difference of the $D^{*\pm}$ meson and the D^0 meson is 145.4 MeV and the D^0 meson has a 3.8% branching ratio to $K^-\pi^+$. Describe the experimental signature that could be used to identify charm production using the tracking system.

   (a) In the rest frame of the D^0 meson, what is the magnitude of the momentum of the K^- from the $K^-\pi^+$ two-body decay? Assume that the decay angle of the two-body decay in the rest frame is orthogonal to momentum direction of the D^0 in the lab frame. If the momentum of the D^0 meson in the lab frame is $p = 3$ GeV/c, what are the magnitudes of the K^- and π^+ momenta in the lab frame?

   (b) What is the magnitude of the momentum of the soft pion from $D^{*\pm}$ decay in the lab frame, assuming the pion is produced at rest in the $D^{*\pm}$ rest frame and the D^0 meson has a momentum of $p = 3$ GeV/c in the lab frame?

   (c) The time-of-flight system is used to distinguish the decay of the D^0 to the $K^-\pi^+$ final state as opposed to the doubly Cabbibo suppressed decay $K^+\pi^-$. What is the required timing resolution (Gaussian standard deviation) of the time-of-flight system required to get three standard deviations of separation of kaons and pions over a flight distance of $L = 2$ m for a D^0-meson decay as described in part (a)?

10. The cross section for a charged kaon to interact with a proton is typically 20% lower than for a charged pion to interact with a proton. If the fraction of charged pions that punch through the hadron calorimeter is 10^{-4}, what is the corresponding fraction for charged kaons assuming the same momentum spectrum?

11. The cross section for a 1 GeV/c K^- to interact with a proton or neutron is over a factor of 2 larger than that for a K^+. This phenomenon leads to more K^+ mesons

punching through the calorimeter and therefore an excess of positively charged muon candidates detected by the muon spectrometers. What final states are available to $K^- + N$ scattering interactions, where N is a nucleon, that have no analogue for $K^+ + N$ scattering?

12. The ratio of the hadronic calorimeter energy response for 2 GeV π^+ compared to π^- is found to be 110% on average in a noncompensating calorimeter. What subset of interactions is causing this asymmetry? Assume the calorimeter is made of material with heavy nuclei.

13. The number of interactions per bunch crossing of a proton-proton collider at high luminosity can be 20 events or more. Assuming these are *minimum bias* events which typically have $\sum E_T = 35$ GeV each, what is the nominal E_x^{miss} resolution for $S' = 45\%$ and a constant term, $C' = 2\%$. If, in addition to the 20 minimum bias events, a W boson is produced at rest and decays leptonically, what is the number of sigma separation of the E_T^{miss} from zero?

14. Assume a muon stops in the center of an $R = 10$ m water Cherenkov detector. If the electron from muon decay carries away $m_\mu c^2/2$ of energy, over what path length will the electron Cherenkov radiate? The $dE/dx \approx 2$ MeV/cm for MeV-scale electrons in water and the index of refraction in water is $n = 1.33$. Repeat the same calculation for muons with 500 MeV/c of kinetic energy. The $dE/dx \approx 1.5$ MeV/cm for muons in water with 100 MeV to 1 GeV of kinetic energy.

15. Assume 3 eV is required to produce an electron-hole pair in a solid-state germanium detector at low temperature. Estimate the Gaussian energy resolution on the measurement of a 4 keV X-ray based on electron-hole pair counting statistics alone. Repeat the same calculation for an $NaI(Tl)$ scintillating crystal detector that yields 38 photons per keV.

16. Cold dark matter. Our Milky Way has a nearly spherical halo of dark matter that comprises most of its mass. The dark matter is thought to consist of stable, neutral elementary particles. If the particles interact only gravitationally, then simulations suggest that their density should rise sharply as one approaches the galactic center. However, astronomical observations suggest that this is not the case. The dark matter density is nearly constant (about 0.4 GeV/c^2/cm^3) inside a radius of 2×10^{20} m, about the distance of the sun from the galactic center. Within this radius, particles travel at speeds of about 300 km/s.

The proposed explanation is that the dark matter particles may scatter elastically off one another. The scattering ensures that they cannot pile up in the center of the galaxy and they are more evenly distributed within the galactic core. A few interactions per particle over the lifetime of the universe (15 billion years) suffices to alter their distribution.

(a) If the dark matter has mass m, estimate the lower bound on the elastic scattering cross section required to alter their distribution.

(b) Compare the scattering cross section of part (a) to an estimate of the scattering cross section for WIMPs, hypothetical weakly interacting particles with $m \approx 100\ \text{GeV}/c^2$. For example, one could consider WIMPs to have the same quantum numbers as left-handed neutrinos and that the elastic scatttering proceeds through the neutral electroweak current.

Suppose the dark matter particles scatter elastically from baryons with the same cross section with which they scatter from themselves. Then, existing dark matter detectors on the surface of Earth or deep underground could, in principle, detect them. However, the particles would first have to penetrate the atmosphere. Assume that the atmosphere consists mostly of nitrogen.

(c) Using the lower bound for the cross section found in part (a) and assuming $m = 100\ \text{GeV}/c^2$, estimate the interaction length (mean free path) in meters due to scattering in the atmosphere.

(d) If the particle scatters from a nucleus in the detector, the recoil energy can be detected provided it exceeds 10 eV. Estimate the energy of the particle when it reaches a detector at sea level. Will the particle be detected?

5.12 References and Further Reading

A selection of reference texts and review articles in experimental particle physics can be found here: [5, 6, 7, 8, 9, 10, 11, 12, 13, 14, 15, 16].

A selection of introductory texts on nuclear physics and related methods can be found here: [17, 18, 19].

A selection of Web sites of related experiments can be found here: [20, 21, 22, 23, 24, 25].

A useful tool for looking at experimental data can be found here: [26]. A simulation tool for tracking particles through different detector materials and geometries is found here: [27].

[1] K. Nakamura et al. (Particle Data Group). *J. Phys. G* **37** (2010). http://pdg.lbl.gov.

[2] I. Tamm. *J. Phys. U.S.S.R.* **1** (1939) 439.

[3] CMS Experiment Public Results. https://twiki.cern.ch/twiki/bin/view/CMSPublic/Physics Results.

[4] Matteo Cacciari, Gavin P. Salam, and Greg Soyez. *J. High Energy Phys.*, **04:063** (2008). doi: 10.1088/1126-6708/2008/04/063.

[5] Dan Green, editor. *At the Leading Edge*. World Scientific Publishing, 2010. ISBN 978-981-4304-67-2.

[6] Claus Grupen and Boris Shwartz. *Particle Detectors*. Cambridge University Press, 2008. ISBN 978-0-521-84006-4.

[7] Dan Green. *The Physics of Particle Detectors*. Cambridge University Press, 2000. ISBN 0-521-66226-5.

[8] Glenn F. Knoll. *Radiation Detection and Measurement*. Wiley, 2000. ISBN 978-0-471-07338-3.

[9] Richard Wigmans. *Calorimetry*. Oxford University Press, 2000. ISBN 0-19-850296-6.

[10] Konrad Kleinknecht. *Detectors for Particle Radiation*. Cambridge University Press, 1998. ISBN 0-521-64854-8.

[11] R. A. Dunlap. *Experimental Physics*. Oxford University Press, 1988. ISBN 0-19-504949-7.

[12] Veljko Radeka. *Ann. Rev. Nucl. Particle Sci.* **38** (1988) 217–277. doi:10.1146/annurev.ns.38.120188.001245.

[13] Thomas Ferbel, editor. *Experimental Techniques in High Energy Physics*. Addison-Wesley, 1987. ISBN 0-201-11487-9.

[14] William R. Leo. *Techniques for Nuclear and Particle Physics Experiments*. Springer-Verlag, 1987. ISBN 0-387-17386-2.

[15] Richard C. Fernow. *Introduction to Experimental Particle Physics*. Cambridge University Press, 1986. ISBN 0-521-30170-X.

[16] Bruno Rossi. *High-Energy Particles*. Prentice-Hall, 1952. ISBN 978-0133873245.

[17] N. A. Jelley. *Fundamentals of Nuclear Physics*. Cambridge University Press, 1990. ISBN 0-521-26994-6.

[18] Kenneth Krane. *Introductory Nuclear Physics*. Wiley, 1988. ISBN 0-471-80553-X.

[19] Emilio Segrè. *Nuclei and Particles*. W. A. Benjamin Publishers, 1977. ISBN 0-8053-8601-7.

[20] The IMB Experiment (1979–1989). http://www-personal.umich.edu/~jcv/imb/imb.html.

[21] Kamiokande Experiment (1982–1990). http://www-sk.icrr.u-tokyo.ac.jp/kam/index.html.

[22] CDMS-II Experiment. http://cdms.berkeley.edu.

[23] Edelweiss-II Experiment. http://edelweiss.in2p3.fr.

[24] WARP Experiment. http://warp.lngs.infn.it.

[25] XENON Experiment. http://xenon.astro.columbia.edu.

[26] ROOT Analysis Tool. http://root.cern.ch.

[27] GEANT Simulation Package. http://geant4.cern.ch.

6 Neutrino Oscillations and CKM Measurements

The quark and lepton mass eigenstates result from the diagonalization of the Higgs-fermion Yukawa couplings between the three mass generations of $SU(2)_L$ doublets and right-handed singlets. The gauge interactions, on the other hand, connect only left-to-left- and right-to-right-handed chirality fermion components. Therefore, gauge interactions that are diagonal in flavor will be unaffected by the phases associated with the unitary rotations for mass diagonalization, as $(U_R^u)^\dagger U_R^u = (U_R^d)^\dagger U_R^d = 1$ and $(U_L^u)^\dagger U_L^u = (U_L^d)^\dagger U_L^d = 1$. However, the charged-current interaction is affected by these rotations, resulting in two separate mixing matrices for quarks and leptons,

$$V_{UD} = U_L^u (U_L^d)^\dagger, \qquad V' = U_L^\nu (U_L^\ell)^\dagger, \tag{6.1}$$

where U_L^u and U_L^d are 3×3 matrices for up-type and down-type left-handed chirality quarks, respectively, and similarly, U_L^ν and U_L^ℓ are 3×3 matrices for neutral and charged left-handed chirality leptons, respectively.

In the case of the quark sector, the u-quark is the only absolutely stable particle. The d-quark through baryon number conservation is stable if within the bound state of a proton. All other quarks decay through the charged-current interaction into the u-quark either directly or via cascade decays. With the exception of the top quark, the quarks interact via the flavor-diagonal QCD interaction on a time scale much shorter than their weak decays. Hence, the flavor of the quark is effectively projected out by the hadronic mass eigenstate that is formed shortly after quark production. The quark-mixing matrix elements act as vertex coefficients for setting the probability to project into a particular quark flavor. Once the quark is in a hadronic state, then the hadronic state may oscillate into other hadronic states if there are degeneracies or near degeneracies in the masses of hadrons and a transition amplitude that connects the states. In particular, hadronic states connected through a transition amplitude involving the weak charged current will be nearly degenerate due to the relative strength of the weak interaction. This is precisely the situation for neutral mesons whose particle and antiparticle states are separate, namely, the K^0–$\bar{K}^0$, D^0–$\bar{D}^0$, B_d^0–$\bar{B}_d^0$, and B_s^0–$\bar{B}_s^0$ systems.

The lepton sector on the other hand has four stable particles, the electron and the three neutrino families. The neutrinos do not interact rapidly after being produced and remain insulated from Standard Model interactions over a long time scale. In addition, the neutrino masses are much more degenerate than the quark masses, so mass selection does not occur as rapidly. Therefore, the neutrino is not rapidly projected into a mass eigenstate following production, but rather is treated as being created in a pure flavor eigenstate tagged by the associated charged lepton in charged-current production. The neutrino then oscillates in flavor during free-particle propagation according to the squared mass differences between nearly degenerate neutrino mass eigenstates. Each mass eigenstate has a nondiagonal flavor composition. The neutrino flavor is then projected out after a finite propagation length through the flavor of the charged lepton created at detection time. Therefore, the natural choice of a mixing matrix in the lepton sector is not V' but rather $(U_L^{\nu})^{\dagger}$, which transforms the neutrino mass eigenstates $\nu = (\nu_1, \nu_2, \nu_3)$ into the neutrino flavor eigenstates $\nu' = (\nu_e, \nu_\mu, \nu_\tau)$, i.e., $\nu' = (U_L^{\nu})^{\dagger}\nu$. The exception to the vacuum oscillation description of the neutrino flavor mixing is in the case where the neutrino is produced in the center of the sun where coherent forward-scattering can occur and where energetic neutrinos will adiabatically fall into mass eigenstates before leaving the solar mass. This exception will be discussed in the section on solar neutrinos.

What is common to both quark and lepton sectors is that the mismatch between weak interaction states and mass eigenstates can give rise to the phenomenon of oscillations. Two types of systems are known to exhibit measurable oscillations: the neutrinos that mix in flavor and the neutral mesons that mix particle and antiparticle states.

6.1 Neutrino Oscillations

The existence of observable interference and oscillations in the propagation of the states requires an uncertainty in energy and momentum of the source of the particles, related to the localization in space and time of the particle production. A quantitative estimate can be taken from the energy uncertainty of neutrinos produced from charged pions decaying to muons. In the π^+ rest frame, the decay kinematics are

$$p_\mu^2 = m_\mu^2 = (p_\pi - p_\nu)^2 = m_\pi^2 + m_\nu^2 - 2p_\pi \cdot p_\nu = m_\pi^2 + m_\nu^2 - 2m_\pi E_\nu. \tag{6.2}$$

This gives

$$E_\nu = \frac{m_\pi^2 + m_\nu^2 - m_\mu^2}{2m_\pi}. \tag{6.3}$$

Therefore, if we consider two possible neutrino mass eigenstates ν_1 and ν_2, then the difference in the produced neutrino energy is related to the difference in the squared neutrino masses

$$E_{\nu_1} - E_{\nu_2} = \frac{m_{\nu_1}^2 - m_{\nu_2}^2}{2m_\pi}. \tag{6.4}$$

The π^+ width is $\Gamma_\pi \simeq 2.5 \cdot 10^{-8}$ eV, which can compared directly to the energy difference of the produced neutrinos $|E_{\nu_1} - E_{\nu_2}|$. Therefore, coherent oscillations will not be present in neutrino mass eigenstates from charged pion decay for squared mass differences greater than $\approx 7\ (\text{eV})^2$.

A neutrino produced in a charged-current decay is flavor tagged at creation according to the charged lepton type. The neutrino flavor eigenstates will be denoted as $|\nu_\alpha\rangle$ and the mass eigenstates as $|\nu_i\rangle$ with masses m_i. Therefore, at $t = 0$ the state $|\nu_\alpha\rangle$ can be expanded in terms of $|\nu_i\rangle$

$$|\nu_\alpha\rangle = \sum_i U^*_{\alpha i}|\nu_i\rangle \tag{6.5}$$

where $U^*_{\alpha i}$ is known as the Maki-Nakagawa-Sakata (MNS) matrix (3.195). At finite t, the expression for the time evolution of a momentum eigenstate is given by

$$|\nu(t)\rangle = \sum_i U^*_{\alpha i}\exp(-iE_i t + i\mathbf{p}\cdot\mathbf{x})|\nu_i\rangle. \tag{6.6}$$

If at time t the neutrino state interacts weakly, the probability that this interaction results in a flavor tag $|\nu_\beta\rangle$ is

$$P(\nu_\alpha \to \nu_\beta) = |\langle\nu_\beta|\nu(t)\rangle|^2 = \left|\sum_i U^*_{\alpha i}U_{\beta i}\exp(-iE_i t)\right|^2. \tag{6.7}$$

In the ultrarelativistic limit, we can expand the energy-momentum relationship in the leading mass term

$$E_i = p + \frac{m_i^2}{2p} \tag{6.8}$$

so that

$$P(\nu_\alpha \to \nu_\beta) = |\langle\nu_\beta|\nu(t)\rangle|^2 = \left|\sum_i U^*_{\alpha i}U_{\beta i}\exp\left(-i\frac{m_i^2 t}{2p}\right)\right|^2 \tag{6.9}$$

for neutrinos propagating in vacuum. In the case of two generations, the one-parameter rotation matrix is given by

$$U = \begin{pmatrix} \cos\theta & \sin\theta \\ -\sin\theta & \cos\theta \end{pmatrix}. \tag{6.10}$$

For an electron-flavor initial state $\alpha = e$, we get the probability of $\nu_e \to \nu_\mu$ oscillation to be

$$\begin{aligned}
P(\nu_e \to \nu_\mu) &= \left|U^*_{e1}U_{\mu 1}\exp\left(-i\frac{m_1^2 t}{2p}\right) + U^*_{e2}U_{\mu 2}\exp\left(-i\frac{m_2^2 t}{2p}\right)\right| \\
&= \left|-\cos\theta\sin\theta\exp\left(-i\frac{(m_1^2 - m_2^2)t}{2p}\right) + \sin\theta\cos\theta\right|^2 \\
&= 2\cos^2\theta\sin^2\theta \\
&\quad -\cos^2\theta\sin^2\theta\left(\exp\left(-i\frac{(m_1^2 - m_2^2)t}{2p}\right) + \exp\left(i\frac{(m_1^2 - m_2^2)t}{2p}\right)\right) \\
&= 2\cos^2\theta\sin^2\theta\left(1 - \cos\left(\frac{(m_1^2 - m_2^2)t}{2p}\right)\right) \\
&= \sin^2 2\theta\sin^2\left(\frac{1.27\Delta m^2 L}{E}\right)
\end{aligned} \tag{6.11}$$

where in the last expression L is in m, E is in MeV, and Δm^2 is in eV^2, giving the resulting coefficient $(10^{-6})^2/(197 \text{ MeV fm})/4 = 1.27 \text{ MeV (eV)}^{-2} \text{ m}^{-1}$. The general expression for a neutrino flavor oscillation for three mass generations $i(j) = 1, 2, 3$ (1, 2, 3) is given by

$$P(\nu_\alpha \to \nu_\beta) = \delta_{\alpha\beta} - 4\sum_{i>j} \mathcal{R}e\{U^*_{\alpha i} U_{\alpha j} U_{\beta i} U^*_{\beta j}\} \sin^2\left(\frac{1.27\Delta m^2_{ij} L}{E}\right) + 2\sum_{i>j} \mathcal{I}m\{U^*_{\alpha i} U_{\alpha j} U_{\beta i} U^*_{\beta j}\} \sin\left(\frac{2(1.27)\Delta m^2_{ij} L}{E}\right) \quad (6.12)$$

where $\Delta m^2_{ij} = m^2_i - m^2_j$. In equation (6.11), there are three important ranges of the argument $\Delta m^2 L/E$:

$$\begin{aligned} &\Delta m^2 L/E \ll 1 \quad && \text{no oscillations, } P(\nu_e \to \nu_\mu) \approx 0, \\ &\Delta m^2 L/E \sim 1 && \theta \text{ and } |\Delta m^2| \text{ can be determined,} \\ &\Delta m^2 L/E \gg 1 && \text{rapid oscillations, } \bar{P}(\nu_e \to \nu_\mu) = \frac{1}{2}\sin^2 2\theta, \end{aligned} \quad (6.13)$$

where $\bar{P}$ is the time-integrated average oscillation probability.

6.1.1 *Atmospheric and Accelerator-Based Neutrino Experiments*

Atmospheric neutrinos come from the leptonic decays of mesons and heavy leptons produced in cosmic ray showers. When a high-energy proton impinges on the gaseous atmosphere of Earth, a multiplicity of pions can be produced when the proton scatters via the strong interaction with a nucleon in the atmosphere. The longitude and latitude of the proton impact point on Earth's atmosphere defines an approximate distance L between the location of the cosmic ray shower and the position of a neutrino detector located at some fixed position underground. Here, L will vary from on order 10–100 km for cosmic ray showers produced in the atmosphere directly above the detector to the diameter of Earth, or approximately 12,500–13,000 km for cosmic ray showers on the opposite side of the planet. Atmospheric neutrinos contain a mix of ν_μ and ν_e and their antiparticles from charged pion decay $\pi^\pm \to \mu^\pm + \nu_\mu(\bar{\nu}_\mu)$ and subsequent muon decay $\mu^\pm \to e^\pm + \nu_e(\bar{\nu}_e) + \bar{\nu}_\mu(\nu_\mu)$. Neutrinos passing through a detector will have a small but finite probability to scatter via the neutral- and charged-current weak interactions. The neutral-current weak interactions, Z-boson exchange, is lepton-flavor independent. The charged-current weak interaction will tag the flavor of the lepton with some flavor bias from processes involving atomic electrons in the scattering material. In the atmospheric data there is a clear depletion of ν_μ, whereas the ν_e flux appears constant as a function of propagation length L from the origin of the cosmic ray shower as determined from the angular information of the particles scattered by the interacting neutrino within the detector. The data from the SuperK experiment is shown in figure 6.1 as a function of L/E for electron-type and muon-type neutrino scattering events [1].

The peak of the cosmic muon neutrino energy spectrum is ~2 GeV, which, given Earth's diameter, is ideally suited for oscillation measurements for squared mass differences of

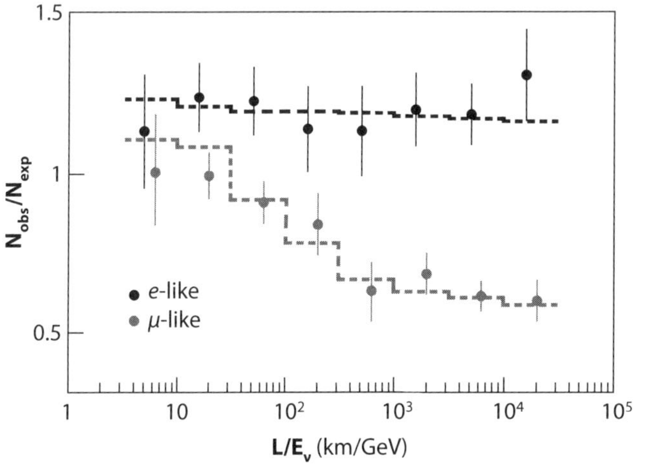

FIGURE 6.1. Observed flux of electron-type and muon-type scattering events from the Super-Kamiokande atmospheric neutrino data as a function of L/E and normalized to the expected flux from a Monte Carlo simulation with no oscillations. The dotted lines show the best fit for neutrino oscillation parameters describing this data. (Reprinted figure from [1] with permission from the APS).

$\Delta m^2 \approx 10^{-3}$–$10^{-4}$ (eV)^2. Fits to the atmospheric (atm) data give the following parameters for two generations of flavor oscillations [2]:

$$|\Delta m^2_{\text{atm}}| \simeq 2.4 \times 10^{-3}\ \text{eV}^2 \quad \text{and} \quad \sin^2\theta_{\text{atm}} \simeq 0.5. \tag{6.14}$$

Although the first measurements of neutrino oscillations were obtained from atmospheric data, the parameters are within the range that can be tested with accelerator-based neutrino sources. The neutrino oscillations seen in atmospheric data were confirmed in muon disappearance measurements by the K2K experiment in Japan and the MINOS experiment, an accelerator-based experiment with neutrino beams produced from a proton beam hitting a fixed target at Fermilab in Chicago and directed to a detector in Soudan, Minnesota, 700 km away. Furthermore, not only are there accelerator-based measurements for muon diasspearance, there is now direct evidence for τ-lepton appearance, measured by the OPERA experiment in the CERN-Gran Sasso beam, as predicted by three-flavor neutrino oscillation parameters described below.

6.1.2 Reactor Neutrino Experiments

A nuclear reactor is an abundant source of antineutrinos. The number of fissions per second of ^{235}U in a commerical reactor is a few times 10^{19} with approximately two detectable antineutrinos produced per fission. The antineutrinos come from a cascade of β decays from neutron-rich fission products. Reactor antineutrinos, $\bar{\nu}_e$, typically have energies of a few MeV, well above the threshold to produce electrons in the final state, and are detected through the reaction $\bar{\nu}_e + p \to e^+ + n$. In the rapid oscillation scenario in equation (6.13)

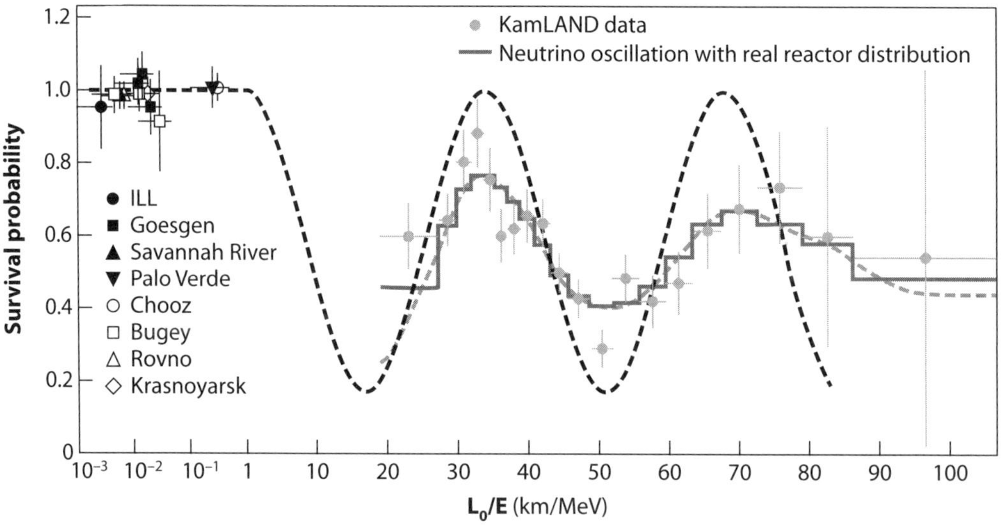

FIGURE 6.2. Survival probability of electron antineutrinos as a function of the effective L_0/E with an average reactor distance of $L_0 = 180$ km (Credit: KamLAND) [3]. The dotted line shows the best-fit oscillation parameters to the KamLAND data showing that the data are collected in a region of rapid oscillations. The solid line is a fit to the measured data including finite smearing effects from real reactor distributions.

for a fixed L, the expectation is to observe a drop in the rate of antineutrino interactions in the detector. The KamLAND long-baseline reactor experiment (average reactor distance is 180 km) measured an oscillation signature with a detected $\bar{\nu}_e$ rate which dips to ~60% of the predicted no oscillation rate, giving a best fit of

$$|\Delta m^2_{\text{reactor}}| \simeq 7.6 \times 10^{-5}\ \text{eV}^2 \quad \text{and} \quad \sin^2\theta_{\text{reactor}} \simeq 0.3 \tag{6.15}$$

when interpreted in terms of two-flavor neutrino oscillations [2]. The KamLAND data and data compiled for short-baseline reactor neutrino experiments are shown in figure 6.2 [3].

6.1.3 Solar Neutrino Data

In the case of the solar neutrinos, the effect of neutrinos propagating through the dense material of the sun has an important effect on the oscillation measurements. Electron neutrinos propagating through matter experience coherent forward scattering due to t-channel charged-current scattering. Matter effects were predicted by Mikheyev, Smirnov, and Wolfenstein (MSW) and are generally referred to as *MSW effects*. As a result of the interaction with atomic electrons, the electron neutrino flavor is constantly being regenerated by the weak interaction. As the weak interaction has a finite mass W propagator, the cross section for neutrino scattering grows with increasing neutrino energy for scattering momentum transfers small compared to M_W. Therefore, the matter effects increase with neutrino momentum. A relatively wide range of neutrino energies is emitted by the sun, as shown in figure 6.3 [4]. The various solar neutrino experiments have also had a range of sensitivies to these fluxes with the gallium experiments reaching the lowest thresholds

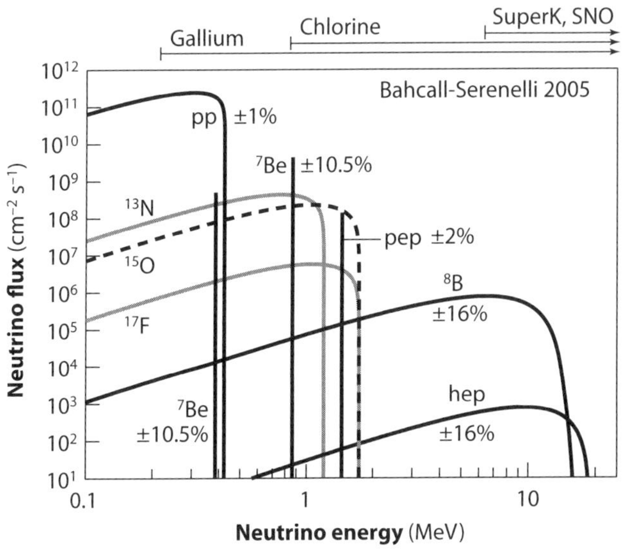

FIGURE 6.3. Solar neutrino energy spectrum predicted by the Standard Solar Model (2005). The uncertainties on the fluxes are quoted, and the detection thresholds for various solar neutrino experiments are shown at the top. (Reproduced by permission of the AAS) [4].

and the water Cherenkov experiments (SuperK) at the high end. The Sudbury Neutrino Observatory (SNO) developed a technique for flavor-independent detection of the neutrino flux from the sun where they measured the processes

$$\begin{aligned} &\nu_e + d \to p + p + e^- \qquad (CC), \\ &\nu_x + d \to p + n + \nu_x \qquad (NC), \\ &\nu_x + e^- \to \nu_x + e^- \qquad (ES), \end{aligned} \tag{6.16}$$

and with this data accurately verified the total flux predictions with the Standard Solar Model in the neutrino energy region corresponding to the SuperK measurements, as shown in figure 6.4. The combination of SNO and SuperK data makes an astounding confirmation of the flux prediction and the vacuum oscillation parameter θ_{12} when corrected for matter (MSW) effects described below.

The effect on the propagation of neutrinos through matter is the addition of an effective index of refraction n given by

$$\begin{aligned} \exp(i(npz - Et)) &= \exp(i(pz - \sqrt{2}\,G_F N_e z - Et)) \\ &= \exp(iE(z - t))\exp\left(-i\left(\frac{m^2}{2p} + \sqrt{2}\,G_F N_e\right)z\right) \end{aligned} \tag{6.17}$$

for an electron neutrino propagating along the z-direction. If we define $m_0^2 = 2\sqrt{2}G_F N_e p$, then the effect on the mixing angle in matter θ'_{12} compared to the vacuum mixing angle θ_{12} is given by

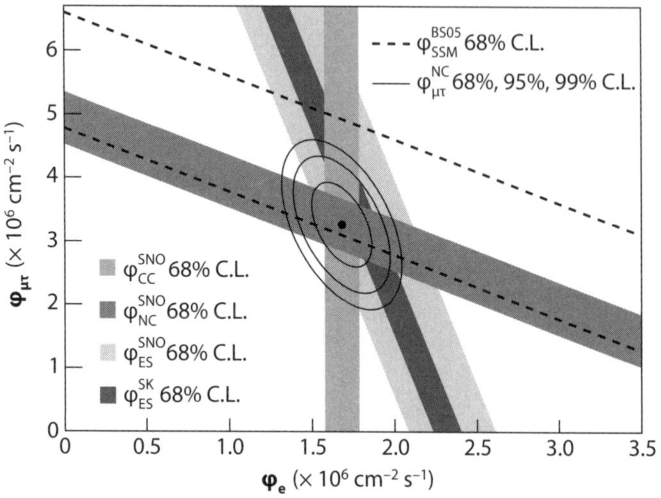

FIGURE 6.4. Comparison of charged-current (CC), elastic scattering (ES) and neutral-current (NC) neutrino fluxes measured by the SNO experiment and including Super-Kamiokande data (Credit: PDG) [2]. The neutral current is sensitive to all neutrino flavors and confirms the total predicted flux from the Standard Solar Model from figure 6.3.

$$\sin^2 2\theta'_{12} = \frac{\sin^2 2\theta_{12}}{\left[\left(\frac{m_0^2}{m_2^2 - m_1^2} - \cos 2\theta_{12}\right)^2 + \sin^2 2\theta_{12}\right]} \tag{6.18}$$

which for $m_0^2 \gg |m_2^2 - m_1^2|$, $\theta'_{12} \to 0$ and neutrinos fall into mass eigenstates while for $m_0^2 \ll |m_2^2 - m_1^2|$, $\theta'_{12} \to \theta_{12}$ and neutrinos undergo vacuum oscillations. Therefore, for the solar neutrino oscillation measurements, only the lowest energy electron neutrinos, those detected by ^{71}Ga experiments, are unaffected by matter effects for the bulk of the detected flux. The ^{71}Ga solar neutrino data give a value of θ_{12} in agreement with the reactor data. For ^{8}B neutrinos, the matter effects are large and the electron neutrinos will remain as an electron-neutrino weak eigenstate during regeneration and then adiabatically follow one of the mass eigenstates as the neutrinos propagate out of the sun and the density slowly decreases. The emerging neutrinos will remain in the mass eigenstate until they propagate to Earth and are detected. The ^{8}B neutrino measurements are, therefore, a static measurement of fixed neutrino mass eigenstates projected onto weak eigenstates. An important property of the denominator of equation (6.18) is the possibility to determine the sign of the squared mass difference $\Delta m_{21}^2 = m_2^2 - m_1^2$ for nonzero $\cos 2\theta_{12}$. For $\Delta m_{21}^2 > 0$ there is a cancellation in the denominator and, therefore, an enhancement in the matter oscillations relative to the vacuum $\sin^2 2\theta'_{12} > \sin^2 2\theta_{12}$. As both the gallium and reactor data set a baseline for the vacuum oscillation probability, the matter enhancement seen in the ^{8}B neutrinos fixes the sign of Δm_{21}^2 as positive where the lighter of the pair of mass states is more "electronlike" while the heavier is found to be a nearly equal admixture of all three neutrino flavors. Correcting for the solar MSW effect, the vacuum θ_{12} value agrees with the chlorine and Cherenkov detector solar neutrino data. The combined fit of solar and reactor data gives $\tan^2\theta_{12} = 0.47 \pm 0.05$ with $\Delta m_{21}^2 = +(7.6 \pm 0.2) \times 10^{-5}$ eV^2 [2].

6.1.4 Neutrino Mass Hierarchy, Flavor Content, and Potential for CP Violation

In a three-neutrino system, we write the general mixing matrix $U \equiv (U_L^\nu)^\dagger$ as

$$U = \begin{pmatrix} c_{12}c_{13} & s_{12}c_{13} & s_{13}\exp(-i\delta) \\ -s_{12}c_{23} - c_{12}s_{23}s_{13}\exp(i\delta) & c_{12}c_{23} - s_{12}s_{23}s_{13}\exp(i\delta) & s_{23}c_{13} \\ s_{12}s_{23} - c_{12}c_{23}s_{13}\exp(i\delta) & -c_{12}c_{23} - s_{12}c_{23}s_{13}\exp(i\delta) & c_{23}c_{13} \end{pmatrix} \quad (6.19)$$

where $s_{ij} = \sin\theta_{ij}$ and $c_{ij} = \cos\theta_{ij}$ with $j > i$. The matrix (6.19) decomposes into the matrix product

$$\begin{aligned} U &= U_{12} \times U_{23} \times U_{13} \\ &= \begin{pmatrix} c_{12} & s_{12} & 0 \\ -s_{12} & c_{12} & 0 \\ 0 & 0 & 1 \end{pmatrix} \begin{pmatrix} 1 & 0 & 0 \\ 0 & c_{23} & s_{23} \\ 0 & -s_{23} & c_{23} \end{pmatrix} \begin{pmatrix} c_{13} & 0 & s_{13}\exp(-i\delta) \\ 0 & 1 & 0 \\ -s_{13}\exp(i\delta) & 0 & c_{13} \end{pmatrix} \end{aligned} \quad (6.20)$$

showing explicitly the individual two-flavor mixing matrices. In the limit where θ_{13} is small, the last mixing matrix is approximately the identity matrix $U_{13} \approx I_3$ and we recover the two-flavor reactor $U_{12} \approx U_{\text{reactor}}$ and atmospheric $U_{23} \approx U_{\text{atm}}$ mixing matrices. This approximation is verified below.

The small squared mass difference describing the reactor oscillation data and the factor of 30 larger squared mass difference found in atmospheric oscillation data imply that Nature has chosen a set of mass splittings such that $|\Delta m_{13}^2| \simeq |\Delta m_{23}^2| \gg |\Delta m_{12}^2|$. Two possible neutrino mass hierarchy scenarios based on these parameters are shown in figure 6.5, labeled arbitrarily as the normal and inverted hierarchies. In this situation, the oscillation formula (6.9) simplifies, for $\alpha \neq \beta$, to

$$P(\nu_\alpha \to \nu_\beta) = 4\left|U_{\alpha 3}^* U_{\beta 3}\right|^2 \sin^2\left(\frac{\Delta m_{23}^2 t}{4p}\right) \quad (6.21)$$

using the unitarity of the three-generation U matrix and assuming that $\Delta m_{12}^2 t/p$ is small compared to unity. This results in the following expected oscillation probabilities:

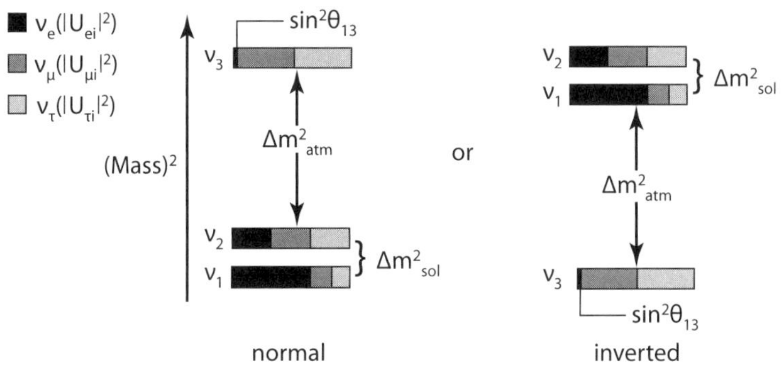

FIGURE 6.5. Two possible neutrino mass hierarchy scenarios, arbitrarily labeled as the *normal* and *inverted hierarchies* (Credit: B. Kayser).

$$P(\nu_e \to \nu_\mu) = \sin^2\theta_{23}\sin^2 2\theta_{13}\sin^2\left(\frac{\Delta m_{23}^2 t}{4p}\right),$$
$$P(\nu_e \to \nu_\tau) = \cos^2\theta_{23}\sin^2 2\theta_{13}\sin^2\left(\frac{\Delta m_{23}^2 t}{4p}\right), \quad (6.22)$$
$$P(\nu_\mu \to \nu_\tau) = \cos^2\theta_{13}\sin^2 2\theta_{23}\sin^2\left(\frac{\Delta m_{23}^2 t}{4p}\right).$$

However, the observed nonoscillation of the atmospheric electron neutrinos indicates that the $P(\nu_e \to \nu_\mu)$ and $P(\nu_e \to \nu_\tau)$ transition probabilities are suppressed. The atmospheric data indicate from muon neutrino disappearance measurements that [2]

$$|\Delta m_{23}^2| \simeq 2.4 \times 10^{-3}\ \mathrm{eV}^2,\ \cos^2\theta_{13}\sin^2 2\theta_{23} \simeq 1. \quad (6.23)$$

The combination of this data points to a small value of θ_{13} giving a $\sin^2 2\theta_{13}$ suppression of $P(\nu_e \to \nu_\mu)$ and $P(\nu_e \to \nu_\tau)$ while $\theta_{23} \simeq 45°$. For θ_{13} small, $\theta_{13} < 10°$, the U matrix can be simplified to

$$U = \begin{pmatrix} c_{12} & s_{12} & \theta_{13}\exp(-i\delta) \\ -s_{12}c_{23} & c_{12}c_{23} & s_{23} \\ s_{12}s_{23} & -c_{12}c_{23} & c_{23} \end{pmatrix}. \quad (6.24)$$

Figure 6.5 shows the flavor composition based on U of the three neutrino mass eigenstates, indicating that each mass eigenstate behaves differently with respect to lepton flavor conservation with one "electronlike," one μ-τ biflavor, and one all-flavor mixture. As shown in the figure, the electronlike neutrino is known to be lighter than the all-flavor neutrino mass eigenstate due to the enhancement in the mixing probability from the MSW effect observed in the solar neutrino data.

To measure *CP* violation in the neutrino sector, as evidenced by the possible nonzero complex phase δ in equation (6.24), requires a measurement to determine that θ_{13} is nonzero, as this is the linear coefficient to the leading *CP*-violating term. Future experiments, such as Daya Bay, with more precise reactor neutrino data on $\bar{\nu}_e$ disappearance provide a promising avenue to place further constraints on or measure θ_{13}. Similarly, electron neutrino appearance experiments, such as NOνA and T2K, aim to detect accelerator-based muon neutrinos oscillating into electron neutrinos. The expected sensitivities for these experiments reach values as low as $\sin^2\theta_{13} = 0.01$, where current constraints from the MNS matrix fits are at the level of $\sin^2\theta_{13} < 0.04$ at the 90% confidence level [2].

6.2 CKM Parameterizations and *CP* Violation

The three u_L^i quarks are linked with a unitary rotation of the triplet of d_L^i quarks

$$V_{UD} = U_L^u (U_L^d)^\dagger. \quad (6.25)$$

The matrix V_{UD} is known as the *Cabibbo-Kobayashi-Maskawa* (CKM) mixing matrix. The values of the matrix indicate how much of the charged-current coupling is shared between the three generations of up-type and down-type quarks, where

$$j^{\mu}_{CC} = \frac{1}{\sqrt{2}} \bar{\Psi}^{U}_{L} \gamma^{\mu} V_{UD} \Psi^{D}_{L} + \frac{1}{\sqrt{2}} \bar{\Psi}^{D}_{L} \gamma^{\mu} V^{*}_{UD} \Psi^{U}_{L} \tag{6.26}$$

with

$$V_{UD} = \begin{pmatrix} V_{ud} & V_{us} & V_{ub} \\ V_{cd} & V_{cs} & V_{cb} \\ V_{td} & V_{ts} & V_{tb} \end{pmatrix}. \tag{6.27}$$

If we consider for the moment only two mass generations, then the general unitary matrix has four parameters and can generally be written in the form

$$V^{(2)}_{UD} = \begin{pmatrix} \exp(i\phi_1)\cos\theta & \exp(i(\phi_1 + \phi_2))\sin\theta \\ -\exp(i\phi_3)\sin\theta & \exp(i(\phi_2 + \phi_3))\cos\theta \end{pmatrix}. \tag{6.28}$$

We can rewrite the 2×2 matrix as

$$\begin{pmatrix} \exp(i\phi_1) & 0 \\ 0 & \exp(i\phi_3) \end{pmatrix} \begin{pmatrix} \cos\theta & \sin\theta \\ -\sin\theta & \cos\theta \end{pmatrix} \begin{pmatrix} 1 & 0 \\ 0 & \exp(i\phi_2) \end{pmatrix}. \tag{6.29}$$

Notice that the matrices involving ϕ_1, ϕ_2, and ϕ_3 are diagonal. These phases simply multiply the individual quark wave functions and by redefinition of the wave functions can be absorbed without changing any physical quantity. The rotation angle θ, however, cannot be removed by redefinition of the quark phases and is the one physical parameter of the two-generation CKM matrix. This angle is denoted θ_C, the Cabibbo angle. The numerical values of the $V^{(2)}_{UD}$ elements are [2]

$$\sin\theta_C = 0.225 \pm 0.001, \quad \text{implying that} \quad \cos\theta_C = 0.975. \tag{6.30}$$

Thus, weak decay rates involving a $u \leftrightarrow s$ quark transition are suppressed by a factor $\sin^2\theta_C = 0.05$, or 5%. The Cabibbo-angle two-generation description is a good estimate (better than 1%) for light-quark transitions.

For three generations of quark families, the three up-type-quark and three down-type-quark wave functions can absorb $2 \cdot 3 - 1 = 5$ relative phases of the 3^2 CKM parameters with quark wave function redefinitions. The remaining four physical parameters can be represented by three rotation angles θ_{12}, θ_{13}, θ_{23} and a phase δ. These can be written in the general form

$$V_{UD} = \begin{pmatrix} c_{12}c_{13} & s_{12}c_{13} & s_{13}\exp(-i\delta) \\ -s_{12}c_{23} - c_{12}s_{23}s_{13}\exp(i\delta) & c_{12}c_{23} - s_{12}s_{23}s_{13}\exp(i\delta) & s_{23}c_{13} \\ s_{12}s_{23} - c_{12}c_{23}s_{13}\exp(i\delta) & -c_{12}c_{23} - s_{12}c_{23}s_{13}\exp(i\delta) & c_{23}c_{13} \end{pmatrix}. \tag{6.31}$$

An accurate approximation of V_{UD} can be written in a form that explicitly shows the suppression of the off-diagonal elements progressively with each mass generation, as we know must occur based on the accuracy of the Cabibbo angle description for the light-quark flavors. This is achieved by defining

$$\begin{aligned} s_{12} &= \sin\theta_C = \lambda,\ s_{23} = A\lambda^2 \ll s_{12}, \\ s_{13}\exp(-i\delta) &= A\lambda^3(\rho - i\eta) \ll s_{23} \text{ in magnitude}, \\ V_{UD} &= \begin{pmatrix} 1-\lambda^2/2 & \lambda & A\lambda^3(\rho-i\eta) \\ -\lambda & 1-\lambda^2/2 & A\lambda^2 \\ A\lambda^3(1-\rho-i\eta) & -A\lambda^2 & 1 \end{pmatrix}. \end{aligned} \tag{6.32}$$

Equation (6.32) is known as the *Wolfenstein parameterization* of the CMS matrix. The terms in the matrix are measured in various weak decay processes. The term $V_{cb} = A\lambda^2$ is measured in $B^{\pm}$ meson decay with a $b \to c$ transition. This gives a value of $|V_{cb}| = 0.0415 \pm 0.001$, which for $\lambda = 0.225 \pm 0.001$ gives $A = 0.81 \pm 0.02$ [2]. The measurement of V_{ub} is determined from the $3^{rd} \to 1^{st}$ generation $b \to u$ transition, neutral B meson mixing, and CP-violating asymmetries and provides a measurement of ρ and η [2]:

$$\rho = 0.14 \pm 0.03 \quad \text{and} \quad \eta = 0.35 \pm 0.02. \tag{6.33}$$

The values of λ, A, ρ, and η are constrained by several measurements. A compact representation for testing the self-consistency of the CKM matrix element measurements is in the form of the unitarity triangle. There are, in fact, six possible unitarity triangles corresponding to the off-diagonal zeros of the unitarity condition $V_{UD} V_{UD}^{\dagger} = 1$, namely,

$$\sum_i V_{ij} V_{ik}^* = 0 \quad \text{for nonidentical columns } (j \neq k), \tag{6.34}$$

and

$$\sum_i V_{ji} V_{ki}^* = 0 \quad \text{for nonidentical rows } (j \neq k), \tag{6.35}$$

where the V_{ij} are the elements of the CKM matrix. A particularly symmetric configuration of legs in the triangle comes from choosing the product terms $V_{ij}V_{ik}^*$ to be of the same order in powers of λ. There are two triangles with all legs with a length of order λ^3 and these are given by

$$\begin{aligned} V_{ud}V_{ub}^* + V_{cd}V_{cb}^* + V_{td}V_{tb}^* &= 0, \\ V_{ud}V_{td}^* + V_{us}V_{ts}^* + V_{ub}V_{tb}^* &= 0. \end{aligned} \tag{6.36}$$

In the study of B-meson decays, the V_{cb} term is particularly important and, therefore, the first of these triangles is commonly known as the *unitarity triangle*. The legs of the unitarity triangle can be normalized by the middle term (V_{cb} containing) of equation (6.36) to give an apex at $(\rho, i\eta)$, as shown in figure 6.6.

The connection between the η parameter and CP violation can be seen by comparing terms in the electroweak Lagrangian under discrete transformations. We can, therefore, examine the terms generated by equation (6.26) in weak charged-current processes. The term $\bar{\Psi}_L^U \gamma^\mu V_{UD} \Psi_L^D W_\mu^+$ are interactions with a V_{UD} coefficient describing a d-quark emitting or absorbing a W boson and rotating into a u-quark, an anti-u-quark emitting or

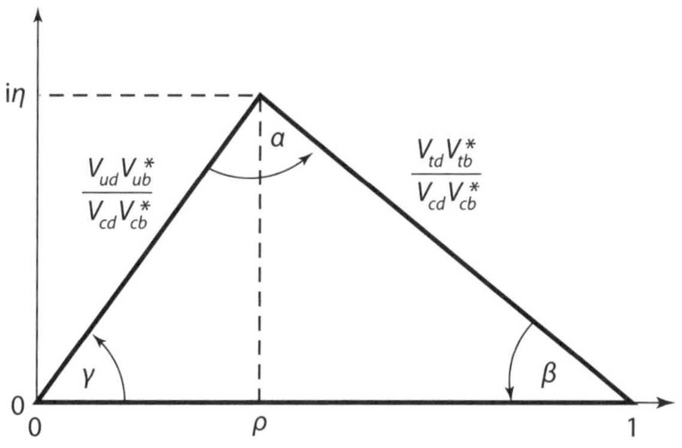

FIGURE 6.6. The unitarity triangle.

absorbing a W boson and rotating into an anti-d-quark, a d-quark annihilating with an anti-u-quark and the pair production of a u-quark and an anti-d-quark. All of these interactions involve elements of V_{UD} (not V_{UD}^*). The corresponding Hermitian conjugate terms (taking up-type quarks to down-type quarks) involve elements of V_{UD}^*. Returning to the properties of the weak charged-current interaction under discrete symmetry operations, the operation of C takes the particle $SU(2)_{\text{flavor}}$ doublet to the antiparticle doublet and maps $u_L \to -\bar{d}_R$ and $d_L \to \bar{u}_R$, i.e., the $SU(2)_{\text{flavor}}$ doublet transforms as

$$\begin{pmatrix} u \\ d \end{pmatrix} \xrightarrow{C} \begin{pmatrix} -\bar{d} \\ \bar{u} \end{pmatrix}, \tag{6.37}$$

but in weak isospin the left-handed chirality u_L quarks are part of $SU(2)_L$ doublets while the right-handed chirality $\bar{d}_R$ quarks are singlets. Therefore, the separate operations of C or P change the chiral gauge charges and generate terms that do not exist in the $SU(2)_L \times U(1)_Y$-invariant electroweak Lagrangian. Applying the combined operation of CP to the term $\bar{\Psi}_L^U \gamma^\mu V_{UD} \Psi_L^D$ yields the CP-transformed processes with V_{UD} coefficients. However, the Hermitian conjugate processes in the Lagrangian have V_{UD}^* coefficients. Therefore, if V_{UD} contains imaginary terms, i.e., $V_{UD} \neq V_{UD}^*$, then CP symmetry is violated in the $SU(2)_L \times U(1)_Y$ interaction with corresponding violations in time-reversed processes.

6.3 Box Diagrams and the GIM Mechanism

The off-diagonal elements of the CKM matrix correspond to rare transitions. The study of rare decays is, therefore, one means of evaluating the couplings to heavy mass states. This effect was used by Glashow, Iliopoulos, and Maiani (GIM) to predict the mass of the charm quark before its discovery. This analysis was first applied to the decay branching fraction of $K_L^0 \to \mu^+\mu^-$, which was measured to be 7.3×10^{-9}. The K_L^0 is nearly CP odd and, therefore, can mix with the light pseudoscalars π^0, η and η' through diagrams with one internal W boson propagator. As with the light $J^P = 0^-$ states, the K_L^0 has a small nonzero decay rate to $\gamma\gamma$. In order for the K_L^0 to decay to dimuons, the $\gamma\gamma$ final state must rescatter to give $K_L^0 \to 2\gamma \to \mu^+\mu^-$. This decay is sufficiently rare to be comparable to second-order charged-current weak processes, i.e., two internal W boson propagators. The second-order weak diagram for this decay is called a *box diagram* with $\bar{K}^0(s\bar{d}) \to W^{+*}W^{-*} \to \mu^+\mu^-$. If the diagram is computed using virtual u-quark exchange alone, then the amplitude for the decay is considerably larger than the observed decay rate. The CKM factor for the u-quark is $V_{us}V_{ud}^*$. However, since the intermediate quark is virtual, we need to include intermediate states of larger mass, where the state with the largest coupling is the c-quark. The c-quark CKM factor is $V_{cs}V_{cd}^*$. In the two-generation approximation,

$$V_{us}V_{ud}^* + V_{cs}V_{cd}^* = \sin\theta_C\cos\theta_C - \cos\theta_C\sin\theta_C = 0, \tag{6.38}$$

by unitarity of the matrix. Thus, the amplitude from the box diagram is suppressed due to a cancellation between u and c quarks. This is known as the *GIM mechanism*. The t-quark CKM factors are very small and do not contribute significantly to the amplitude. However, for box diagrams involving initial b-quarks instead of s-quarks, the CKM factors are much larger for the virtual t-quark contribution. As a result, virtual t-quarks typically provide the dominant amplitude, compared to u and c exchange, for box diagrams in the B-meson system.

6.4 Neutral Meson Oscillations and *CP* Violation

The mixing of particle and antiparticle neutral meson states comes from second-order weak transitions that connect the two states through common virtual or real intermediate states, as shown in figure 6.7. The virtual states contribute to the mass of the mesons while the real states contribute to the partial width according to the complex mass matrix

$$i\frac{\partial}{\partial t}\begin{pmatrix} B^0 \\ \bar{B}^0 \end{pmatrix} = \left(\mathbf{M} - \frac{i}{2}\mathbf{\Gamma}\right)\begin{pmatrix} B^0 \\ \bar{B}^0 \end{pmatrix} = \begin{pmatrix} m - \frac{i}{2}\Gamma & \Delta m_{12} - \frac{i}{2}\Gamma_{12} \\ \Delta m_{12}^* - \frac{i}{2}\Gamma_{12}^* & m - \frac{i}{2}\Gamma \end{pmatrix}\begin{pmatrix} B^0 \\ \bar{B}^0 \end{pmatrix}. \tag{6.39}$$

The diagonal terms describe the decay of the neutral B mesons with m being the mass of the flavor eigenstates B^0 and $\bar{B}^0$ and Γ their common decay width before the weak interaction is turned on. The second-order weak interaction introduces off-diagonal terms in the complex mass matrix (6.39) and is responsible for B^0–$\bar{B}^0$ transitions. Thus, the weak interaction results in a new set of mass eigenstates and total widths, changing m and Γ for the physical states. The heavy (H) and light (L) states have mass and width splittings given by

$$\begin{aligned} \Delta m &\equiv M_H - M_L = 2\mathcal{R}e\sqrt{\left(\Delta m_{12} - \frac{i}{2}\Gamma_{12}\right)\left(\Delta m_{12}^* - \frac{i}{2}\Gamma_{12}^*\right)}, \\ \frac{\Delta\Gamma}{2} &\equiv \frac{\Gamma_H}{2} - \frac{\Gamma_L}{2} = -2\mathcal{I}m\sqrt{\left(\Delta m_{12} - \frac{i}{2}\Gamma_{12}\right)\left(\Delta m_{12}^* - \frac{i}{2}\Gamma_{12}^*\right)}. \end{aligned} \tag{6.40}$$

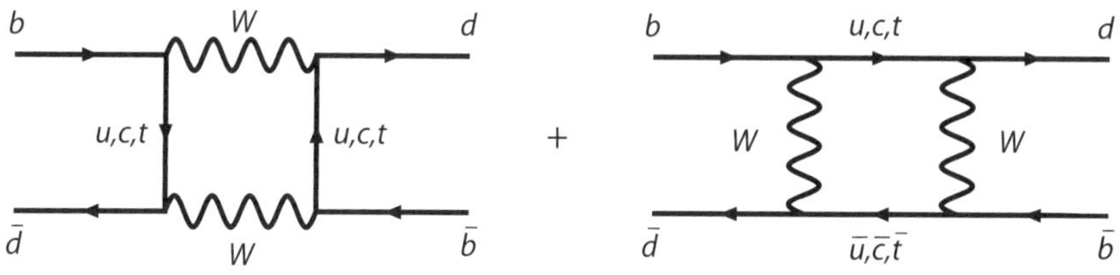

FIGURE 6.7. Second-order weak transitions contributing to neutral B_d-meson oscillations.

The eigenstates of matrix (6.39) have the general form

$$\begin{aligned} |P_L\rangle &= p|P^0\rangle + q|\bar{P}^0\rangle, \\ |P_H\rangle &= p|P^0\rangle - q|\bar{P}^0\rangle \end{aligned} \tag{6.41}$$

where p and q are complex constants that satisfy $|p|^2+|q|^2=1$. The ratio of q/p is found by diagonalizing (6.39) and is given by

$$\frac{q}{p} = \sqrt{\frac{\Delta m_{12}^* - i\Gamma_{12}^*/2}{\Delta m_{12} - i\Gamma_{12}/2}}. \tag{6.42}$$

The operation of CP gives

$$CP|P^0\rangle \to \eta_{CP}|\bar{P}^0\rangle \tag{6.43}$$

where we can choose $\eta_{CP} = -1$. In the absence of CP violation, CP eigenstates are formed and the coefficients are equal and given by

$$p = q = \frac{1}{\sqrt{2}}, \tag{6.44}$$

corresponding to

$$|P^0_{CP=\pm1}\rangle = \frac{1}{\sqrt{2}}\left(|P^0\rangle \mp |\bar{P}^0\rangle\right). \tag{6.45}$$

The physical states in the presence of the weak interaction are, therefore, an admixture of particle and antiparticle neutral meson flavor eigenstates. A similar mass matrix occurs in the K^0–$\bar{K}^0$ system where the short-lived (S) state decays primarily to the CP-even 2π final state and the long-lived state (L) decays to the CP-odd 3π final state. The lifetime difference in the physical kaon states is mainly due to the size of the phase space for CP-conserving $n\pi$ decays, namely, $K_S^0 \to 2\pi$ and $K_L^0 \to 3\pi$. The first evidence for CP violation was observed in 1964 by Fitch, Cronin, et al. in the decay $K_L^0 \to \pi^+\pi^-$ far downstream the beam line from the region of $K_S^0 \to \pi^+\pi^-$ decays. The magnitude of the ratio of the amplitudes for K_L^0 and K_S^0 decay to $\pi^+\pi^-$ is approximately given by [1]

$$|\epsilon_K| = \left|\frac{p-q}{p+q}\right| \approx \left|\frac{A(K_L^0 \to \pi^+\pi^-)}{A(K_S^0 \to \pi^+\pi^-)}\right| = (2.28 \pm 0.011) \times 10^{-3}. \tag{6.46}$$

The dominant contribution to the CP violation in the kaon sector comes from $\Delta S = 2$ mixing amplitudes, corresponding to the two units of strangeness (S) change in the mixing process $s \to \bar{s}$. However, the weak $\Delta S = 1$ decays through so-called *penguin diagrams* have also been found to violate CP *directly*, as measured by the parameter ϵ' where the magnitude squared of the double ratio is given by [2]

$$\left| \frac{A(K_L^0 \to \pi^0\pi^0)}{A(K_S^0 \to \pi^0\pi^0)} \bigg/ \frac{A(K_L^0 \to \pi^+\pi^-)}{A(K_S^0 \to \pi^+\pi^-)} \right|^2 \approx 1 - 6\epsilon'/\epsilon \tag{6.47}$$

with

$$\epsilon'/\epsilon = (1.65 \pm 0.26) \times 10^{-3}. \tag{6.48}$$

In the neutral B-meson systems, there are many possibilities for the observation of CP violation. Direct CP violation, for example, is observed in the decay rate asymmetry of $B_d^0 \to K^+\pi^-$ as compared to $\bar{B}_d^0 \to K^-\pi^+$ at the level of 10%, almost two orders of magnitude larger than in the $\pi\pi$ decays of kaons. However, the branching fraction of the B_d^0 meson to $K^+\pi^-$ is approximately 2×10^{-5} and therefore the production of 10^8 or more neutral B mesons is needed to study this decay mode. Proton-proton and proton-antiproton colliders are copious sources of B mesons, inclusively produced in hadronic interactions. Hadron colliders are described in more detail in chapter 8. Another source of B mesons is from the correlated B–$\bar{B}$ decays of the $\Upsilon(4S)$ and $\Upsilon(5S)$ resonances, where the time evolution of the correlated pair introduces additional methods to measure the CKM parameters. In particular, B-meson pairs from $\Upsilon(4S)$ decay have been studied extensively at two asymmetric e^+e^- collider experiments, the BaBar experiment at PEP-II and the Belle experiment at KEKB.

A neutral B-meson oscillation experiment begins by creating a correlated pair of B^0–$\bar{B}^0$ mesons from the decay of an upsilon resonance, $\Upsilon(4S) \to B^0\bar{B}^0$. In an asymmetric e^+e^- collider, the electron and positron beam energies are unequal and, therefore, the Υ resonance is boosted in the lab frame. This boost allows the experiment to measure the time difference between the decaying B mesons, as shown in figure 6.8. The first meson to decay is "flavor" tagged according to its decay products. At this reference time, the second B meson is known to be in either a pure B^0 or $\bar{B}^0$ state. The difference in decay length along the beam direction is then translated into a proper time difference in the rest frame of the second B meson. Thus, the time evolution of the probability to find the second meson in the same state or the opposite state is given by (with $|q/p| = 1$)

$$|f_\pm(t)|^2 = \frac{1}{2}\exp(-\Gamma t)\left(\cosh\left(\frac{\Delta\Gamma}{2}t\right) \pm \cos(\Delta m t)\right) \tag{6.49}$$

where the plus sign indicates the same state and the minus sign the opposite state, according to

$$\begin{aligned} |B^0(t)\rangle &= f_+(t)|B^0\rangle + f_-(t)\frac{q}{p}|\bar{B}^0\rangle, \\ |\bar{B}^0(t)\rangle &= f_+(t)|\bar{B}^0\rangle + f_-(t)\frac{p}{q}|B^0\rangle. \end{aligned} \tag{6.50}$$

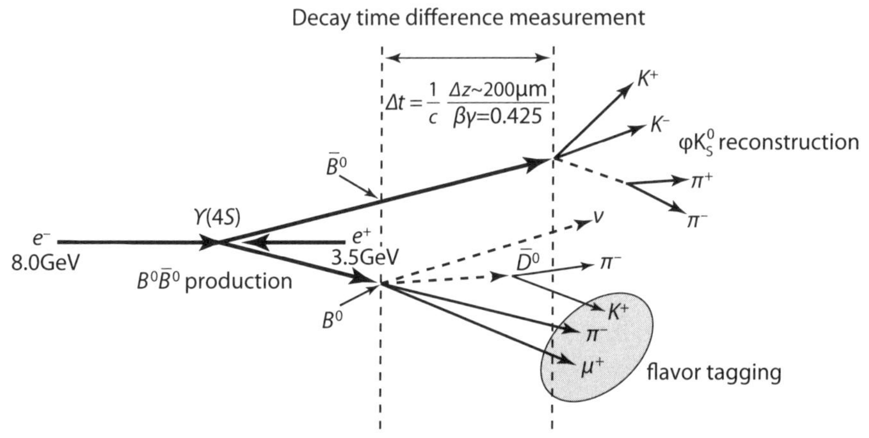

FIGURE 6.8. Neutral *B*-meson oscillation and decay measurements at an asymmetric *B*-meson factory (Credit: BABAR).

Equation (6.50) shows the explicit dependence of the mass difference Δm and the decay width difference $\Delta\Gamma$ of the heavy and light states.

If *CP* is conserved, then $\Delta m_{12} = \Delta m^*_{12}$ and $\Gamma_{12} = \Gamma^*_{12}$ and therefore $q/p = 1$. Deviations of q/p from unity indicate *CP* violation. If the magnitudes of $|q|$ and $|p|$ differ, then this is called *direct CP violation* and can be determined from the ratio of the decay amplitudes as in the ϵ'/ϵ measurement in the kaon system (6.47). It is more common to detect a phase difference between q and p and in order to be sensitive to a phase, two interfering amplitudes must be present. The most common technique is to use the interference of direct decay with oscillation plus decay for common decay modes, as sketched in figure 6.9. This method of measuring *CP* violation is called *indirect CP-violation.*

Unlike the K^0–$\bar{K}^0$ system, the B^0_d and $\bar{B}^0_d$ decays share few real transitions (decay modes). Common decay modes are Cabibbo suppressed, such as $b \to c\,(\bar{c}d)$ where the (W^-) decay is Cabibbo suppressed and similarly for $\bar{b} \to \bar{c}\,(c\bar{d})$. This implies that $\Delta\Gamma_{B_d} \ll \bar{\Gamma}_{B_d}$ where $\bar{\Gamma}_{B_d}$ is the average of the heavy and light B_d meson widths. There is a larger effect on the width

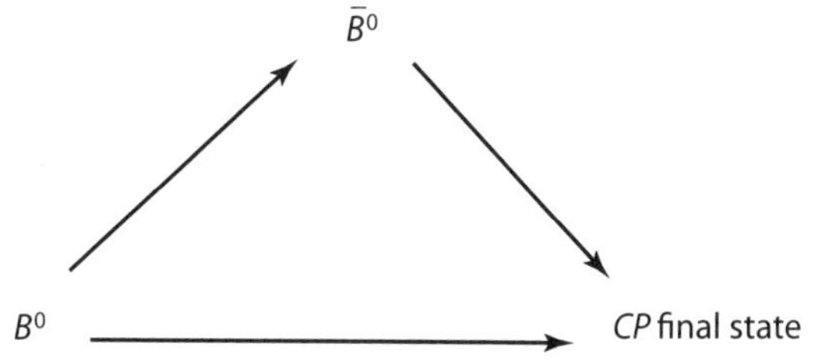

FIGURE 6.9. The direct decay amplitude of a neutral *B* meson to a *CP* eigenstate will interfere with a neutral meson that first oscillates to a $\bar{B}$ meson and then subsequently decays to the same final state.

in the B_s^0–$\bar{B}_s^0$ system due to Cabibbo-favored decays with $\Delta\Gamma_{B_s}/\bar{\Gamma}_{B_s} \approx 12\%$. The dominant contribution from virtual states in the B^0–$\bar{B}^0$ system involve the t-quark. The diagram for $\bar{B}_d^0 \to B_d^0$ will proceed through $(b\bar{d}) \to t\bar{t} \to (d\bar{b})$ or $(b\bar{d}) \to W^{+*}W^{-*} \to (d\bar{b})$ with amplitudes proportional to $V_{tb}^2(V_{td}^*)^2$ and hence sensitive to CP violation as it involves a $3^{\text{rd}} \to 1^{\text{st}}$ mass generation transition. The corresponding diagrams for B_s^0–$\bar{B}_s^0$ have amplitudes proportional to $V_{tb}^2(V_{ts}^*)^2$.

The measurement of oscillations in the B_d^0–$\bar{B}_d^0$ system gives a mass difference equal to [2]

$$\Delta m_d = 0.507 \pm 0.005 \text{ ps}^{-1} \quad \text{or, equivalently,} \quad \Delta m_d^2 \approx 1.1 \times 10^{-7} \text{ eV}^2, \tag{6.51}$$

which in terms of the natural width is given by

$$x_d = \frac{\Delta m_d}{\bar{\Gamma}_{B_d}} = 0.774 \pm 0.008. \tag{6.52}$$

The CDF and DØ experiments observed B_s^0–$\bar{B}_s^0$ mixing and due to the rapid oscillations of the states, the mass difference is given by [2]

$$\Delta m_s = 17.8 \pm 0.1 \text{ ps}^{-1} \quad \text{or, equivalently,} \quad \Delta m_s^2 \approx 1.4 \times 10^{-4} \text{ eV}^2, \tag{6.53}$$

also written as

$$x_s = \frac{\Delta m_s}{\bar{\Gamma}_{B_s}} = 26.2 \pm 0.5. \tag{6.54}$$

The value of x_s is predicted to be

$$\frac{|V_{ts}|^2}{|V_{td}|^2} = f_{SU(3)}^{-1} \frac{\Delta m_s}{\Delta m_d} \approx 25 \tag{6.55}$$

using an $SU(3)_{\text{flavor}}$ violation factor $f_{SU(3)} \approx 1.4$ that affects the comparison of the mixing parameters of the B_s and B_d states. The value of x_s is in agreement with a self-consistent CKM matrix description of quark mixing.

Oscillations have been observed in the neutral charm mesons D^0–$\bar{D}^0$ at the B-factories. The large rate of $D^{*\pm}$ production from B-hadron decay is used to tag the flavor $(D^0/\bar{D}^0)$ of the neutral D meson through the slow pion decay $D^{*\pm} \to \pi^{\pm} D^0$. The shape of the time-dependent distribution of the doubly Cabbibo–suppressed (DCS) decay $D^0 \to K^+\pi^-$ is analyzed for interference contributions from D^0–$\bar{D}^0$ mixing followed by the Cabbibo-favored (CF) decay $\bar{D}^0 \to K^+\pi^-$. The first indications are that the mass difference between the states is less than ~1% of the average width and that the mixing probability has a larger contribution from for the difference in widths

$$y = \frac{\Delta\Gamma}{2\Gamma} \approx 1\%. \tag{6.56}$$

Oscillations are less pronounced in the D^0–$\bar{D}^0$ system due to the short lifetime of the favored weak decay (increasing Γ) and the corresponding suppresion from CKM elements in the second-order weak transition (decreasing Δm).

6.5 Constraints on the Unitarity Triangle

In the general analysis of the B-meson oscillation and decay, one can construct the quantity Λ given by

$$\Lambda = \left(\sqrt{\frac{M^*}{M}}\right)_{B_x^0} \frac{\bar{D}}{D} \left(\sqrt{\frac{M^*}{M}}\right)_{K^0} \tag{6.57}$$

where the first term is the square root of the ratio of the coupling terms for the mixing transition $\bar{B}_x^0 \rightarrow B_x^0$ to $B_x^0 \rightarrow \bar{B}_x^0$, the second is the ratio of the decay transition coupling terms for $\bar{B}_x^0 \rightarrow X$ to $B_x^0 \rightarrow X$, and the third term is an additional mixing term for final states which contain a K_s^0 meson. Decays with no K_s^0 in the final state only have the first two product terms. For example, the complex phase of Λ for the process $B_d^0 \rightarrow \Psi K_s^0$ is given by

$$\begin{aligned}\Lambda(B_d^0 \rightarrow \Psi K_s^0) &= \left(\frac{V_{td}^* V_{tb}}{V_{td} V_{tb}^*}\right)\left(\frac{V_{cs}^* V_{cb}}{V_{cs} V_{cb}^*}\right)\left(\frac{V_{cd}^* V_{cs}}{V_{cd} V_{cs}^*}\right) \\ &= \exp\left(i2\tan^{-1}\left(\frac{\eta}{1-\rho}\right)\right) = \exp(i2\beta).\end{aligned} \tag{6.58}$$

For the process $B_d^0 \rightarrow \pi^+\pi^-$, the phase of Λ is given by

$$\begin{aligned}\Lambda(B_d^0 \rightarrow \pi^+\pi^-) &= \left(\frac{V_{td}^* V_{tb}}{V_{td} V_{tb}^*}\right)\left(\frac{V_{ud}^* V_{ub}}{V_{ud} V_{ub}^*}\right) \\ &= \exp\left(i2\tan^{-1}\left(\frac{\eta}{1-\rho}\right) - 2\tan^{-1}\left(\frac{\eta}{\rho}\right)\right) \\ &= \exp(i2(\beta-\gamma)) = \exp(i2(\alpha-\pi)) = \exp(-i2\alpha).\end{aligned} \tag{6.59}$$

The measurement of the third angle, γ, is most directly determined from B_s^0–$\bar{B}_s^0$ mixing and decay. The angles α, β, and γ are the internal angles of the unitarity triangle, as shown in figure 6.6. A nonzero area of the unitarity triangle indicates the presence of CP violation, i.e., the triangle collapses when the complex phase $\eta = 0$. In fact, the area of the unitarity triangle, denoted J, is approximately 25 standard deviations from zero and is invariant under the choice of triangle from the six possible unitarity conditions of the CKM matrix from equations (6.34) and (6.35).

In addition to constraints on the angles of the unitarity triangle, a number of measurements restrict the lengths of the legs of the triangle. These are ϵ_K from the kaon system, the $3^{\text{rd}} \rightarrow 1^{\text{st}}$ generation decays $|V_{ub}|$, and the mixing frequencies of the $B_x^0 - \bar{B}_x^0$ systems. The current fit to the unitarity triangle describing the self-consistency of the measurements of the CKM matrix elements is shown in figure 6.10, where relative to figure 6.6 we define

$$\bar{\rho} = \rho(1-\lambda^2/2), \quad \bar{\eta} = \eta(1-\lambda^2/2). \tag{6.60}$$

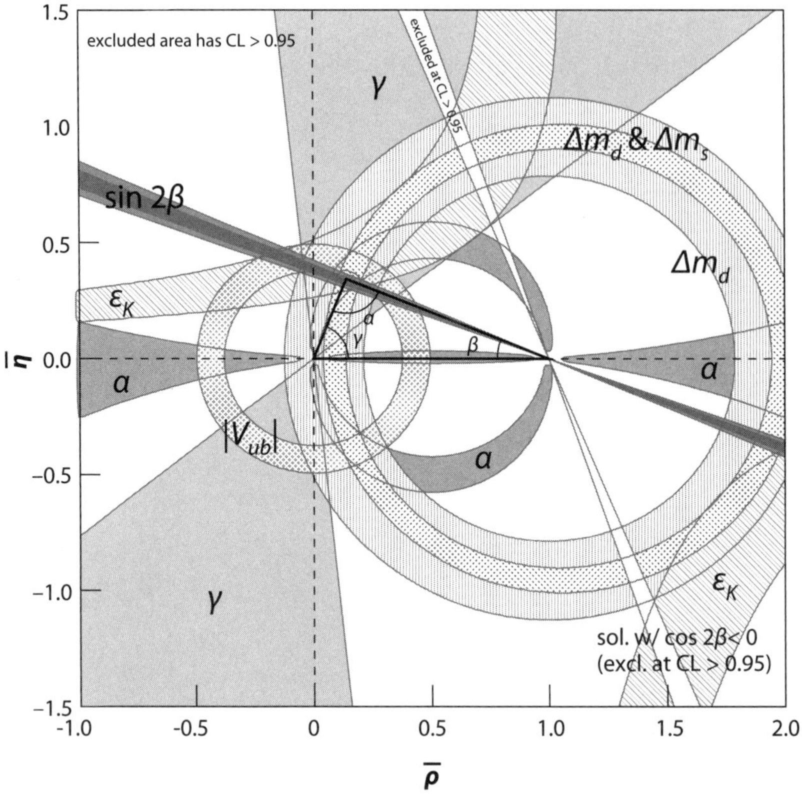

FIGURE 6.10. Fit to the unitarity triangle in the $(\bar{\eta}, \bar{\rho})$-plane indicating measurement constraints from α, $\sin 2\beta$, γ, ϵ_K, $|V_{ub}|$, Δm_d and Δm_s (Credit: CKMfitter) [5].

6.6 Exercises

1. **SuperK.** The SuperKamiokande experiment sees evidence for flavor oscillations in neutrinos produced in the atmosphere by cosmic rays. One interpretation is that the deficit in observed muon neutrinos is the result of $\nu_\mu \to \nu_\tau$ mixing. Using a two-generation mixing model, the $\nu_\mu \to \nu_\tau$ mixing probability is given by

$$P(\nu_\mu \to \nu_\tau) = \sin^2(2\theta)\sin^2\left(\frac{\Delta m^2 L}{4E}\right)$$

where θ is the mixing angle between flavor and mass eigenstates, Δm^2 is the difference between the masses squared of the mass eigenstates, E is the neutrino energy, and L is the distance from the source, in units where $\hbar = c = 1$. The term $\sin^2(2\theta)$ is referred to as the amplitude of the oscillations. SuperK's results, interpreted in this model, give $\Delta m^2 \approx 2.4 \times 10^{-3}$ eV^2, $\sin^2(2\theta) \approx 1$.

However, there are three known generations of neutrinos. Thus, in general, mixing will be governed by a 3×3 mixing matrix

$$U = \begin{pmatrix} U_{e1} & U_{e2} & U_{e3} \\ U_{\mu 1} & U_{\mu 2} & U_{\mu 3} \\ U_{\tau 1} & U_{\tau 2} & U_{\tau 3} \end{pmatrix}.$$

This general case requires a great deal of experimental information to pin down, but we can make inferences about specific models. Including solar neutrino results suggests a model where two of the neutrino masses are very close together, and the third is larger:

$$m_3 = M > m = m_1 = m_2,$$

with SuperK's $\Delta m^2 = M^2 - m^2$. In this problem, assume this mass spectrum and also assume that there is no *CP* violation.

(a) What does the assumption of no *CP* violation tell us about the matrix U?

(b) Derive the probability of $\nu_\mu \to \nu_\tau$ as a function of distance from the source. (*Hint:* It helps to put in the mass model early in the calculation.)

(c) When interpreted as two-generation mixing, SuperK measures $\sin^2(2\theta) \approx 1$. In our model of three-generation mixing, what is this a measurement of? To allow a definite result, assume that SuperK measures $\sin^2(2\theta) = 0.98$ and that $U_{\tau 3} = 1/\sqrt{2}$. What is the amplitude of ν_μ oscillations to electron neutrinos?

2. **MSW effect.** Just as with light, the propagation of neutrinos through a material medium can be characterized by an index of refraction. This neutrino index of refraction plays an important role in the MSW hypothesis concerning the shortage of electron-type neutrinos coming from the sun. As with light, the index of refraction n can be related to the neutrino energy $\hbar\omega$, the number density N of scatters in the medium, and the scattering amplitude f for *forward* neutrino scattering:

$$n = 1 + \frac{2\pi N}{k^2} f$$

where $k = \omega/c$.

(a) According to the Standard Model, electron-type and muon-type neutrinos are expected to have different scattering amplitudes, hence different indices of refraction. Explain qualitatively why this is so, referring to appropriate Feynman diagrams as necessary. (This difference is at the heart of the MSW mechanism.)

(b) We are concerned here with neutrino energies small compared to the rest energy of the intermediate vector bosons but large compared to electron binding energies of atoms. Suppose that the medium is composed of neutral atoms of number density N, with Z electrons and $A - Z$ neutrons per atom. Let G_F be the standard "four-fermion" coupling constant of ordinary β-decay. For the purposes of this question, treat the neutrinos as massless.

Obtain an explicit expression for the difference, $n_e - n_\mu$, of the indices of refraction of electron-type and muon-type neutrinos of energy $\hbar\omega$.

Evaluate this difference numerically for neutrinos of energy 1 MeV in a medium of hydrogen atoms with number density $N = 5 \times 10^{25}$ atoms/cm^3. The Fermi constant is $G_F \approx 1.166 \times 10^{-5}$ GeV^{-2}.

3. **Double-bang τ-neutrino detection.** Recent experimental measurements suggest that high-energy electron neutrinos produced at astronomical distances oscillate to τ neutrinos before reaching Earth. High-energy neutrino observatories seek to measure the flux of τ neutrinos through neutrino interactions with the polar ice cap in the southern hemisphere. A volume of 1 km^3 of ice is instrumented with phototubes to detect Cherenkov light. The following information may be useful: mass of τ lepton $m_\tau = 1.8$ GeV/c^2, mass of muon $m_\mu = 0.1$ GeV/c^2, mass of electron $m_e = 0.5$ MeV/c^2, and the lifetime of a τ lepton $\tau_\tau = 0.3$ ps.

   (a) Sketch and label the tree-level Feynman diagram of a τ neutrino interacting with the ice to produce a charged tau lepton in the final state.

   Assume for the following parts of the problem that the τ lepton carries away 75% of the initial neutrino energy.

   (b) For a τ neutrino with an energy of 10^7 GeV, estimate the distance in ice that the τ lepton will travel before decaying.

   (c) List four of the possible decay modes of the τ lepton.

   (d) What is the experimental signature of a 10^7 GeV τ neutrino interacting with the polar ice cap? Make a rough plot of the relative amount (not absolute rates) of Cherenkov light seen at a phototube as a function of time. Label in the plot the effect from the τ decay length.

   (e) Several events are observed in the ice that have a single high-energy shower with energies very close to 6.4×10^6 GeV. What flavor of neutrino is responsible for this, and is it the particle or antiparticle which is interacting? Describe what the interaction is and why 6.4×10^6 GeV is special for this detector.

4. **β-Beams.** Design an Earth-based experiment that utilizes ν_e neutrinos from the β-decay of an energetic beam of ^{13}N ions to detect muon appearance. Assume the ^{13}N ion beam circulates in a beamline consisting of two long, straight sections, one pointing toward the center of Earth and the other pointing up toward the sky. A near detector, as shown in figure 6.11, is located just on top of the accelerator, and a far detector is located on the other side of the planet. The energy of the ν_e from the decay $^{13}\text{N} \rightarrow {}^{13}\text{Ce}^+\nu_e$ in the rest frame of the ^{13}N ion is 2.22 MeV.

   (a) What is the minimum energy of the ν_e in the lab frame that is kinematically required to produce an upward-going μ^- from a weak charged-current interaction with a neutron in a target nucleus in the far detector, assuming that $\nu_e \rightarrow \nu_\mu$ oscillations occur on this length scale? What is the corresponding required beam momentum of the ^{13}N ions in the straight section of the accelerator, also known as a *β-beam*?

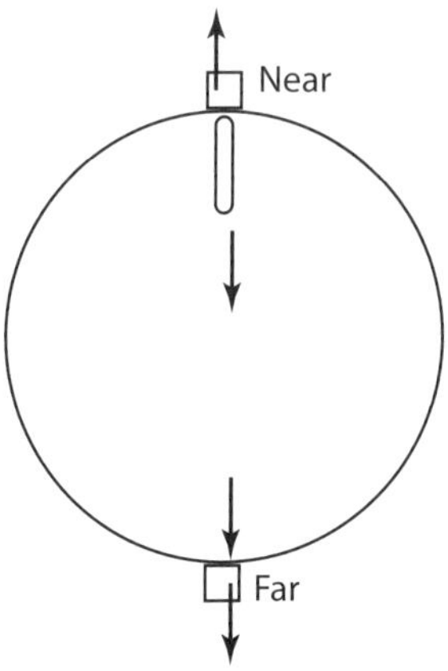

FIGURE 6.11. Vertical ^{13}N ion β-beam with a near detector located above the accelerator and a far detector on the opposite side of the Earth (*not drawn to scale*).

(b) Using the minimum ν_e energy for muon production, the diameter of Earth, and the MNS matrix parameters, what is the probability that the β-decay electron neutrinos have oscillated to muon neutrinos $\nu_e \to \nu_\mu$ by the time they reach the far detector?

(c) If Earth was of sufficiently high density, would it be possible for this experiment to observe matter oscillation enhancement of the mixing? If so, would this effect determine the sign of Δm^2_{31}? Explain.

(d) For an actual experiment, what would be the function of the near detector?

5. **Momentum uncertainty in antineutrinos from a nuclear reactor.** The KamLAND experiment measured neutrino oscillations in antineutrinos coming from nuclear reactors. Show that the momentum uncertainty of antineutrinos coming from free neutron β-decay is compatible with the squared mass difference in antineutrino mass eigenstates from the oscillation signal. Repeat the same analysis for the β-decay of ^{93}Br with a half-life of $T_{1/2} = 102$ ms and for each of the β-decays in the following chain of unstable of fission fragments:

$$^{93}\text{Br} \to {}^{93}\text{Kr} \to {}^{93}\text{Rb} \to {}^{93}\text{Sr} \to {}^{93}\text{Y} \to {}^{93}\text{Kr} \to {}^{93}\text{Nb}. \tag{6.61}$$

6. **Phases of elements of V_{CKM}.** In the Standard Model, the charged-current interactions of the quarks take the form

$$\mathcal{L}_{\text{CC}} = -\frac{g}{\sqrt{2}}(\bar{u}_L \quad \bar{c}_L \quad \bar{t}_L)\gamma^\mu V_{\text{CKM}} \begin{pmatrix} d_L \\ s_L \\ b_L \end{pmatrix} W^+_\mu + h.c.,$$

where the W_μ fields correspond to the W bosons and V_{CKM} is a unitary matrix giving the amplitudes for weak transitions between the different flavors of quark. A more explicit representation of V_{CKM} is

$$V_{\mathrm{CKM}} = \begin{pmatrix} V_{ud} & V_{us} & V_{ub} \\ V_{cd} & V_{cs} & V_{cb} \\ V_{td} & V_{ts} & V_{tb} \end{pmatrix},$$

where the V_{ab} are complex numbers subject, by virtue of the unitarity of V_{CKM}, to constraints such as

$$V_{cd}V_{cb}^{*} + V_{td}V_{tb}^{*} + V_{ud}V_{ub}^{*} = 0.$$

This constraint can be graphically summarized in "unitary triangles" as in figure 6.6. The phases of the V_{ab} are not individually meaningful: they can be changed by redefining the phases of the quark fields (viz. $u \to \exp(i\alpha)u$). If there is no way to choose the phases of the quarks such that all the V_{ab} are real, then CP (and T) are broken with all the consequences that follow. We will now explore some methods for determining the phases of elements of V_{CKM}.

(a) Consider the neutral mesons $B^0 = \bar{b}d$ and $\bar{B}^0 = b\bar{d}$. The states $|B^0\rangle$ and $|\bar{B}^0\rangle$ are not mass eigenstates because they mix through the diagrams in figure 6.7. To a good approximation (this has to do with the GIM mechanism), these diagrams are dominated by those in which the internal lines are all top quarks. Show that both diagrams are proportional to $(V_{td}V_{tb}^{*})^2$. Assuming that all other contributions to the amplitude are real, the overall phase is determined by this CKM factor.

(b) Neither B^0 nor $\bar{B}^0$ are mass eigenstates. To find the mass eigenstates, we must diagonalize a 2×2 Hamiltonian whose off-diagonal elements are given by the mixing amplitude. Show that the energy eigenstates are of the form

$$|B_{\pm}\rangle = \frac{1}{\sqrt{2}}\left(\exp(i\phi_M)|B^0\rangle \pm \exp(-i\phi_M)|\bar{B}^0\rangle\right)$$

and show that ϕ_M is related to the phase of the $V_{td}V_{tb}^{*}$ leg of the unitary triangle in figure 6.6.

(c) We observe mixing by watching decays to a channel open to both B^0 and $\bar{B}^0$, such as $B^0(\bar{B}^0) \to J/\Psi\pi^0$, which we assume to proceed via the quark diagrams (we are simplifying reality here) in figure 6.12. Identify the CKM factors associated with these diagrams and show that they correspond to another leg of the unitarity triangle, shown in figure 6.6.

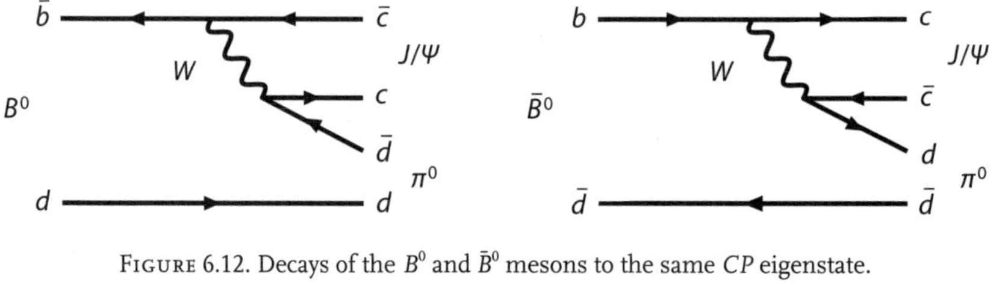

FIGURE 6.12. Decays of the B^0 and $\bar{B}^0$ mesons to the same CP eigenstate.

(d) In what follows we will assume that the complex phase ϕ_W of the decay amplitude comes entirely from the CKM factor. Now suppose that a B^0 (or $\bar{B}^0$) is created at time $t = 0$. It will evolve into a linear combination of B^0 and $\bar{B}^0$ and the rate for decay to the $f = J/\Psi\pi^0$ final state at later times t will have time-dependent interference effects. Construct the time-dependent asymmetry

$$A(t) = \frac{\Gamma(\bar{B}^0 \to f) - \Gamma(B^0 \to f)}{\Gamma(\bar{B}^0 \to f) + \Gamma(B^0 \to f)}$$

and show that by measuring it, one can determine a combination of the phases ϕ_W and ϕ_M. Show that this combined phase is physically meaningful, i.e., is independent of redefinitions of the phases of the quark fields. To what angle of the unitarity triangle does it correspond?

7. Give an example of mixing and decay in the $B_s^0 - \bar{B}_s^0$ system that measures the angle γ in the unitarity triangle.

8. **D^0–$\bar{D}^0$ Mixing**. An abundant source of $D^{*\pm}$ production is from B-hadron decays at B-factories. The flavor of a neutral D meson ($D^0/\bar{D}^0$) coming from $D^{*\pm}$ decay can be tagged by the charge of the soft pion. Assume that the mass difference of the heavy and light neutral D mesons is negligible compared to the average width, i.e., $x = \Delta m/\Gamma \approx 0$. Let the difference in width between the heavy and light neutral D mesons, compared to twice their average width, be given by $y = \Delta\Gamma/(2\Gamma) \approx 1\%$.

(a) Show that the decay $D^0 \to K^-\pi^+$ is Cabbibo-favored (CF) while the decay $D^0 \to K^+\pi^-$ is doubly Cabbibo suppressed (DCS).

(b) If the D^0 meson is produced with a momentum of $p = 3$ GeV/c from the decay of a $D^{*\pm}$, what is the typical decay length of the D^0 meson in the lab frame?

(c) Assume that the $D^0 \to K^+\pi^-$ decay only proceeds through D^0–$\bar{D}^0$ mixing followed by the CF decay $\bar{D}^0 \to K^+\pi^-$. What is the time dependence of the decay rate for small times (i.e., small decay distances in the lab frame)?

(d) Assume the direct DCS decay $D^0 \to K^+\pi^-$ is two orders of magnitude larger than the probability for D^0–$\bar{D}^0$ mixing followed by the CF decay $\bar{D}^0 \to K^+\pi^-$ in the characteristic time scale of the neutral D-meson lifetime in the lab frame. What is the approximate magnitude of the deviation of the observed $D^0 \to K^+\pi^-$ decay rate from a pure DCS decay rate and how does it depend on time (i.e., decay distance in the lab frame) for small times?

6.7 References and Further Reading

Further information on neutrinos and astroparticle physics can be found here: [6, 7, 8, 9, 10].

A selection of Web sites of related neutrino experiments can be found here: [11, 12, 13, 14, 15, 16, 17, 18, 19, 20].

A selection of Web sites of ultra-high-energy cosmic ray observatories can be found here: [21, 22, 23, 24, 25].

A selection of Web sites of related B-factory experiments can be found here: [26, 27, 28, 29].

Compilations of CKM results can be found here: [5].

[1] Super-Kamiokande Collaboration, *Phys. Rev. Lett.* **81** (1998) 1562.

[2] K. Nakamura et al. (Particle Data Group). *J. Phys. G* **37**, 2010. http://pdg.lbl.gov.

[3] KamLAND Experiment. http://www.awa.tohoku.ac.jp/awa/eng/index.html.

[4] Bahcall, Serenelli, and Basu, *ApJ* **621**, L85 (2005).

[5] CKMfitter. http://ckmfitter.in2p3.fr.

[6] Paul Langacker. *The Standard Model and Beyond.* CRC Press, 2010. ISBN 978-1-4200-7906-7.

[7] Abraham Seiden. *Particle Physics.* Addison-Wesley, 2005. ISBN 0-8053-8736-6.

[8] Rabindra N. Mohapatra and Palash B. Pal. *Massive Neutrinos in Physics and Astrophysics.* World Scientific Publishing, 2004. ISBN 981-238-071-X.

[9] K. Grotz and H. V. Klapdor. *The Weak Interaction in Nuclear, Particle and Astrophysics.* IOP Publishing, 1990. ISBN 0-85274-313-0.

[10] P. D. B. Collins, A. D. Martin, and E. J. Squires. *Particle Physics and Cosmology.* Wiley, 1989. ISBN 0-471-60088-1.

[11] Super-K Experiment on the T2K Beamline. http://www-sk.icrr.u-tokyo.ac.jp/sk/index-e.html.

[12] SNO Experiment, (1999–2006). http://www.sno.phy.queensu.ca.

[13] Borexino Experiment. http://borex.lngs.infn.it.

[14] OPERA Experiment on the CNGS Beamline. http://operaweb.lngs.infn.it.

[15] ICARUS Experiment on the CNGS Beamline. http://icarus.lngs.infn.it.

[16] MINOS Experiment on the NUMI Beamline. http://www-numi.fnal.gov.

[17] NOvA Experiment on the NUMI Beamline. http://www-nova.fnal.gov.

[18] K2K Experiment. http://neutrino.kek.jp.

[19] T2K Experiment. http://jnusrv01.kek.jp/public/t2k/index.html.

[20] Daya Bay Experiment. http://dayawane.ihep.ac.cn.

[21] IceCube Experiment. http://icecube.wisc.edu.

[22] ANTARES Experiment. http://antares.in2p3.fr.

[23] Pierre Auger Observatory. http://www.auger.org.

[24] AGASA at the Akeno Observatory. http://www-akeno.icrr.u-tokyo.ac.jp/AGASA.

[25] HiRes Experiment. http://www.cosmic-ray.org.

[26] CLEO Experiment at CESR. http://www.lns.cornell.edu/public/CLEO.

[27] BaBar Experiment at PEP-II (1999–2008). http://www.public.slac.stanford.edu/babar.

[28] Belle Experiment at KEKB (1999–2010). http://belle.kek.jp.

[29] LHCb Experiment at the LHC. http://lhcb-public.web.cern.ch/lhcb.

7 e^+e^- Collider Physics

An extensive number of processes in high-energy e^+e^- collider physics are predicted and measured to high precision. Figure 7.1 shows a spectrum of different final states compared with their measured cross sections from the L3 experiment from the LEP collider 85 GeV $< \sqrt{s} <$ 209 GeV. The Z peak resonance appears at $M_Z \approx 91.2$ GeV and decays into fermion-antifermion (two-fermion) final states, the leptonic final states $\ell^+\ell^-$ and $\nu_\ell \bar{\nu}_\ell$ for the three lepton flavors ($\ell = e, \mu, \tau$), and the hadronic final states $q\bar{q}$ for the five energetically accessible quark flavors ($q = u, d, s, c, b$). These processes also have amplitude contributions from a virtual photon propagator with QED couplings.

The full lineshape above the Z peak is separated into two components, the full collision energy events and the "radiative-return" events where the effective center-of-mass of the collision, known as $\sqrt{s'}$, is lower than the full center-of-mass energy $\sqrt{s} = 2E_{\text{beam}}$. The process of radiative return will be discussed in more detail later in the chapter. Also shown in figure 7.1 are the four-fermion electroweak and QED final states. The process $e^+e^- \to e^+e^- q\bar{q}$ (or similarly for $e^+e^- \to e^+e^- \ell^+\ell^-$) is called a *two-photon interaction* and is described effectively by the collision of two photons, each emitted from the incoming electron and positron, to form a two-fermion final state in addition to the scattered electron and positron. The two-photon cross section grows as a function of $\sqrt{s}$ because the process probes fixed mass scales (or fixed Q^2 photon propagators) and therefore has a larger and larger spectrum of radiated photons to collide for an increase in beam energy.

The four-fermion electroweak production comes through W^+W^- and ZZ pair production and subsequent two-fermion decays of each boson. There are also single Z ($e^+e^- \to e^+e^- Z$) and single W ($e^+e^- \to e^+\nu_e W^-$ or $e^-\bar{\nu}_e W^+$) production mechanisms with four-fermions in the final state. The curve for $e^+e^- \to HZ$ is a reference curve for the search for Higgs production and will be described in chapter 9.

7.1 Fermion Pair Production

The invariant matrix element for the process $e^+e^- \to f\bar{f}$ through single photon exchange is given by

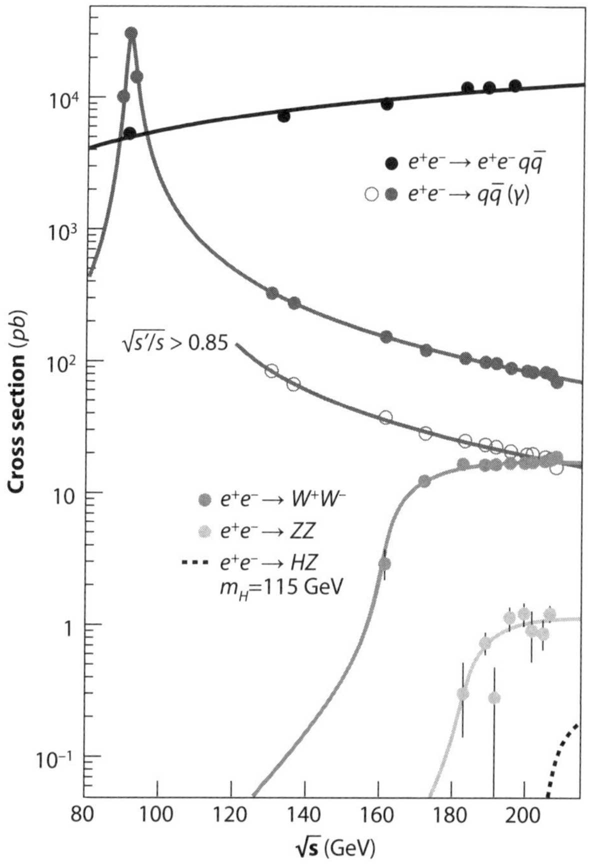

FIGURE 7.1. Cross-section measurements by the L3 experiment at LEP (Credit: L3) [1].

$$\mathcal{M}_\gamma = \frac{ie^2 Q_e Q_f}{s} \bar{v}_2 \gamma_\mu u_1 \bar{u}_3 \gamma^\mu v_4. \tag{7.1}$$

If s is large compared to m_e^2, then we can separate the chiral contributions to the total cross section using the chirality projection operators

$$P_{L,R} = \frac{1 \mp \gamma^5}{2}. \tag{7.2}$$

Given that γ^μ anticommutes with γ^5, the vector nature of electromagnetic current forces annihilating fermions to have the same chirality, i.e., pushing the chirality projection operator past a γ^μ flips the sign of the γ^5 term, giving $\gamma^\mu P_L = P_R \gamma^\mu$, which when acting to the right projects out the left-handed chirality component of $\bar{\Psi}$. A similar result occurs for the chiralities of the pair-produced fermions. Therefore, one can parameterize the four nonzero contributions to the scattering diagram as follows:

$$\mathcal{M}_\gamma = \frac{ie^2 Q_e Q_f}{s} \bar{v}_2 \gamma_\mu \left(\frac{1 + h_i \gamma^5}{2}\right) u_1 \bar{u}_3 \gamma^\mu \left(\frac{1 + h_f \gamma^5}{2}\right) v_4 \tag{7.3}$$

where in terms of helicity $h_i = 1$ corresponds to $h(e^-) = -h(e^+) = 1$ and $h_i = -1$ corresponds to $h(e^-) = -h(e^+) = -1$ and likewise for $h_f = \pm 1$ for the final-state fermions. The resulting differential cross section is

$$\begin{aligned}\frac{d\sigma}{d\Omega} &= \frac{2\alpha^2 Q_e^2 Q_f^2}{s^3}\left[(1+h_i h_f)(p_2\cdot p_3)(p_1\cdot p_4) + (1-h_i h_f)(p_2\cdot p_4)(p_1\cdot p_3)\right] \\ &= \frac{\alpha^2 Q_e^2 Q_f^2}{8s}\left[(1+h_i h_f)(1+\cos\theta)^2 + (1-h_i h_f)(1-\cos\theta)^2\right] \\ &= \frac{\alpha^2 Q_e^2 Q_f^2}{4s}\left[(1+\cos^2\theta) + 2h_i h_f \cos\theta\right].\end{aligned} \tag{7.4}$$

If we sum over the four possible helicity configurations, corresponding to the total cross section for unpolarized beams, the term linear in $\cos\theta$ disappears. The "$1+\cos^2\theta$" angular dependence is a universal signature of unpolarized fermion pair production from a vector particle. Equation (7.4) shows that if both the initial- and final-state leptons are polarized, then the resulting differential cross section is asymmetric, with $h_i h_f = \pm 1$ giving $d\sigma/d\Omega \propto (1 \pm \cos\theta)^2$, as shown in figure 7.2. In the standard model, the weak interaction component has different couplings for left- and right-handed chirality fermions, hence the angular distribution of the final-state fermions provides a test of the model. If one now includes the contribution to the process $e^+e^- \to f\bar{f}$ from the Z boson, then the total matrix element is given by ($f \neq e$)

$$\begin{aligned}\mathcal{M}(e^+e^- \to f\bar{f}) &= \mathcal{M}_\gamma + \mathcal{M}_Z \\ &= \frac{ie^2 Q_e Q_f}{s}\bar{v}_2\gamma_\mu u_1 \bar{u}_3\gamma^\mu v_4 \\ &\quad + \frac{ig^2}{\cos^2\theta_W (s - M_Z^2)}\bar{v}_2\gamma_\mu\left(\frac{g_V^e - g_A^e\gamma^5}{2}\right)u_1\bar{u}_3\gamma^\mu\left(\frac{g_V^f - g_A^f\gamma^5}{2}\right)v_4\end{aligned} \tag{7.5}$$

where $g_V = C_L + C_R$ and $g_A = C_L - C_R$ and the $q^\mu q^\nu/M_Z^2$ term of the Z propagator is neglected. Thus $|\mathcal{M}|^2$ is composed of the following three terms:

$$|\mathcal{M}|^2 = |\mathcal{M}_\gamma|^2 + |\mathcal{M}_Z|^2 + (\mathcal{M}_\gamma\mathcal{M}_Z^* + \mathcal{M}_Z\mathcal{M}_\gamma^*). \tag{7.6}$$

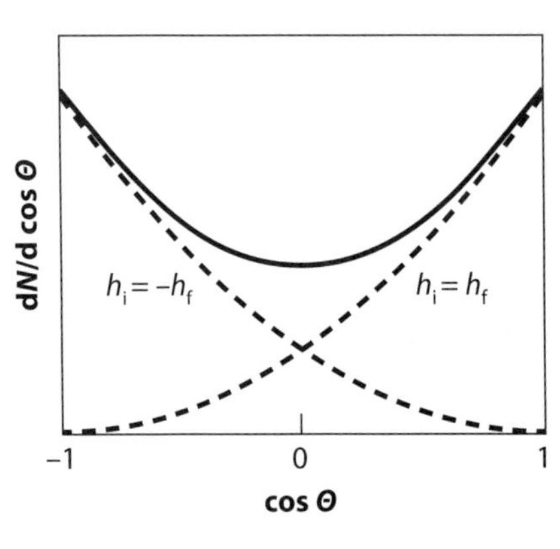

FIGURE 7.2. Angular dependence of outgoing fermion f in the process $e^+e^- \to f\bar{f}$. The polarized contributions from $h_i = h_f$ and $h_i = -h_f$ are shown.

The resulting differential cross section for unpolarized e^+e^- beams has the form

$$\frac{d\sigma}{d\Omega} = \frac{\alpha^2}{4s}\left[C_1(1+\cos^2\theta) + C_2\cos\theta\right] \tag{7.7}$$

where θ is the outgoing polar angle of fermion f with respect to the incoming e^- momentum. The contributions to C_1 and C_2 from $|\mathcal{M}_\gamma|^2$ and $|\mathcal{M}_Z|^2$ are

$$\begin{aligned} C_1' &= Q_e^2 Q_f^2 + \frac{F^2}{16}\left[(g_V^e)^2 + (g_A^e)^2\right]\left[(g_V^f)^2 + (g_A^f)^2\right], \\ C_2' &= \frac{F^2}{2} g_V^e g_A^e g_V^f g_A^f, \end{aligned} \tag{7.8}$$

where

$$F = \frac{g^2}{e^2\cos^2\theta_W}\frac{s}{s-M_Z^2} = \frac{\sqrt{2}G_F}{\pi\alpha}\frac{sM_Z^2}{(s-M_Z^2)}. \tag{7.9}$$

Note that since $M_W = M_Z\cos\theta_W$, the limit as $M_Z \to \Delta$ is well-defined to be $F = -\sqrt{2}G_F s/(\pi\alpha)$. If $\sqrt{s}$ is in the continuum region 10 GeV $< \sqrt{s} <$ 60 GeV, then the contributions from F^2 are negligible. However, the interference term $(\mathcal{M}_\gamma\mathcal{M}_Z^* + \mathcal{M}_Z\mathcal{M}_\gamma^*)$ has measurable effects: the large $\mathcal{M}_\gamma$ term amplifies the effect of the high-mass resonance giving terms linear in F

$$\begin{aligned} C_1^{\text{int}} &= \frac{F}{2} Q_e Q_f g_V^e g_V^f, \\ C_2^{\text{int}} &= F Q_e Q_f g_A^e g_A^f. \end{aligned} \tag{7.10}$$

The *forward-backward asymmetry*, defined as $A_{FB} = (N_F - N_B)/(N_F + N_B)$, is related to C_1 and C_2 by

$$A_{FB} = \frac{3C_2}{8C_1} \simeq \frac{3Fg_A^e g_A^f}{8Q_eQ_f}. \tag{7.11}$$

Data from the PEP and PETRA e^+e^- colliders are plotted in figure 7.3 for muon pair production. One can therefore see the onset of the Z boson through interference terms, even though the center-of-mass energy is subthreshold for direct production. Figure 7.3 shows measurements of $A_{FB}^{\mu\mu}$ up to $\sqrt{s} = 45$ GeV. Notice that the Z boson never *decouples* from the photon even in the limit of infinite Z mass. The nondecoupling of weak boson masses is a recurring property of the electroweak theory.

On the Z peak, the *forward-backward asymmetry* A_{FB} is given by

$$A_{FB} = \frac{3C_2'}{8C_1'} = \frac{3g_V^e g_A^e g_V^f g_A^f}{((g_V^e)^2 + (g_A^e)^2)((g_V^f)^2 + (g_A^f)^2)} = \frac{3}{4}A_e A_f. \tag{7.12}$$

The nonzero value of the forward-backward asymmetry comes from the $SU(2)_L$ contribution to the couplings of the Z boson to fermions. There is a preference to produce Z bosons from left-handed e^- (matter) and right-handed e^+ (antimatter) and, similarly, the outgoing fermions from Z decay are also polarized. The net final-state polarization prefers to align with the initial-state polarization and, hence, the nonzero value of A_{FB}.

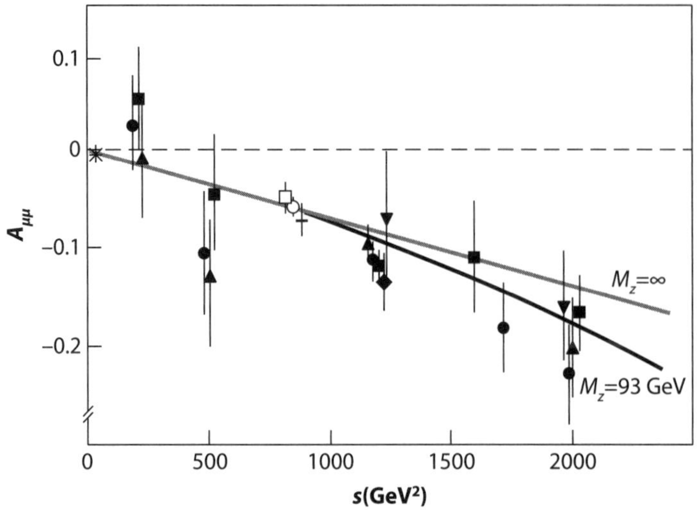

FIGURE 7.3. Measurements of A_{FB} from dimuon data at $\sqrt{s} < 45$ GeV (Credit: PEP and PETRA experiments).

This is most clearly seen in the τ-leptons described in the section on polarized τ-lepton decays. Enhancement of the initial-state Z boson polarization can be achieved with polarized beams, as described in the section on the left-right asymmetry A_{LR}.

7.2 W Boson Decay

The decay of the W^- boson with four-momentum p and polarization vector ϵ^λ to a final-state fermion f_a and antifermion $\bar{f}_b$

$$W^-(p,\epsilon^\lambda) \to f_a(p_a,s_a) + \bar{f}_b(p_b,s_b) \tag{7.13}$$

is given by

$$\mathcal{M} = \frac{-ig}{\sqrt{2}} C\epsilon^\lambda_\mu \bar{u}_a \gamma^\mu \frac{(1-\gamma^5)}{2} v_b \tag{7.14}$$

where $C = 1$ for a leptonic final-state $(f_a, \bar{f}_b) = (\ell^-, \bar{\nu}_\ell)$ and for a quark-antiquark final state we have the corresponding CKM matrix element $C = V_{q_a q_b}$ for $(f_a, \bar{f}_b) = (q_a, \bar{q}_b)$ and q_a an up-type quark. For a down-type q_a, the factor is $C = V^*_{q_b q_a}$. The square of the invariant amplitude is given by

$$|\mathcal{M}|^2 = \frac{g^2|C|^2}{2} \mathrm{tr}\left\{u_a\bar{u}_a\gamma^\mu\left(\frac{1-\gamma^5}{2}\right)v_b\bar{v}_b\gamma^\mu\left(\frac{1-\gamma^5}{2}\right)\right\}\epsilon^\lambda_\mu\epsilon^{\lambda*}_\nu. \tag{7.15}$$

Averaging over the three polarization states of the W boson and summing over the final spins yields

$$
\begin{aligned}
\overline{|\mathcal{M}|^2} &= \frac{g^2|C|^2}{3(2)}\,\mathrm{tr}\left\{(\not{p}_a + m_a)\gamma^\mu\left(\frac{1-\gamma^5}{2}\right)(\not{p}_b + m_b)\gamma^\nu\left(\frac{1-\gamma^5}{2}\right)\right\}\sum_\lambda \epsilon^\lambda_\mu \epsilon^{\lambda *}_\nu \\
&= \frac{g^2|C|^2}{6}\,\mathrm{tr}\left\{\not{p}_a\gamma^\mu \not{p}_b\gamma^\nu\left(\frac{1-\gamma^5}{2}\right)^2\right\}\sum_\lambda \epsilon^\lambda_\mu \epsilon^{\lambda *}_\nu \\
&= \frac{g^2|C|^2}{12}\,\mathrm{tr}\left\{\not{p}_a\gamma^\mu \not{p}_b\gamma^\nu (1-\gamma^5)\right\}\sum_\lambda \epsilon^\lambda_\mu \epsilon^{\lambda *}_\nu
\end{aligned}
\tag{7.16}
$$

where the mass terms dropped out using $(\frac{1+\gamma^5}{2})(\frac{1-\gamma^5}{2}) = 0$ and simplifying the square using $(\gamma^5)^2 = 1$. Equation (7.16) can be further reduced by using the completeness relation for the sum over polarizations

$$
\sum_\lambda \epsilon^\lambda_\mu \epsilon^{\lambda *}_\nu = -g_{\mu\nu} + p_\mu p_\nu / M_W^2
\tag{7.17}
$$

and noting that the γ^5 term will be the contraction of an antisymmetric tensor with the symmetric term in equation (7.17) and thus vanish. This gives

$$
\begin{aligned}
\overline{|\mathcal{M}|^2} &= \frac{g^2|C|^2}{3}\left[p_a^\mu p_b^\nu + p_a^\nu p_b^\mu - g^{\mu\nu}(p_a\cdot p_b)\right]\left(-g_{\mu\nu} + \frac{p_\mu p_\nu}{M_W^2}\right) \\
&= \frac{g^2|C|^2}{3}\left[p_a\cdot p_b + 2\frac{(p\cdot p_a)(p\cdot p_b)}{M_W^2}\right] \\
&\simeq \frac{g^2|C|^2}{3} M_W^2
\end{aligned}
\tag{7.18}
$$

evaluated in the W rest frame where

$$
p = (M_W, 0,0,0),\ p_a = (M_W/2, 0,0, M_W/2),\ \text{and}\ p_b = (M_W/2, 0,0, -M_W/2),
\tag{7.19}
$$

taking the z-axis along the direction of the decay. We also take m_a and m_b small compared to M_W. The partial decay width of the W boson is given by

$$
\begin{aligned}
\Gamma_{W\to f_a\bar{f}_b} &= \frac{1}{2M_W}\int \overline{|\mathcal{M}|^2}\frac{d^3p_a}{(2\pi)^3 2E_a}\frac{d^3p_b}{(2\pi)^3 2E_b}(2\pi)^4\delta^{(4)}(p-p_a-p_b) \\
&= \int \frac{\overline{|\mathcal{M}|^2}}{(2\pi)^2 8M_W E_a E_b}(4\pi p_a^2 dp_a)\,\delta(M_W - E_a - E_b) \\
&= \int \frac{\overline{|\mathcal{M}|^2}}{8\pi M_W E_b}(p_a dE_a)\,\delta(M_W - 2E_a) \\
&= \frac{\overline{|\mathcal{M}|^2} p_a}{8\pi M_W^2} = \frac{g^2|C|^2}{48\pi}M_W \approx 0.23\ \text{GeV}
\end{aligned}
\tag{7.20}
$$

where we have used $E_a = E_b$ in the W rest frame to complete the δ-function integration and substituted in equation (7.18) in the last line. The possibility of gluon radiation requires an additional correction for the hadronic W decays, and this increases the hadronic partial width by the QCD multiplicative factor of $(1 + \alpha_s(M_W^2)/\pi)$. To a good approximation, the W decays to left-handed doublets independent of flavor, and therefore

$$\Gamma_W \approx 3\Gamma_{\ell\bar{\nu}} + N_c(2\Gamma_{\ell\bar{\nu}}) = 9\Gamma_{\ell\bar{\nu}} \simeq 2.1 \text{ GeV} \qquad (N_c = 3) \tag{7.21}$$

corresponding to 11% branching fractions to each of the lepton families $(e\bar{\nu}_e)$, $(\mu\bar{\nu}_\mu)$ and $(\tau\bar{\nu}_\tau)$ and 33% branching fractions to each of $(d\bar{u})$ and $(s\bar{c})$. In detail, one must account for all hadronic terms according to the CKM matrix and the QCD multiplicative factor. This reduces the leptonic branching fraction $\Gamma_{\ell\bar{\nu}}/\Gamma_W$ to 10.8%.

The contributions to the differential decay distribution from the three polarization states can be obtained from equation (7.15). The polarization states are labeled as Left (L), Right (R), and Scalar (S) and are defined as follows:

$$\epsilon^\mu_{L,R} = \frac{1}{\sqrt{2}}(0, 1, \pm i, 0), \qquad \epsilon^\mu_S = \frac{1}{M_W}(|\mathbf{p}|, 0, 0, E). \tag{7.22}$$

Note that these polarization vectors satisfy $\epsilon^\lambda_\mu \cdot p^\mu = 0$, as required, and $\epsilon^\lambda_\mu \epsilon^{\mu\lambda} = -1$ for each polarization λ. Defining $p^\mu_a = M_W/2 \cdot (1, \sin\theta\cos\phi, \sin\theta\sin\phi, \cos\theta)$ and applying the trace theorems to equation (7.15), we obtain

$$\begin{aligned} |\mathcal{M}_{L,R}|^2 &= \frac{g^2|C|^2 M_W^2}{4}(1 \mp \cos\theta)^2, \\ |\mathcal{M}_S|^2 &= \frac{g^2|C|^2 M_W^2}{2}\sin^2\theta. \end{aligned} \tag{7.23}$$

where the $(1 \mp \cos\theta)^2$ behavior matches that of the single-photon propagator. However, the scalar or longitudinal polarization has a $\sin^2\theta$ angular dependence as is characteristic of a two-body decay of a spin-0 particle. The production and decay of a single W^+ boson through an up-type and anti-down-type fermion pair annihilation and fermion pair production, respectively, will produce fully polarized initial and final states and, therefore, have a contribution only from the $|\mathcal{M}_L|^2$ portion of the decay distribution in the high-energy limit. Single W boson production is not a resonant process at e^+e^- colliders. However, this process is a leading process for proton-antiproton collisions and the differential cross section is plotted in figure 7.4 for W bosons produced at the $Sp\bar{p}S$ collider at CERN and measured by the UA1 experiment showing a clear $(1 + \cos\theta^*_e)^2$ dependence for the final-state electron from $W \to e\nu$ decay in the rest frame of the W boson [2].

7.3 Resonance Production

The Z boson can be resonantly produced by colliding e^+e^- at a center-of-mass at or near the "pole" mass of the propagator, i.e., at $\sqrt{s} = M_Z$ corresponding to the pole in the denominator of the relativistic propagator $(s - M_Z^2)^{-1}$. The pole in the propagator assumes a zero-width state; however, the W and Z bosons are unstable corresponding to the lifetime computed from the inverse of the total decay width $\tau_Z \approx 2.5 \times 10^{-25}$ s. The result is that we can replace the pole mass M by $M - i\Gamma/2$ in the propagator.

The total width of the Z boson is computed summing over the contributions from the three neutrino families $(\nu_e, \nu_\mu, \nu_\tau)$, three charged lepton families (e, μ, τ), and five quark flavors (u, d, c, s, b). Each fermion-antifermion decay has a partial decay width of

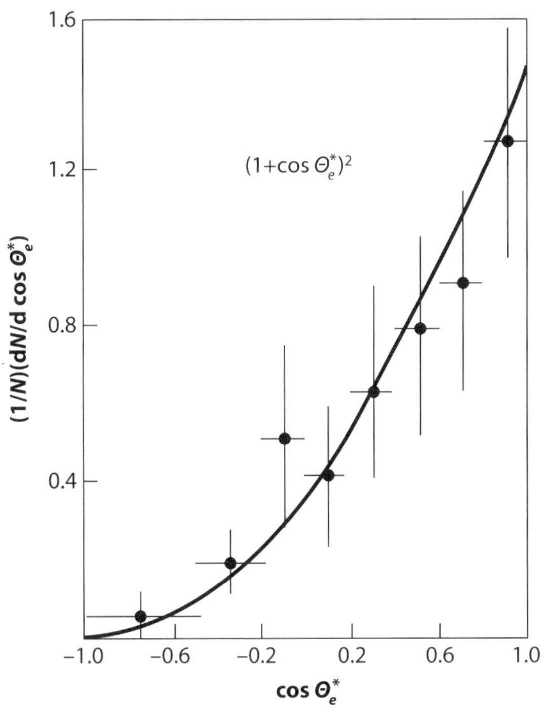

FIGURE 7.4. Differential angular distribution of electrons (positrons) from W boson decay relative to the direction of the proton (antiproton) momentum in the rest frame of the W boson. The data were collected by the UA1 experiment in proton-antiproton collisions at CERN (Credit: UA1) [2].

$$\Gamma_{Z\to f\bar{f}} \simeq \frac{N_c g^2 M_Z}{24\pi\cos^2\theta_W}\left(|C_L|^2 + |C_R|^2\right) \tag{7.24}$$

with N_c the number of colors and

$$\begin{aligned} C_L &= T^3 - Q\sin^2\theta_W, \\ C_R &= -Q\sin^2\theta_W. \end{aligned} \tag{7.25}$$

This gives

$$\begin{aligned} \Gamma_Z &\simeq 3\Gamma_{\ell^+\ell^-} + 3\Gamma_{\nu\bar{\nu}} + N_c(3\Gamma_{d\bar{d}}) + N_c(2\Gamma_{u\bar{u}}) \qquad (N_c = 3) \\ &= \frac{g^2 M_Z}{8\pi\cos^2\theta_W}\Big[(-\tfrac{1}{2} + \sin^2\theta_W)^2 + (\sin^2\theta_W)^2 + (\tfrac{1}{2})^2 \\ &\quad +3(-\tfrac{1}{2} + \tfrac{1}{3}\sin^2\theta_W)^2 + 3(\tfrac{1}{3}\sin^2\theta_W)^2 \\ &\quad +2(\tfrac{1}{2} - \tfrac{2}{3}\sin^2\theta_W)^2 + 2(-\tfrac{2}{3}\sin^2\theta_W)^2\Big] \\ &= \frac{g^2 M_Z}{8\pi\cos^2\theta_W}\left(\tfrac{7}{4} - \tfrac{10}{3}\sin^2\theta_W + \tfrac{40}{9}\sin^4\theta_W\right). \end{aligned} \tag{7.26}$$

Note that the hadronic partial width must be corrected by the QCD multiplicative factor of $(1 + \alpha_s(M_Z^2)/\pi)$ accounting for the possibility of gluon emission in the final state.Using

$\sin^2\theta_W \simeq 0.232$, the value of expression (7.26) is in good agreement with the measured width of the Z boson, $\Gamma_Z = 2.495 \pm 0.002$ GeV.

The estimation of the total cross section for Z boson production can be simplified near the resonant pole. We can factorize the matrix element into the form

$$\mathcal{M} \simeq \mathcal{M}_{12}\left(\frac{1}{s-(M_Z - i\Gamma_Z/2)^2}\right)\mathcal{M}_{34} \tag{7.27}$$

where $(1\bar{2})$ is the incoming electron and positron and $(3\bar{4})$ is the outgoing fermion-antifermion pair. The square of the factorized invariant amplitude is

$$|\mathcal{M}|^2 \simeq |\mathcal{M}_{12}|^2\left(\frac{1}{(s-M_Z^2)^2 + M_Z^2\Gamma_Z^2}\right)|\mathcal{M}_{34}|^2 \tag{7.28}$$

where a shift of $-\Gamma_Z^2/4$ in the pole mass squared M_Z^2 is neglected. The total cross section is given by

$$\begin{aligned}\sigma_{\text{tot}}(s) = \int &\frac{|\mathcal{M}_{12}|^2}{4(2s)[(s-M_Z^2)^2 + M_Z^2\Gamma_Z^2]} \\ &\times\left\{(2\pi)^4\delta^{(4)}(p_f - p_i)|\mathcal{M}_{34}|^2 \frac{d^3p_3 d^3p_4}{(2\pi)^6 2E_3 2E_4}\right\}\end{aligned} \tag{7.29}$$

where the fermion masses have been neglected, the factor of 4 is from the spin average over the (assumed) unpolarized beams, and the flux plus incident wave function normalization term is $4[(p_1 \cdot p_2)^2 - m_1^2 m_2^2]^{1/2} \simeq 2s$. The expression in curly brackets can be identified with $2M_Z\Gamma_{34}$, where Γ_{34} is the partial decay width to $(3\bar{4})$. The analogous calculation as in equation (7.20) shows that

$$\Gamma_{12} = \frac{|\mathcal{M}_{12}|^2}{(2J+1)16\pi M_Z} \tag{7.30}$$

where J is the total angular momentum of the resonance, i.e., $J = 1$ for the Z boson. Therefore, combining the above terms, the resonant cross section can be written in *Breit-Wigner* form

$$\sigma_{\text{tot}}(s) = \frac{4\pi(2J+1)\Gamma_{12}\Gamma_{34}}{(s-M_Z^2)^2 + M_Z^2\Gamma_Z^2}. \tag{7.31}$$

Figure 7.5 shows the measured lineshape of the Z boson.

Figure 7.6 shows a fit to the Z lineshape for the number of neutrino families. An additional light neutrino family would broaden the total width by $\Gamma_{\nu\bar{\nu}} \simeq 0.167$ GeV. Notice also that by setting $s = M_Z^2$ in equation (7.31), the peak cross section is

$$\sigma_{\text{tot}}^0(M_Z^2) = \frac{12\pi\Gamma_{e^+e^-}\Gamma_{\text{had}}}{M_Z^2\Gamma_Z^2}. \tag{7.32}$$

In the Z lineshape measurement, only the hadronic decays of the Z boson are counted (and only for e^+e^- production). Therefore, the peak cross section of the observed lineshape also changes with the number of neutrino families, decreasing for an increase in the number of neutrino families from the reduction in $\Gamma_{e^+e^-}/\Gamma_Z$ and $\Gamma_{\text{had}}/\Gamma_Z$.

7.3.1 Radiative Return

The cross section for e^+e^- annihilation has a pole at the Z mass giving resonant production. For center-of-mass energies at and above the Z peak, there are significant corrections to the total cross section from initial-state radiation from QED. The effect on the ideal Breit-Wigner of the Z peak is shown in figure 7.5. Far above the Z peak, the radiation of a single or multiple energetic photons from the electron or positron immediately before collision has the effect of bringing the effective center-of-mass energy $\sqrt{s'}$ down to the Z pole $\sqrt{s'} = M_Z$. This causes the observed Z lineshape to be asymmetric, increasing the total cross section above the Z peak. The Z bosons produced in this process are on-shell and recoil off the radiated photon in the lab frame.

This process is known as *radiative return* and has a particularly important role in the verification in the number of neutrino families. The technique for neutrino counting described in the previous section compares the visible width of the Z boson, decays to charged lepton pairs and hadrons, to that of the total width Γ_Z from the center-of-mass scan

$$\Gamma_Z = \Gamma_{\text{inv}} + \Gamma_{\text{vis}}. \tag{7.33}$$

The number of neutrinos can be determined from the invisible width of the Z boson using

$$\Gamma_{\text{inv}} = \Gamma_Z - \Gamma_{\text{had}} - \Gamma_{\text{lept}}, \qquad N_\nu = \frac{\Gamma_{\text{inv}}}{\Gamma_{\nu\nu}^{\text{SM}}} \tag{7.34}$$

where $\Gamma_{\nu\nu}^{\text{SM}}$ for a single light neutrino flavor is predicted from the Standard Model theory. The predictions for the visible Z lineshapes are shown in figure 7.6 for multiple numbers of light neutrino families ($N = 2, 3, 4$). In direct neutrino counting experiments, one can detect the photon radiated from the initial-state electron or positron as an indication that

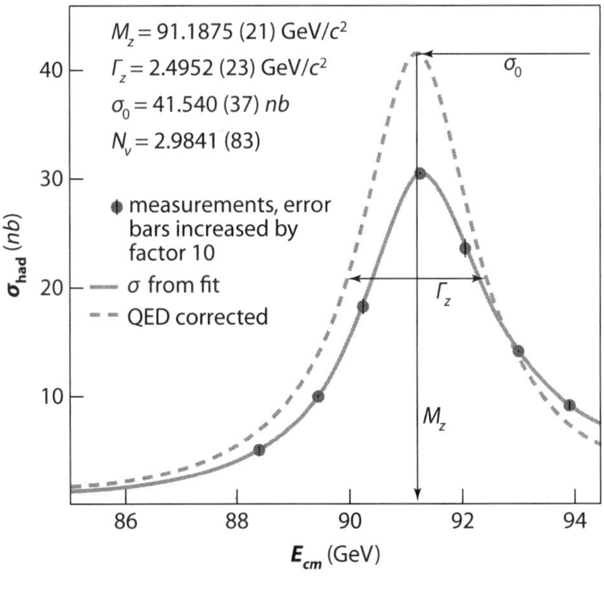

FIGURE 7.5. Measured lineshape of the Z boson decaying to hadrons showing the effect of QED radiative corrections to the Breit-Wigner form (Credit: LEP EWWG) [3].

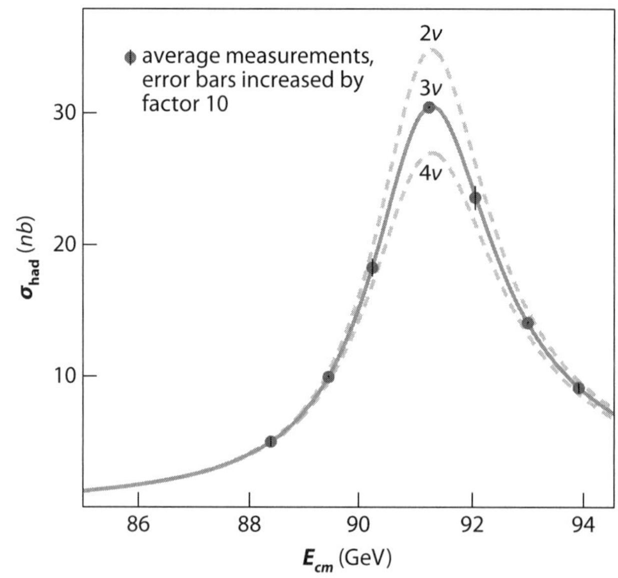

FIGURE 7.6. The measurements of the Z lineshape decaying into hadrons compared against the prediction for 2, 3 and 4 light neutrino families (Credit: LEP EWWG) [3].

a high-energy collision has occurred. One then counts the number of events which detect single photons and nothing else, indicating that the Z boson decayed to neutrinos. The invariant mass recoiling against a single or, in general, multiple photons is given by

$$M_{\text{recoil}} = \sqrt{(\sqrt{s} - \sum_i E_{\gamma_i})^2 - \left|\sum_i p_{\gamma_i}\right|^2} \tag{7.35}$$

where i is an index over the number of radiative photons measured in the detector. For a single photon, equation (7.35) reduces to

$$M_{\text{recoil}} = \sqrt{s - 2\sqrt{s}\,E_\gamma}\,. \tag{7.36}$$

The recoil mass for the LEP collider data taken far above the Z peak, $\sqrt{s} \geq 183$ GeV, peaks at the Z boson mass, as shown in figure 7.7. As the number of neutrino families increases, the predicted cross section of radiative-return events increases. The measured rate of these photons can therefore be used to directly measure the partial width to neutrinos $N_\nu \Gamma_{\nu\nu}^{\text{SM}}$. The recoil mass spectrum and corresponding cross-section measurements as a function of center-of-mass energy are shown in figures 7.7 and 7.8, respectively [4].

A general approach to incorporating radiative-return effects from a leading-order s-channel cross section σ_0 is to introduce a radiator function $H(x_\gamma, \theta_\gamma; s)$:

$$H = \frac{\alpha_{\text{QED}}}{2\pi x_\gamma}\left[2\,\frac{1 + (1 - x_\gamma)^2}{\sin^2\theta_\gamma + 4m_e^2/s} - x_\gamma^2\right] \tag{7.37}$$

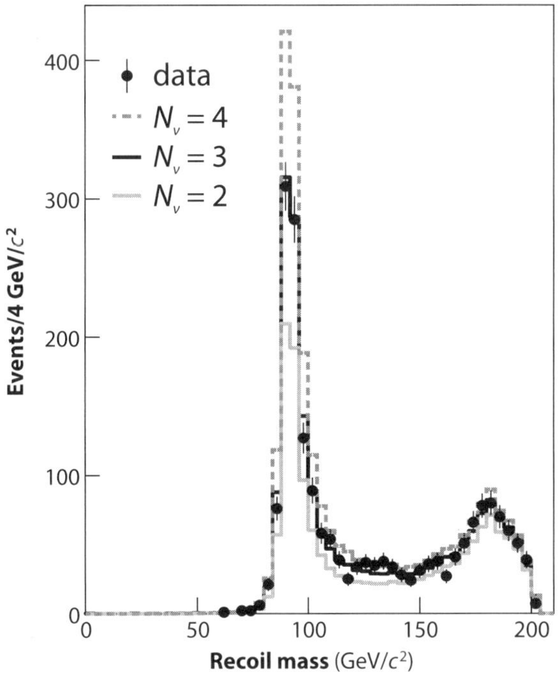

FIGURE 7.7. Recoil mass spectrum of radiative-return events measured by the L3 experiment at the LEP e^+e^- collider for $\sqrt{s} = 183$–209 GeV (Credit: L3) [4].

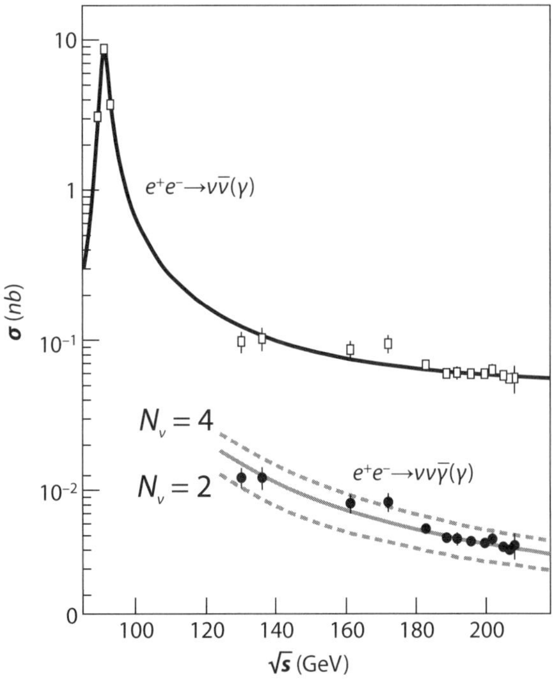

FIGURE 7.8. Measured cross section of radiative-return events by the L3 experiment at LEP compared to the rates of 2, 3, and 4 light neutrino families (Credit: L3) [4].

where $x_\gamma = E_\gamma / E_{\text{beam}}$ and m_e is the electron mass. The differential cross section of the reaction $e^+e^- \to \nu\bar{\nu}\gamma$ is then given by

$$\frac{d^2\sigma}{dx_\gamma d\cos\theta_\gamma} = \sigma_0([1 - x_\gamma] \cdot s) \cdot H(x_\gamma, \theta_\gamma; s). \tag{7.38}$$

7.4 Measurement of A_{LR}

The most direct avenue for probing the chirality of the neutral-current interaction is to collide longitudinally polarized beams at the center-of-mass of the Z resonance. In particular, electrons from polarized sources, such as electron emission from pulsed polarized laser light hitting the surface of a semiconductor, can be linearly accelerated and brought into collision with an unpolarized positron beam. This technique, employed by the SLC collider at SLAC, provided a unique sample of highly polarized Z bosons with enhanced sensitivity to $\sin^2\theta_W$ through the left-handed and right-handed chiral couplings (7.25) as they relate to the Z boson production cross section and the decay angular distributions.

For any given $e^+e^- \to Z$ interaction, the Z boson is produced with its total spin pointing either parallel or antiparallel to the beam axis. We can compute the dependence of the number of right-handed Z bosons N_R produced with a polarization pointing along the incoming e^- momentum direction in terms of the number of right-handed chirality e^- in the electron beam N_R^-, the number of right-handed chirality e^+ in the positron beam N_L^+ and the square of the right-handed coupling of the Z boson to the electron $(C_R^e)^2$ as given by equation (7.25):

$$N_R \propto N_R^- N_L^+ (C_R^e)^2 \propto (1 + P^-)(1 - P^+)(C_R^e)^2, \tag{7.39}$$

where $P^\pm$ is the polarization of the $e^\pm$ beam:

$$P^\pm = \frac{N_R^\pm - N_L^\pm}{N_R^\pm + N_L^\pm}. \tag{7.40}$$

Using a similar calculation for the number of left-handed Z bosons N_L, the net polarization of the ensemble of Z bosons produced from polarized beams is given by P_Z:

$$\begin{aligned} N_L &\propto N_L^- N_R^+ (C_L^e)^2 \propto (1 - P^-)(1 + P^+)(C_L^e)^2, \\ P_Z &= \frac{N_R - N_L}{N_R + N_L} = \frac{(1 + P^-)(1 - P^+)(C_R^e)^2 - (1 - P^-)(1 + P^+)(C_L^e)^2}{(1 + P^-)(1 - P^+)(C_R^e)^2 + (1 - P^-)(1 + P^+)(C_L^e)^2}. \end{aligned} \tag{7.41}$$

Recall that for unpolarized beams, $P^\pm = 0$, the Z boson has a nonzero polarization given by

$$P_Z = \frac{(C_R^e)^2 - (C_L^e)^2}{(C_R^e)^2 + (C_L^e)^2} = -\frac{2g_V^e g_A^e}{(g_V^e)^2 + (g_A^e)^2} = -A_e \approx -15\%. \tag{7.42}$$

The forward-backward asymmetry measured from beam-polarized Z bosons on the Z peak is therefore modified relative to equation (7.12) to the more general form

$$A_{FB} = -\frac{3}{4} P_Z A_f, \tag{7.43}$$

which for beam-polarized $|P_Z|$ values substantially greater than $|A_e|$ provides an enhancement in the sensitivity to $\sin^2\theta_W$. Similarly, if one computes the cross sections to produce left- and right-handed Z bosons, given by σ_L and σ_R, respectively, an asymmetry parameter

$$A_{LR} = \frac{\sigma_L - \sigma_R}{\sigma_L + \sigma_R} = A_e, \tag{7.44}$$

known as the *left-right asymmetry*, can be formed. The measurement of A_{LR} is therefore a direct determination of minus the Z boson polarization from unpolarized beams in equation (7.42). Ideally, the left- and right-handed cross sections are measured using $\mp 100\%$ polarized electron beams. In practice, cross-section measurements from any two well-determined beam polarization values are sufficient to solve for $(C_R^e)^2$ and $(C_L^e)^2$ independently using matrix methods. The advantage of the A_{LR} ratio is the cancellation of common uncertainties entering into the overall normalization of the cross-section measurements.

7.5 Polarized τ-Lepton Decays

The τ-lepton has several low-multiplicity decay modes owing to the unique placement of its mass, $m_\tau \approx 1.777$ GeV, relative to the leptons and low-mass hadrons. The lifetime of the τ-lepton is short, but finite, and can be measured in collider experiments giving a decay length of approximately 2 mm for τ leptons produced in Z decay. The τ-lepton flavor is conserved in QED and QCD and therefore decays through the weak charged-current interaction.The main decay modes are listed in table 5.1.

The single charged-track decay modes, categorized as "1-prong" decays, comprise 85.3% of all τ-lepton decays. As the charm quark is too heavy to contribute to τ-lepton decays, the strange-quark decay modes are coming primarily from the CKM matrix, i.e., the Cabbibo angle decomposition of (u, d'). The chirality properties of the weak charged-current decays allow the τ-lepton to act as a polarimeter in the angular and energy distributions of its decay products. The angular dependence of the decay $\tau^- \to \pi^- \nu_\tau$ has the following form for a given τ-lepton longitudinal polarization P_τ:

$$|\mathcal{M}|^2 \propto 1 + P_\tau \cos\theta_\pi^*, \tag{7.45}$$

where θ_π^* is the angle made by the π^- with respect to the reference axis direction for the spin polarization of the τ^- in the τ^- rest frame. The π^- energy spectrum in the lab frame is also sensitive to P_τ. Neglecting small masses and defining $x_\pi = 2E_\pi/\sqrt{s} \simeq (1 + \cos\theta_\pi)/2$, we get the following differential rate as a function of x_π:

$$\frac{dN}{dx_\pi} = 1 + P_\tau(2x_\pi - 1). \tag{7.46}$$

The corresponding distribution for the three-body leptonic decay $\tau^- \to \ell^- \bar{\nu}_\ell \nu_\tau$ is given by

$$\frac{dN}{dx_\ell} = \frac{1}{3}\left[5 - 9x_\ell^2 + 4x_\ell^3 + P_\tau\left(1 - 9x_\ell^2 + 8x_\ell^3\right)\right]. \tag{7.47}$$

Similar distributions can be computed for the other τ-lepton decay modes.

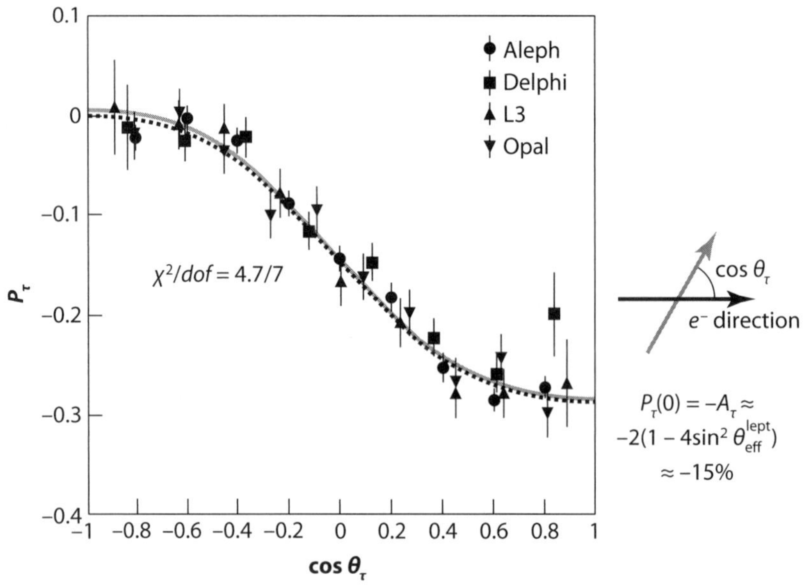

FIGURE 7.9. Angular dependence of τ-lepton polarization in Z boson decay (Credit: LEP EWWG) [3].

7.5.1 *Measurement of* $\sin^2\theta_W$ *from* τ *Decays*

The most precise single measurement of $\sin^2\theta_W$ comes from the measurement of the angular dependence of the τ polarization in Z-boson decays. The τ-lepton polarization is determined from fits of dN/dx as a function of polar angle θ_τ relative to the incoming electron beam direction. The intrinsic polarization imparted to the τ lepton comes from A_τ and originates from the preference of the Z boson to couple to left-handed fermions and right-handed antifermions. The value of A_τ is given by

$$A_\tau = \frac{2g_V^\tau g_A^\tau}{(g_V^\tau)^2 + (g_A^\tau)^2} = \frac{2(1-4\sin^2\theta_W)}{1+(1-4\sin^2\theta_W)^2} \approx 15\%. \tag{7.48}$$

The A_τ parameter is extracted from a fit to the dependence of P_τ as a function of $\cos\theta_\tau$ in Z-boson decays with $P_\tau(\cos\theta_\tau = 0) = -A_\tau$. The data are plotted in figure 7.9. The τ polarization P_τ has an angular dependence due to the initial polarization of the Z boson from the electron couplings. Notice how P_τ goes to zero at $\cos\theta_\tau = -1$ as expected since the Z couplings to the electron and τ are equal and in this configuration exactly cancel.

7.6 W^+W^- Pair Production

Perhaps one of the most telling results on the high-energy behavior of the electroweak interaction is the cross section for W^+W^- pair production. Theories with massive vector

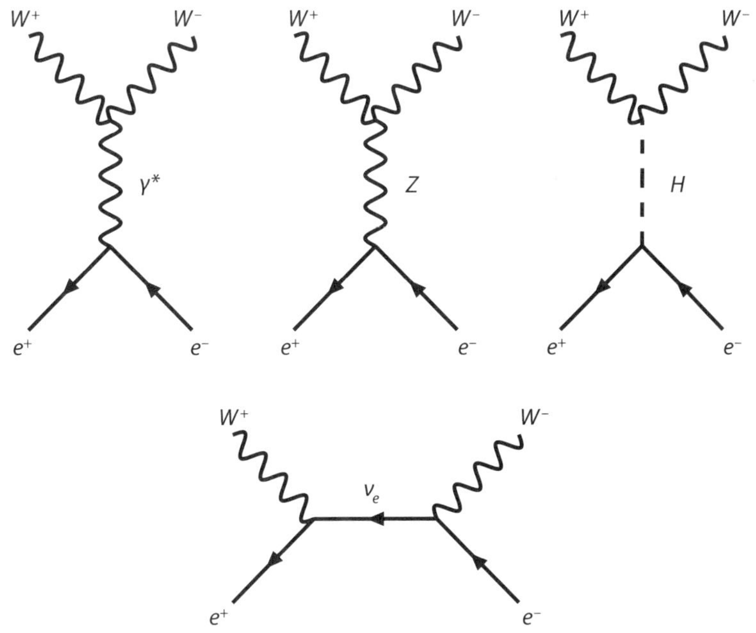

FIGURE 7.10. W^+W^- pair production diagrams for an e^+e^- collider.

bosons are subject to growing longitudinal polarizations at high energy. Therefore, the e^+e^- annihilation cross section to two longitudinally polarized W bosons is a test of the self-consistency of the Standard Model predictions.

There are four diagrams for W^+W^- pair production, as shown in figure 7.10. The Higgs s-channel diagram is suppressed by a factor m_e/M_W and thus completely negligible. At threshold, the orbital angular momentum of the final-state W^+W^- is zero and based on symmetry arguments, the total spin angular momentum must be $J=2$ for the two-boson system. The $J=2$ final state is accessible only via the t-channel neutrino exchange diagram and hence is the dominant production process at the W^+W^- threshold. The matrix element for this process is given by

$$-i\mathcal{M} = \left(\frac{ig}{\sqrt{2}}\right)^2 \bar{v}_L(p_+)\gamma^\mu \frac{i(\not{p}_- - \not{k}_-)}{(p_- - k_-)^2}\gamma^\nu u_L(p_-)\epsilon^*_\mu(k_+)\epsilon^*_\nu(k_-) \tag{7.49}$$

using the left-handed chiral projection operators

$$\frac{1}{2}(1-\gamma_5)u = u_L. \tag{7.50}$$

Hence, only left-handed electrons contribute to this process in the high-energy limit. There are two interesting limits of the t-channel neutrino process, one of which is the threshold production. The total differential cross section for unpolarized beams is given by

$$\frac{d\sigma}{d\Omega} = \frac{\beta}{64\pi^2 s} \sum_{s_+, s_-, \lambda_+, \lambda_-} \frac{1}{4} |\mathcal{M}|^2 \tag{7.51}$$

where $\beta = \sqrt{1 - M_W^2/E^2}$, the velocity of the W bosons. The threshold production cross section using the t-channel matrix element (7.49) for $\beta \ll 1$ is therefore

$$\sigma(\sqrt{s} \approx 2M_W) \approx \frac{\pi\alpha_{\text{QED}}^2}{s} \frac{\beta}{4\sin^2\theta_W} \approx 2\text{–}6 \text{ pb} \tag{7.52}$$

as shown in figure 7.1. The mass of the W boson can be precisely determined from the direct β-dependence of the threshold cross section.

The other interesting limit is the high-energy behavior of the t-channel matrix element (7.49), which can be approximated by substituting in an approximation for the longitudinal polarization of the W boson

$$\epsilon_L^\mu(k) = \left(\frac{k}{M_W}, 0, 0, \frac{E}{M_W}\right) \approx \frac{k^\mu}{M_W} + \mathcal{O}(M_W/E). \tag{7.53}$$

This gives

$$-i\mathcal{M}_{LL} \approx -i\frac{g^2}{2} \bar{v}_L(p_+) \frac{\not{k}_+}{M_W} \frac{(\not{p}_- - \not{k}_-)}{(p_- - k_-)^2} \frac{\not{k}_-}{M_W} u_L(p_-). \tag{7.54}$$

We can further simplify this expression using $\not{p}_- u_L(p_-) = 0$ from the Dirac equation in the high-energy limit,

$$(\not{p}_- - \not{k}_-)\not{k}_- u_L(p_-) = -(\not{p}_- - \not{k}_-)(\not{p}_- - \not{k}_-) u_L(p_-) = -(\not{p}_- - \not{k}_-)^2 u_L(p_-), \tag{7.55}$$

which, substituting into (7.54), gives

$$-i\mathcal{M}_{LL} = i\frac{g^2}{2} \bar{v}_L(p_+) \frac{\not{k}_+}{M_W^2} u_L(p_-) = i\frac{g^2}{4M_W^2} \bar{v}_L(p_+)(\not{k}_+ - \not{k}_-) u_L(p_-) \tag{7.56}$$

by symmetry. Using equation (7.51), the contribution from $\mathcal{M}_{LL}$ would diverge at high energy due to the growing longitudinal polarization of the W bosons. This divergent contribution is canceled, however, by the $SU(2)_L$ part of the s-channel Z boson diagram, as already indicated by the plateau in the W^+W^- production cross-section measurements in figure 7.1. This cancellation shows that the underlying local gauge invariance of $SU(2)_L$ preserved under spontaneous symmetry breaking is a stabilizing symmetry that prevents divergent predictions in the theory.

7.7 Exercises

1. **Tau leptonic decays to pseudoscalar mesons.** A charged pion $\pi^-(d\bar{u})$ is a spin-0 particle with negative intrinsic parity (pseudoscalar meson). The d- and anti-u-quarks can annihilate through the charged current interaction to produce a lepton and lepton antineutrino in the final state. As the incoming d- and anti-u-quarks originate from a bound state, the free-quark current $j_q^\mu = \bar{v}\gamma^\mu(1-\gamma^5)u$ is replaced by an effective

hadronic current j^μ_π constructed out of the only available four-vector describing the initial state, the pion four-momentum q^μ. The pion four-momentum is related by energy-momentum conservation to the final-state lepton p^μ_1 and antineutrino p^μ_2 four-momenta, $q^\mu = p^\mu_1 + p^\mu_2$. Therefore, we parameterize the pion hadronic current

$$j^\mu_\pi = if_\pi V_{ud} q^\mu = if_\pi V_{ud}(p^\mu_1 + p^\mu_2) \tag{7.57}$$

where $f_\pi \approx 90$ MeV is called the *pion decay constant* and V_{ud} is the CKM matrix element for a $d\bar{u}$ charged current interaction.

(a) Calculate the decay rate for $\tau^- \to \pi^- \nu_\tau$ in terms of the pion decay constant f_π. Assume the neutrino is massless. Use $V_{ud} \approx \cos\theta_C = 0.975$, $m_\tau = 1777$ MeV/c^2, $m_\pi =$ 139.6 MeV/c^2.

(b) The charged kaon $K^-(s\bar{u})$ is also a pseudoscalar meson. Approximate the kaon decay constant f_K to be equal to the pion decay constant $f_K \approx f_\pi$ (using $SU(3)$ flavor symmetry). Compute the ratio of the partial decay widths of $\tau^- \to \pi^- \nu_\tau$ to $\tau^- \to K^- \nu_\tau$. Use $V_{us} \approx \sin\theta_C = 0.221$, $m_K = 494$ MeV/c^2.

(c) Consider the τ lepton decays as being of the general form $\tau^- \to W^{*-}\nu_\tau$, where the virtual W boson (W^{*-}) can propagate to any kinematically allowed W-decay mode. Assuming all the fermion masses in the final state are negligible and that the quarks from W^* decay are free (not influenced by hadronic resonances), show that

$$Br(\tau^- \to e^- \bar{\nu}_e \nu_\tau) = Br(\tau^- \to \mu^- \bar{\nu}_\mu \nu_\tau) \simeq \frac{1}{5}. \tag{7.58}$$

2. **WW pair production at e^+e^- colliders.** An experiment at the LEP2 collider at CERN studies e^+e^- annihilation at a center-of-mass energy of 172 GeV. The experiment reports a cross section $\sigma(e^+e^- \to WW \to q_1\bar{q}_2q_3\bar{q}_4) = 5.1 \pm 1.0$ pb, corresponding to the observation of 59 events of this type found in their data.

(a) How much (integrated) luminosity did the accelerator deliver to the experiment?

(b) Events of the type $WW \to q\bar{q}'\mu\nu$ are identified with approximately the same degree of efficiency as events of the type $WW \to q_1\bar{q}_2q_3\bar{q}_4$. However, events of type $WW \to q\bar{q}'\tau\nu$ are more difficult to identify. What is the main reason for this?

(c) Estimate how many events of the type $WW \to q\bar{q}'\mu\nu$ are seen by the experiment.

(d) The experiment reports a number for $\sigma(e^+e^- \to WW)$. Estimate this number.

(e) How might the W mass be extracted from the $WW \to q_1\bar{q}_2q_3\bar{q}_4$ event sample?

(f) Draw the dominant tree-level diagram that contributes to Standard Model Higgs production at LEP2. Draw another diagram which will also contribute to Higgs production.

3. **Neutrino production at an e^+e^- collider.** Compute the matrix elements that contribute to neutrino production $\sum_i \mathcal{M}(e^+e^- \to \nu_i\bar{\nu}_i)$ for $(i = e, \mu, \tau)$ at an e^+e^- collider. (*Hint:*

There are two leading-order diagrams that produce a neutrino-antineutrino pair in the final state. No other particles are in the final state.)

(a) Write down the Feynman diagrams and compute $\sum_i \mathcal{M}(e^+e^- \to \nu_i\bar{\nu}_i)$.

(b) Compute $\overline{|\mathcal{M}|^2}$ for unpolarized electron beams and summing over final-state spins. Do not reduce traces beyond inserting completeness relations and/or partial widths.

(c) How does the sign of the interference term in part (b) change with center-of-mass energy?

4. **Searching for narrow resonances.** It's early November 1974. You have heard rumors that a narrow resonance of mass m_J decaying to e^+e^- has been produced in proton fixed-target experiments. Your detector is on an e^+e^- collider with enough center-of-mass energy to scan a wide region of masses (2–5 GeV) to find this resonance. However, the resonance is so narrow that it's impossible to operate within the finite width of the resonance without knowing the mass to better than a few MeV.

   (a) One of the graduate students comes to you and suggests taking all the data at the highest center-of-mass energy. What is he talking about? How can one find a narrow resonance by colliding e^+e^- at a center-of-mass energy above the mass of the resonance? Draw the relevant Feynman diagram from QED that would allow the resonance to be produced for $\sqrt{s} > m_J$. (*Hint:* The resonance has the same quantum numbers as a photon (it's a "heavy" photon).)

   (b) Write the matrix element for an e^+e^- annihilation that produces a final state that includes the massive resonance. Show that the resonance will be produced dominantly at one (nonzero) momentum p_J in the lab frame and compute the momentum p_J of the J particle in terms of the electron/positron beam energy E_b (where $\sqrt{s} = 2E_b$).

7.8 References and Further Reading

A selection of reference texts citing results in e^+e^- physics can be found here: [5, 6, 7, 8, 9].

A selection of Web sites of related experiments is found here: [1, 10, 11, 12, 13].

[1] L3 Experiment at LEP (1989–2000). http://l3www.cern.ch.

[2] UA1 Collaboration, *Phys. Lett.* B **166** (1986) 484–490.

[3] LEP Electroweak Working Group. http://lepewwg.web.cern.ch/LEPEWWG.

[4] L3 Collaboration, *Phys. Lett.* B **587** (2004) 16–32.

[5] Paul Langacker, *The Standard Model and Beyond.* CRC Press, 2010. ISBN 978-1-4200-7906-7.

[6] G. Altarello, T. Sjöstrand, and F. Zwirner. *Physics at LEP2*, vols. 1 and 2 (1996); CERN-96-01-V-1 http://cdsweb.cern.ch/record/300671/files/CERN-96-01_full_document.pdf; CERN-96-01-V-2 http://cdsweb.cern.ch/record/473529/files/CERN-96-01-V-2.pdf.

[7] R. K. Ellis, C. T. Hill, and J. D. Lykken, editors. *Perspectives in the Standard Model*. World Scientific Publishing, 1992. ISBN 981-02-1990-3.

[8] Peter Renton. *Electroweak Interactions*. Cambridge University Press, 1990. ISBN 0-521-36692-5.

[9] G. Altarelli, R. Kleiss and C. Verzegnassi. *Z Physics at LEP1*, vols. 1, 2, and 3. (1989); CERN-89-08-V-1 http://cdsweb.cern.ch/record/116932/files/CERN-89-08-V-1.pdf; CERN-89-08-V-2 http://cdsweb.cern.ch/record/367652/files/CERN-89-08-V-2.pdf; CERN-89-08-V-3 http://cdsweb.cern.ch/record/367653/files/CERN-89-08-V-3.pdf.

[10] ALEPH Experiment at LEP (1989–2000). http://aleph.web.cern.ch/aleph.

[11] DELPHI Experiment at LEP (1989–2000). http://delphiwww.cern.ch.

[12] OPAL Experiment at LEP (1989–2000). http://opal.web.cern.ch/Opal.

[13] SLD Experiment at the SLC (1992–1998). http://www-sld.slac.stanford.edu/sldwww/sld.html.

8 Hadron Colliders

The energy frontier in experimental particle physics is at hadron colliders. Superconducting magnets permit high-momentum protons (and antiprotons) to be accumulated and ramped to the highest energies with negligible energy loss due to synchrotron radiation. The Tevatron proton-antiproton collider at the Fermilab National Laboratory outside of Chicago, IL, USA, has operated for nearly a decade with 0.98 TeV beam energies. In 2010, the energy frontier increased to 3.5 TeV per beam at the Large Hadron Collider at the CERN laboratory outside of Geneva, Switzerland, designed to eventually operate at beam energies of up to 7 TeV. However, the beam energy in hadron colliders sets the range of collision energies, not a precise center-of-mass energy as is the case for s-channel interactions at e^+e^- colliders. Hadrons are composite particles and their collisions probe a wide range of length scales. In fact, the bulk of the total cross section in hadron-hadron collisions arises from soft multiple collisions, called *minimum bias nteractions.* The term "min-bias" means all events that pass the most minimal level of energy thresholds required to record a scattering event, the highest rate of which involve momentum transfers at the scale of Λ_{QCD}. Such processes are outside the scope of perturbative QCD, which requires the presence of at least one large-momentum scale. A small fraction of interactions, however, do involve hard collisions and have cross sections that can be computed as if the collisions were a direct result of quark and gluon interactions. The top quark is the highest mass known elementary particle and has been produced uniquely in hadron colliders. The uniqueness of the top quark has long been speculated to be related to the origin of the electroweak scale, and, hence, top quark physics is a rich area of study, as described below.

8.1 Drell-Yan and the Parton Model

The clearest connection between e^+e^- and hadron collider physics is with the "time-reversed" reaction $q\bar{q}\to\gamma^*\to\ell^+\ell^-$, known as the Drell-Yan process. The Drell-Yan process is computed for continuum $\ell^+\ell^-$ pair production ignoring heavy resonances. In proton-antiproton collisions, the quark and antiquark "beams" are obtained primarily from valence constituents of protons and antiprotons, respectively, or in the case of proton-proton collisions from valence quarks colliding with antiquarks from quantum vacuum $q\bar{q}$ fluctuations, known as "sea" antiquarks. The sea antiquarks typically have a lower probability than the valence quarks to carry a large fraction of the proton momentum.

The *parton model* treats high-energy lepton-hadron, neutrino-hadron, and hadron-hadron scattering in terms of a subprocess involving the direct interaction of free quarks and gluons through Standard Model interactions, neglecting the hadronic state. The distribution of parton momenta within a proton or antiproton is known as the *parton distribution function* (PDF). Thus, an important aspect of hadron collider physics is to understand the proper kinematic variables to describe the collision in the lab frame and to transform the proton-(anti)proton collision luminosity to a parton luminosity. Taking first the example of the Drell-Yan process, the lowest order diagram for hadron-hadron collisions (labeling the initial hadrons as h_1 and h_2) is given by $h_1h_2 \to \gamma^* + X \to \ell^+\ell^- + X$, where X denotes all the remaining particles produced. Thus, X can come from particles produced from the incident hadron inelastic scattering, also known as the *underlying event*. In the h_1h_2 center-of-mass, neglecting the hadron masses, the four-momenta are given by $p_1 = (p, 0, 0, p)$ and $p_2 = (p, 0, 0, -p)$ with $s = 4p^2$. Neglecting parton masses and transverse momenta (transverse momenta of partons within the hadron), the parton momenta can be written $k_1 = x_1p_1$ and $k_2 = x_2p_2$ for the quark and antiquark, respectively. Hence, the virtual photon (or $\ell^+\ell^-$ pair) has four-momentum

$$q_\mu = ((x_1 + x_2)p, 0, 0, (x_1 - x_2)p) \quad \text{with} \quad q^2 = 4x_1x_2p^2 = x_1x_2s. \tag{8.1}$$

The kinematics for the process can thus be fixed by measuring the squared lepton-pair mass $m_{\ell\ell}^2 = q^2$ and the longitudinal momentum q_z of the pair, as described by the Feynman-x variable, $x_F = 2q_z/\sqrt{s} = x_1 - x_2$. The cross section for the parton subprocess is given by

$$\frac{d\hat{\sigma}}{dm_{\ell\ell}^2}(q_i\bar{q}_i \to \ell^+\ell^-) = \frac{4\pi\alpha^2}{3m_{\ell\ell}^2} Q_i^2 \delta(x_1x_2s - m_{\ell\ell}^2). \tag{8.2}$$

Note that integrating both sides over $m_{\ell\ell}^2$ gives the familiar result for the process $e^+e^+ \to \mu^+\mu^-$,

$$\sigma(e^+e^- \to \mu^+\mu^-) = \frac{4\pi\alpha^2}{3s}, \tag{8.3}$$

with s replaced by $\hat{s} = x_1x_2s$. Similarly, in the dilepton rest frame the Drell-Yan leptons have a $1 + \cos^2\theta^*$ angular distribution, as expected for spin-1/2 particles coming from a vector propagator. In order to obtain the overall cross section for Drell-Yan dilepton production, the cross section contribution from (8.2) must be weighted by the appropriate fluxes of quarks and antiquarks, giving

$$\begin{aligned} &\frac{d\sigma}{dm_{\ell\ell}^2}(h_1h_2 \to \ell^+\ell^- + X) \\ &= \frac{1}{3}\sum_i \int dx_1dx_2[f_{q_i}(x_1)f_{\bar{q}_i}(x_2) + f_{\bar{q}_i}(x_1)f_{q_i}(x_2)]\frac{d\hat{\sigma}}{dm_{\ell\ell}^2} \\ &= \frac{4\pi\alpha^2}{9m_{\ell\ell}^2 s}\sum_i Q_i^2 \int dx_1dx_2[f_{q_i}(x_1)f_{\bar{q}_i}(x_2) + f_{\bar{q}_i}(x_1)f_{q_i}(x_2)]\delta(x_1x_2 - \tau) \end{aligned} \tag{8.4}$$

where $\tau = m_{\ell\ell}^2/s$, $f_q(x)$ and $f_{\bar{q}}(x)$ are the parton distribution functions, and the factor of $\frac{1}{3}$ arises because there are three possible colorless $q\bar{q}$ pairings out of nine possible combinations. Equation (8.4) predicts that $m_{\ell\ell}^4 d\sigma/dm_{\ell\ell}^2$ is a function of $\tau_{\ell\ell}$ only, i.e., it scales with $\tau_{\ell\ell}$. The scaling behavior predictions of the parton model were a triumph for early studies of Drell-Yan production. As the precision of experimental measurements improved, small deviations from scaling were observed in the data. Scaling violations were some of the first predictions from the theory of QCD. A more detailed description of the PDFs and of the running of α_s include QCD quantum corrections which depend logarithmically on q^2.

8.1.1 Kinematic Variables and Parton Distribution Functions

The initial kinematics are described by the incoming proton momentum p and partons i with longitudinal momentum fractions $x_i (0 \leq x_i \leq 1)$. The four-momenta and center-of-mass energy are approximated by

$$k_1 = (x_1 p, 0, 0, x_1 p), \quad k_2 = (x_2 p, 0, 0, -x_2 p), \quad \hat{s} = (k_1 + k_2)^2 = x_1 x_2 s = \tau s \tag{8.5}$$

and the *rapidity* y is given by

$$y = \frac{1}{2} \ln \left(\frac{E + p_z}{E - p_z} \right) = \frac{1}{2} \ln \frac{x_1}{x_2} \tag{8.6}$$

with $x_{1,2} = \sqrt{\tau} e^{\pm y}$. Rapidity differences Δy are invariant under longitudinal boosts. This can be shown as follows:

$$E \to E' = \gamma E + \gamma \beta p_z \quad \text{and} \quad p_z' = \gamma \beta E + \gamma p_z. \tag{8.7}$$

Therefore,

$$y' = \frac{1}{2} \ln \left(\frac{\gamma E + \gamma \beta p_z + \gamma \beta E + \gamma p_z}{\gamma E + \gamma \beta p_z - \gamma \beta E - \gamma p_z} \right) = y + \frac{1}{2} \ln \left(\frac{1 + \beta}{1 - \beta} \right). \tag{8.8}$$

Note, in the limit of massless particles

$$\begin{aligned} E \pm p_z &= E(1 \pm p_z/E) = E(1 \pm \beta \cos\theta), \\ y \to \eta &= \frac{1}{2} \ln \left(\frac{1 + \beta \cos\theta}{1 - \beta \cos\theta} \right) \overset{\beta \to 1}{=} -\ln \tan \frac{\theta}{2}, \end{aligned} \tag{8.9}$$

where η is called the *pseudorapidity*. The R-measure in terms of η and ϕ, is given by

$$R = \sqrt{(\eta_1 - \eta_2)^2 + (\phi_1 - \phi_2)^2} \tag{8.10}$$

where the 2π periodicity of ϕ must be explicitly removed by taking ϕ differences in the range $[-\pi, \pi]$. The R-measure is a commonly used angular measure for a wide range of quantities at hadron colliders. For example, a cone with a half-angle of $R = 0.5$ typically contains the core of a jet of particles, as discussed in the previous section on jets and jet algorithms.

A convenient variable transformation of the four-momentum of a particle of mass m is given by

$$\begin{aligned} p^\mu &= (E, p_x, p_y, p_z) \\ &= (m_\mathrm{T}\cosh y, p_\mathrm{T}\sin\phi, p_\mathrm{T}\cos\phi, m_\mathrm{T}\sinh y) \end{aligned} \tag{8.11}$$

where the transverse mass is defined as $m_\mathrm{T} = \sqrt{m^2 + p_\mathrm{T}^2}$ and $p_\mathrm{T}^2 = p_x^2 + p_y^2$. Notice that $m_\mathrm{T}^2 = E^2 - p_z^2$ by construction. Expression (8.11) comes from identifying rapidity y with the rotation angle of Lorentz transformations in the time-space plane (2.92). For computing transverse components of momenta, the relationship between pseudorapidity and polar angle is given by

$$\sin\theta = 1/\cosh\eta \tag{8.12}$$

where equation (8.12) follows the Lorentz transform (8.11) using

$$\tanh y = \frac{p_z}{E} = \beta\cos\theta \overset{\beta\to 1}{=} \cos\theta \tag{8.13}$$

in the massless limit $E = |\boldsymbol{p}|$. The corresponding differential volumes transform as

$$\frac{d^3p}{E} = p_\mathrm{T}dp_\mathrm{T}d\phi dy = \frac{1}{2}dp_\mathrm{T}^2 d\phi dy \equiv d^2p_\mathrm{T}dy \tag{8.14}$$

using

$$\frac{dy}{dp_z} = 1/E. \tag{8.15}$$

In expression (8.14) we define $d^2p_\mathrm{T} \equiv p_\mathrm{T}dp_\mathrm{T}d\phi$.

Parton distribution functions are fits to experimental data and theoretical extrapolations.

An online source for these databases is *http://durpdg.dur.ac.uk/hepdata/pdf3.html* maintained by the Durham high-energy group. An example set of PDFs is shown in figure 8.1 for the proton at a squared momentum scale $Q^2 = 4m_t^2$ corresponding to top quark pair production. The relationship between the parton momentum fraction x, Q^2, and rapidity y is shown in figure 8.2 for LHC beam energies. One can observe in figure 8.2 that for a fixed mass scale, such as Z boson production $Q^2 = M_Z^2$, that only a finite range of parton momenta fractions will produce Z bosons and that this range corresponds to a finite range of Z boson production rapidities.

The relationships (8.5) and (8.6) can be tested against the rapidity distributions of W and Z boson production at the Tevatron. Substituting in the maximum value for x, $x = 1$, the Z boson can be produced for rapidities satisfying

$$1 \geq \frac{m_Z}{1.96\ \mathrm{TeV}} e^y, \tag{8.16}$$

which gives $y < 3$, as shown in figure 8.3.

The W boson rapidity has a charge asymmetry due to the higher probability to find at least one of the two valence up-(anti)quarks at larger x-values in the (anti)proton compared to the sole valence d-(anti)quark. The Tevatron rapidity distributions for W and Z bosons are shown in figure 8.4. At the LHC, Z boson production will be peaked in both

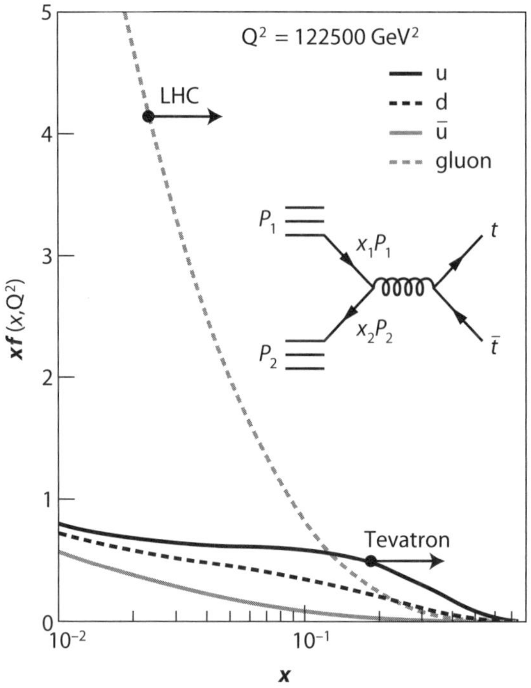

FIGURE 8.1. The PDF distribution for $Q^2 = 4m_t^2$ (Credit: HepData) [1].

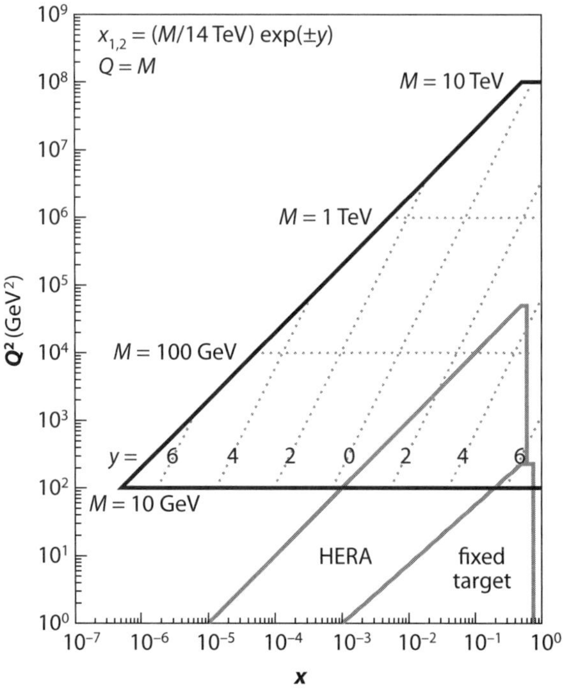

FIGURE 8.2. Regions of Q^2 and x to be probed at the Large Hadron Collider (LHC). PDF regions measured at HERA and in fixed-target experiments are indicated in the plot (Credit: MSTW) [2].

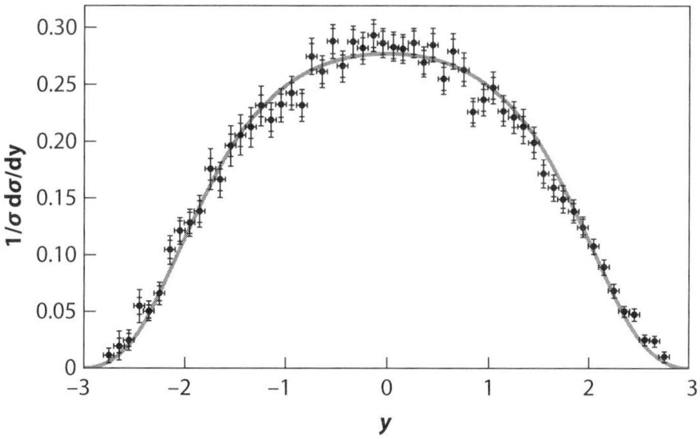

FIGURE 8.3. Rapidity of Z bosons produced at the Tevatron (Credt: D0) [3].

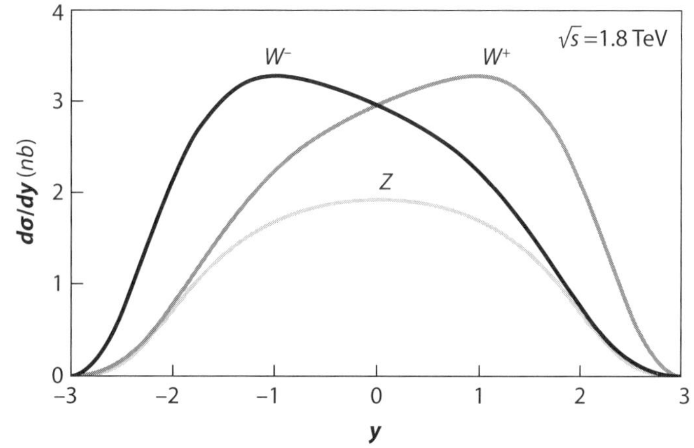

FIGURE 8.4. Charge asymmetry in the rapidity of W bosons produced at the Tevatron. The rapidity distribution of Z bosons is also shown.

the backward and forward rapidity regions, corresponding to which proton supplied the (low x) antiquark for the $q\bar{q}$ annihilation.

8.2 Single Boson Production

The cross section for single W boson production $h_1h_2 \to W^{\pm} + X$ and single Z boson production $h_1h_2 \to Z + X$ are resonant as they are in e^+e^- collisions. However, due to the parton luminosity distributions the single boson will have a distribution of longitudinal momenta p_z. Single boson production is therefore frequently described in terms of the differential rapidity distribution of the boson in the lab frame, as expressed through the Jacobian transformation of integration variables

$$dx_1 dx_2 = \frac{d\hat{s}dy}{s}. \tag{8.17}$$

Resonance production yields a δ-function in the phase space integral for the outgoing boson

$$\int \frac{d^3 p_W}{2E_W} \delta^{(4)}(p_W - p_1 - p_2) = \delta(\hat{s} - M_W^2) \tag{8.18}$$

that will be removed with the $d\hat{s}$ integration. The parton momentum fractions can then be expressed in terms of the rapidity of the outgoing boson

$$x_{1,2} = \frac{M_W}{\sqrt{s}} \exp(\pm y). \tag{8.19}$$

As an example, the distribution of $1/\sigma(d\sigma/dy)$ for single Z boson production at the Tevatron is shown in figure 8.3. One can further analyze single boson production by writing the differential cross section in terms of the transverse momentum p_T of the decay products in the rest frame of the boson. The approach taken here is to first take the known angular distribution of the leptons in the rest frame of the boson and then transform to p_T. For example, for a W boson decaying via $W^+ \to e^+\nu_e$, the angular distribution of the e^+ in the rest frame of the W boson relative to the incoming $\bar{d}$ from $u\bar{d} \to W^+ \to e^+\nu_e$ is given by

$$\frac{d\sigma}{d\cos\theta^*}(u\bar{d} \to W^+ \to e^+\nu_e) \propto (1 + \cos\theta^*)^2. \tag{8.20}$$

The differential cross section in terms of the single boson rapidity y in the lab frame can be written in terms of $\cos\theta^*$ computing the relationship between y and the e^+ rapidity y^* in the W rest frame, given by

$$y^* = \frac{1}{2}\ln\frac{x_1}{x_2} - y. \tag{8.21}$$

Neglecting the e^+ mass, we can relate the differential cross section in y^* to $\cos\theta^*$ with equation (8.13) and using

$$\frac{d\eta}{d\cos\theta^*} = \frac{1}{\sin^2\theta^*} \tag{8.22}$$

and, therefore,

$$\frac{d\sigma}{dy^*} = \sin^2\theta^* \frac{d\sigma}{d\cos\theta^*}. \tag{8.23}$$

Now that all the angular terms have been assembled, the Jacobian transformation to p_T is computed using

$$p_T^2 = \frac{\hat{s}}{4}\sin^2\theta^* \tag{8.24}$$

where the star notation is dropped from p_T. From equation (8.23), this gives

$$\frac{d\sigma}{dy^*} = \frac{4p_T^2}{\hat{s}} \frac{d\sigma}{d\cos\theta^*} \frac{d\cos\theta^*}{dp_T^2} = \frac{8p_T^2}{\hat{s}^2\sqrt{1-\frac{4p_T^2}{\hat{s}}}} \frac{d\sigma}{d\cos\theta^*} \tag{8.25}$$

and for the specific case of the W-boson decay from equation (8.20), we get

$$\frac{d\sigma}{dy^*}(u\bar{d} \to W^+ \to e^+\nu_e) \propto \frac{16p_T^2(1-\frac{2p_T^2}{\hat{s}})}{\hat{s}^2\sqrt{1-\frac{4p_T^2}{\hat{s}}}} \tag{8.26}$$

where terms odd in $\cos\theta^*$ have been dropped since the quantity p_T is forward-backward symmetric. The most important and general feature of two-body decays is the singularity in the Jacobian transformation to p_T shown in equation (8.26). This is known as the *Jacobian peak,* a name derived from the origin of the singularity. A similar singularity occurs for the differential cross section in terms of the W-boson transverse mass, defined by

$$m_T = \sqrt{(|\mathbf{p}_T^e| + |\mathbf{p}_T^\nu|)^2 - (\mathbf{p}_T^e + \mathbf{p}_T^\nu)^2} \tag{8.27}$$

where the singularity in the cross section

$$\frac{d\sigma}{dm_T} \propto (M_W^2 - m_T^2)^{-1/2} \tag{8.28}$$

occurs at $m_T = M_W$ corresponding to the upper edge of the possible range of m_T values $0 \le m_T \le M_W$. The shape of the m_T distribution for W bosons can been seen in figure 8.22, where measurement resolutions have smeared out the sharpness of the upper edge [4].

The cross sections for Drell-Yan and single boson production are subject to QCD correction factors, known as *K*-factors, that multiply the overall cross section. In the case of $p\bar{p} \to W^\pm + X$, the *K*-factor is approximately

$$K \simeq 1 + \frac{8\pi}{9}\alpha_s(M_W^2). \tag{8.29}$$

8.3 QCD Jet Production

In QCD, dijet events result when an incoming parton from one hadron scatters off an incoming parton from the other hadron to produce two high-p_T partons which are observed as jets. From momentum conservation, the two final-state partons are produced with equal and opposite momenta in the subprocess center-of-mass frame. In the lab frame, the two jets will be back-to-back in azimuthal angle and balanced in transverse momentum. Diagrams (and their implied space-time rotated diagrams) contributing to dijet production are given in figure 8.5. As an example, the scattering process $q(p_1)q'(p_2) \to q(p_3)q'(p_4)$ has the matrix element

$$\begin{aligned}\mathcal{M}(qq' \to qq') &= (-ig_s)^2 \frac{(\lambda^a)_{ij}}{2} \frac{(\lambda^b)_{kl}}{2} \bar{u}_3\gamma_\mu u_1 \left(\frac{-i\delta_{ab}g^{\mu\nu}}{\hat{t}}\right) \bar{u}_4\gamma_\nu u_2 \\ &= ig_s^2 \frac{(\lambda^a)_{ij}(\lambda^b)_{kl}}{4\hat{t}} \bar{u}_3\gamma_\mu u_1 \bar{u}_4\gamma^\mu u_2\end{aligned} \tag{8.30}$$

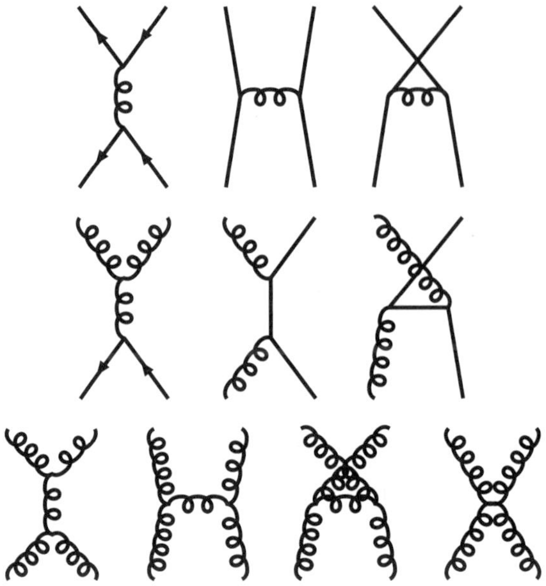

FIGURE 8.5. A subset of QCD 2 → 2 diagrams contributing to dijet production.

where $\{ij\}$ and $\{kl\}$ are incoming and outgoing color indices of the q and q' quarks, respectively. Each of the vertices contributes a color factor $\mathrm{tr}\{\lambda^a \lambda^a\}/4 = \frac{1}{2}$, and summing over the eight possible exchanged gluons a, and averaging over the nine initial color states, gives the overall color factor $\frac{2}{9}$. Thus, summing and averaging over both spins and colors gives

$$\overline{|\mathcal{M}|^2}\,(qq' \to qq') = \frac{4g_s^4}{9}\left(\frac{\hat{s}^2 + \hat{u}^2}{\hat{t}^2}\right) \tag{8.31}$$

where $\hat{s}$, $\hat{t}$, and $\hat{u}$ are the Mandelstam variables (2.268) evaluated in the parton-parton subprocess. In terms of the parton-parton center-of-mass system, we have for massless partons

$$\hat{t} = -\frac{\hat{s}}{2}(1 - \cos\theta^*),\ \hat{u} = -\frac{\hat{s}}{2}(1 + \cos\theta^*). \tag{8.32}$$

The invariant matrix elements squared corresponding to the diagrams in figure 8.5 are given in table 8.1 where the color and spin indices are averaged (summed) over initial (final) states. It is interesting to compare the relative contributions before parton luminosity weighting. This is done by evaluating the matrix elements at $\theta^* = \pi/2$, which corresponds to $\hat{t} = \hat{u} = -\hat{s}/2$, as listed in table 8.1. The $gg \to gg$ scattering contributions dominate, originating from the relatively larger color charge of the gluon as compared to quarks. As the gluons reside in a lower x region of the parton distribution functions, their contribution is strongest for low-p_T dijets.

The short-distance differential cross section $d\hat{\sigma}$ is given by the standard form

$$d\hat{\sigma} = \frac{1}{2\hat{s}} \frac{d^3p_3}{(2\pi)^3 2E_3} \frac{d^3p_4}{(2\pi)^3 2E_4} (2\pi)^4 \delta^{(4)}(p_1 + p_2 - p_3 - p_4) \sum \overline{|\mathcal{M}|^2}. \tag{8.33}$$

Using the phase space element from equation (8.14), the total differential cross section for dijet production can be written as

$$\frac{d^3\sigma}{dp_T^2 dy_3 dy_4}(h_1 h_2 \to jj) = \frac{1}{16\pi^2 s^2} \sum_{ijkl} \frac{f_i(x_1) f_j(x_2)}{x_1 x_2} \overline{|\mathcal{M}|^2}\,(ij \to kl) \tag{8.34}$$

where the sum is over the matrix elements in table 8.1 and the momentum fractions of the initial scattered partons are given in terms of the jet rapidities $y_{3,4}$ in the lab frame

$$\begin{aligned} x_1 &= \tfrac{1}{2} x_T(\exp(y_3) + \exp(y_4)), \\ x_2 &= \tfrac{1}{2} x_T(\exp(-y_3) + \exp(-y_4)), \end{aligned} \tag{8.35}$$

with

$$x_T = 2p_T/\sqrt{s}\,. \tag{8.36}$$

The parton-luminosity-weighted subprocess contributions to the jet E_T distribution and the measured differential cross section at the Tevatron are shown in figures 8.6 and 8.7. Events at large E_T correspond to scattering at the shortest distance scales. This is the most sensitive probe of quark substructure, which, if present, would result in an excess of high-E_T events.

Table 8.1 Comparison of QCD 2 → 2 squared matrix elements evaluated at $\theta^* = \pi/2$.

Subprocess	$\overline{\vert\mathcal{M}\vert^2}/g_s^4$	$\theta^* = \pi/2$
$q\bar{q} \to q'\bar{q}'$	$\frac{4}{9}\frac{\hat{t}^2+\hat{u}^2}{\hat{s}^2}$	0.22
$q\bar{q} \to gg$	$\frac{32}{27}\frac{\hat{t}^2+\hat{u}^2}{\hat{t}\hat{u}} - \frac{8}{3}\frac{\hat{t}^2+\hat{u}^2}{\hat{s}^2}$	1.04
$q\bar{q} \to q\bar{q}$	$\frac{4}{9}\left(\frac{\hat{s}^2+\hat{u}^2}{\hat{t}^2} + \frac{\hat{t}^2+\hat{u}^2}{\hat{s}^2}\right) - \frac{8}{27}\frac{\hat{u}^2}{\hat{s}\hat{t}}$	2.59
$q\bar{q}' \to q\bar{q}'$	$\frac{4}{9}\frac{\hat{s}^2+\hat{u}^2}{\hat{t}^2}$	2.22
$qq' \to qq'$	$\frac{4}{9}\frac{\hat{s}^2+\hat{u}^2}{\hat{t}^2}$	2.22
$qq \to qq$	$\frac{4}{9}\left(\frac{\hat{s}^2+\hat{u}^2}{\hat{t}^2} + \frac{\hat{s}^2+\hat{t}^2}{\hat{u}^2}\right) - \frac{8}{27}\frac{\hat{s}^2}{\hat{u}\hat{t}}$	3.26
$gq \to gq$	$-\frac{4}{9}\frac{\hat{s}^2+\hat{u}^2}{\hat{s}\hat{u}} + \frac{\hat{u}^2+\hat{s}^2}{\hat{t}^2}$	6.11
$gg \to q\bar{q}$	$\frac{1}{6}\frac{\hat{t}^2+\hat{u}^2}{\hat{t}\hat{u}} - \frac{3}{8}\frac{\hat{t}^2+\hat{u}^2}{\hat{s}^2}$	0.15
$gg \to gg$	$\frac{9}{2}\left(3 - \frac{\hat{t}\hat{u}}{\hat{s}^2} - \frac{\hat{s}\hat{u}}{\hat{t}^2} - \frac{\hat{s}\hat{t}}{\hat{u}^2}\right)$	30.4

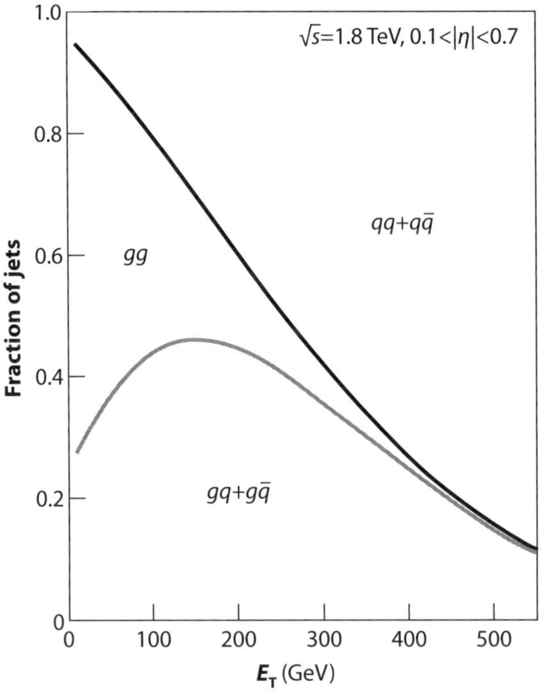

FIGURE 8.6. Contributions to the fraction of the jet E_T distribution from different initial-state parton combinations at the Tevatron $\sqrt{s}$ = 1.8 TeV.

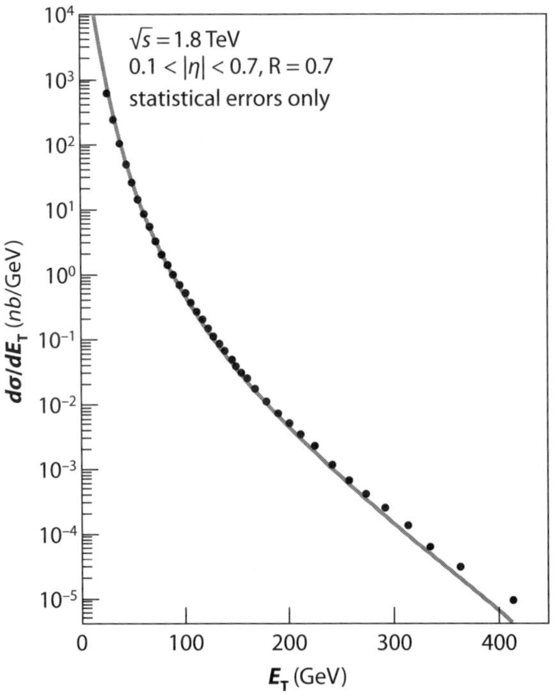

FIGURE 8.7. Measured jet E_T distribution from CDF $\sqrt{s}$ = 1.8 TeV with initial-state parton contributions predicted in figure 8.6 (Credit: CDF) [5].

We can also write the differential cross section for dijet production in terms of the invariant mass of the jet-jet system and the scattering angle θ^* in the rest frame of the dijet system. The laboratory rapidity y of the two-parton system and the equal and opposite rapidities $\pm y^*$ of the two jets in the parton-parton center-of-mass system are given in terms of the observed rapidities $y_{3,4}$ by

$$y = (y_3 + y_4)/2, \qquad y^* = (y_3 - y_4)/2. \tag{8.37}$$

With these variables and equation (8.36), the longitudinal momentum fractions are given by

$$x_{1,2} = x_{\mathrm{T}} \cosh y^* \exp(\pm y). \tag{8.38}$$

It follows from equations (8.38) that the invariant mass of the jet-jet system can be written as

$$m_{jj}^2 = \hat{s} = 4p_{\mathrm{T}}^2 \cosh^2 y^*. \tag{8.39}$$

Thus, using the relationships in equations (8.17), (8.37), and (8.39), the differential element in p_{T}^2 and lab rapidities can be transformed into

$$dp_{\mathrm{T}}^2 dy_3 dy_4 = \frac{1}{2} d\hat{s} dy \frac{dy^*}{\cosh^2 y^*} = \frac{1}{2} s dx_1 dx_2 d\cos\theta^*. \tag{8.40}$$

Therefore, we can now write the differential cross section for dijet production in terms of the squared dijet mass and jet center-of-mass angular distribution by integrating over $dx_1 dx_2$ against a δ-function in $(x_1 x_2 s - m_{jj}^2)$ to collect the parton luminosity contributions corresponding to the same dijet mass

$$\frac{d^2\sigma}{dm_{jj}^2 d\cos\theta^*}(h_1 h_2 \to jj) = \frac{1}{32\pi m_{jj}^2 s} \int dx_1 dx_2 \sum_{ijkl} f_i(x_1) f_j(x_2) \delta(x_1 x_2 - \tau) \overline{|\mathcal{M}|^2}(ij \to kl) \tag{8.41}$$

where $\tau = m_{jj}^2/s$.

8.3.1 Single Boson+Jet

The two leading-order processes of V+jet production with $V = \gamma, Z, W$ are analogous to pair annihilation and Compton scattering from QED. In the γ+jet pair annihilation process $q\bar{q} \to \gamma g$, the photon and gluon are the final-state particles. This diagram has the squared matrix element

$$\overline{|\mathcal{M}|^2}(q_i \bar{q}_i \to \gamma g) = g_s^2 e^2 Q_i^2 \left(\frac{8}{9}\right) \frac{\hat{t}^2 + \hat{u}^2}{\hat{t}\hat{u}}. \tag{8.42}$$

The Compton scattering process $gq \to \gamma q$ has the squared matrix element

$$\overline{|\mathcal{M}|^2}(g q_i \to \gamma q_i) = -g_s^2 e^2 Q_i^2 \left(\frac{1}{3}\right) \frac{\hat{s}^2 + \hat{u}^2}{\hat{s}\hat{u}}. \tag{8.43}$$

Similarly, the corresponding pair annihilation and Compton scattering processes for Z+jet with neutral-current couplings include terms in M_Z^2

$$\overline{|\mathcal{M}|^2}\,(q_i\bar{q}_i \to Zg) = \frac{g_s^2 g^2((g_V^{q_i})^2 + (g_A^{q_i})^2)}{8\cos^2\theta_W}\left(\frac{8}{9}\right)\frac{\hat{t}^2 + \hat{u}^2 + 2\hat{s}M_Z^2}{\hat{t}\hat{u}},$$
$$\overline{|\mathcal{M}|^2}\,(gq_i \to Zq_i) = \frac{-g_s^2 g^2((g_V^{q_i})^2 + (g_A^{q_i})^2)}{8\cos^2\theta_W}\left(\frac{1}{3}\right)\frac{\hat{s}^2 + \hat{u}^2 + 2\hat{t}M_Z^2}{\hat{s}\hat{u}}. \tag{8.44}$$

The γ+jet and Z+jet processes are direct probes of the parton-level kinematics of the substituted gluon diagram. The p_T-balanced two-body final state is also a powerful experimental tool for cross-calibrating the jet response against a well-calibrated electromagnetic calorimeter or muon spectrometer, as in the case of $Z \to \mu^+\mu^-$ decays.

The W+jet processes have the same form of squared matrix element as the Z+jet

$$\overline{|\mathcal{M}|^2}\,(q_i\bar{q}_j \to Wg) = \frac{g_s^2 g^2 |V_{q_i q_j}|^2}{16}\left(\frac{8}{9}\right)\frac{\hat{t}^2 + \hat{u}^2 + 2\hat{s}M_W^2}{\hat{t}\hat{u}},$$
$$\overline{|\mathcal{M}|^2}\,(gq_i \to Wq_j) = \frac{-g_s^2 g^2 |V_{q_i q_j}|^2}{16}\left(\frac{1}{3}\right)\frac{\hat{s}^2 + \hat{u}^2 + 2\hat{t}M_W^2}{\hat{s}\hat{u}}. \tag{8.45}$$

8.3.2 *W+Multijet*

For multijet final states, there are fewer tractable analytical predictions given the number of perturbative contributions that can contribute to the same final-state jet multiplicity. In the case of QCD jet production in association with W boson production, general scaling behavior is observed at the Tevatron as described by *Berends scaling*

$$\frac{\sigma(W + (N+1)_{\text{jets}})}{\sigma(W + N_{\text{jets}})} = \alpha \tag{8.46}$$

where α is a constant for a given set of jet E_T and η requirements, typically $\alpha \approx 0.2$. Equation (8.46) describes an exponential dependence of the cross section for W+jets on the inclusive jet multiplicity N_{jets}. The W+jets process is an important background to studying top quark production. General methods for predicting N_{jet} distributions involve *jet-parton jet matching* algorithms such as those used in the ALPGEN Monte Carlo generator and others listed in section 8.7.

8.4 Top Quark Physics

The fundamental theory of the strong interaction was developed at a time when the heaviest known quark was the strange quark with $m_s \sim 300$ MeV comparable to Λ_{QCD}. At that time, the behavior of the $SU(3)_C$ gauge interaction was extrapolated to high-energy and predicted to be perturbative in this limit. The top quark discovered in 1995 at the Tevatron $p\bar{p}$ collider has a mass that is three orders of magnitude higher than Λ_{QCD}. While not originally foreseen by the authors of QCD, top quark pair production is far into the perturbative regime and is one of the most accurately predicted QCD cross sections (~10% accuracy). The top quark decay proceeds promptly via the weak charged-current interaction and consequently enables the top quark mass to be the most precisely measured (~1% accuracy) of all quark masses. The measurements of top quark pair production, electroweak single top quark production, and the top quark mass and properties are described in the sections that follow.

8.4.1 Pair Production

Diagrams for top quark pair production are given in figure 8.8. The invariant matrix elements squared which result from the diagrams in figure 8.8 are given by

$$\begin{aligned}\sum \overline{|\mathcal{M}|^2}\,(q\bar{q} \to Q\bar{Q}) &= \frac{4g_s^4}{9}\left(\frac{\hat{t}^2+\hat{u}^2}{\hat{s}^2} + \frac{2m^2}{\hat{s}}\right),\\ \sum \overline{|\mathcal{M}|^2}\,(gg \to Q\bar{Q}) &= g_s^4\left(\frac{\hat{s}^2}{6\hat{t}\hat{u}} - \frac{3}{8}\right)\left(\frac{\hat{t}^2+\hat{u}^2}{\hat{s}^2} + \frac{4m^2}{\hat{s}} - \frac{4m^4}{\hat{t}\hat{u}}\right),\end{aligned} \tag{8.47}$$

where relative to table 8.1 the top mass terms have been included. The top quark pair production cross section versus $\sqrt{s}$ is plotted in figure 8.9 for $p\bar{p}$ and pp colliders. The difference in the cross section near threshold originates from the PDFs plotted in figure 8.1. The equality of the $p\bar{p}$ and pp cross sections at large $\sqrt{s}$ comes partly from the dominance of the parton luminosity feeding the $gg \to t\bar{t}$ process at low x.

Given the short lifetimes of the top quark and W boson, the top pair-production cross section is commonly treated as a six-particle decay amplitude including the top quark and W boson propagators in the matrix element for the process $t\bar{t} \to W^+bW^-\bar{b} \to (w_1w_2)b\,(w_1'w_2')\bar{b}$. Neglecting spin correlations, the leading-order matrix element for $q\bar{q}$ s-channel top quark pair production is given by

$$|\mathcal{M}|^2\,(q\bar{q} \to t\bar{t} \to W^+bW^-\bar{b} \to (w_1w_2)b\,(w_1'w_2')\bar{b}) = \frac{g_s^4}{9}F_{w_1w_2b}F_{w_1'w_2'\bar{b}}(2-\beta^2\sin^2\theta^*), \tag{8.48}$$

where g_s is the strong coupling constant, β is the velocity of the top quarks in the $t\bar{t}$ rest frame, and θ^* denotes the angle between the incoming parton and the outgoing top quark in the $t\bar{t}$ rest frame. The expression for $F_{w_1w_2b}$ is given by

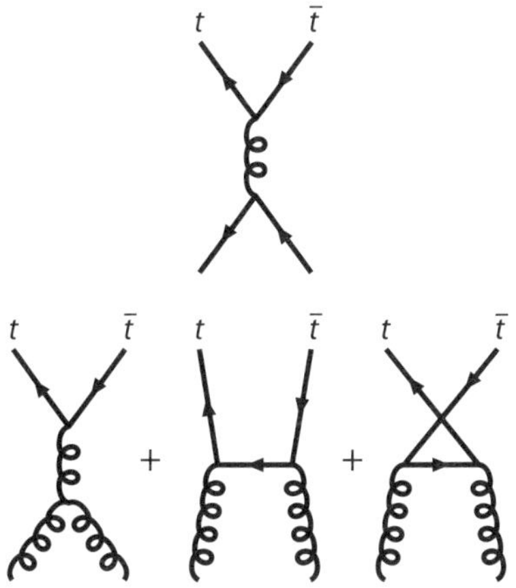

FIGURE 8.8. Feynman diagrams for top quark pair production at a hadron collider.

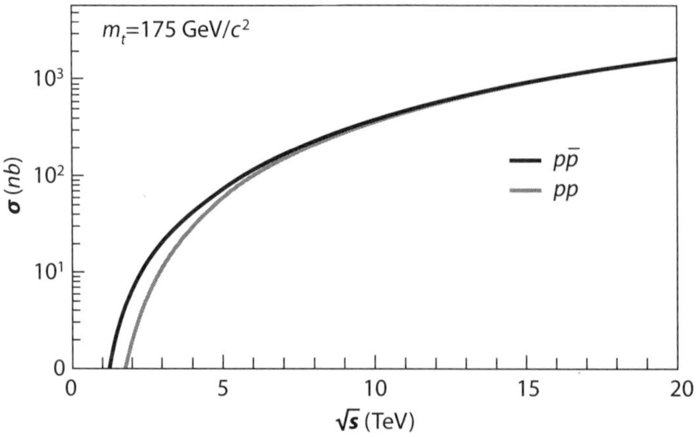

FIGURE 8.9. Comparison of cross sections for top quark pair production versus $\sqrt{s}$ at proton-antiproton and proton-proton colliders for $m_t = 175$ GeV/c^2. The enhancement in the cross section in proton-antiproton colliders near threshold originates from the PDFs plotted in figure 8.1 and thus enabled the Tevatron to discover the top quark at $\sqrt{s} = 1.8$ TeV.

$$F_{w_1 w_2 b} = \frac{g^4}{4}\left(\frac{m^2_{w_1 w_2 b} - m^2_{w_1 w_2}}{\left(m^2_{w_1 w_2 b} - m_t^2\right)^2 + \left(m_t \Gamma_t\right)^2}\right)\left(\frac{m^2_{w_1 w_2 b}\left(1 - \cos^2\hat{\theta}_{bw_1}\right) + m^2_{w_1 w_2}\left(1 + \cos\hat{\theta}_{bw_1}\right)^2}{\left(m^2_{w_1 w_2} - M_W^2\right)^2 + \left(M_W \Gamma_W\right)^2}\right), \quad (8.49)$$

for $F_{w_1' w_2' \bar{b}}$ replace $(w_1 w_2 b) \leftrightarrow (w_1' w_2' \bar{b})$, where $\hat{\theta}_{bw_1}$ is the angle between the b-quark and the charged lepton of the leptonically decaying W^+ in the W boson rest frame. In the case of a hadronically decaying W boson, one cannot easily distinguish the quark and the antiquark jet from W decay and, therefore, it is common to symmetrize $F_{w_1 w_2 b}$ in terms of w_1 and w_2. The result of this symmetrization is to replace the term $(1 + \cos\hat{\theta}_{bw_1})^2$ in equation (8.49) by $1 + \cos^2\hat{\theta}_{bw_1}$ where in the W rest frame $\cos^2\theta_{bw_1} = \cos^2\theta_{bw_2}$.

The leading-order matrix element for $gg \to t\bar{t}$ has several contributing diagrams, giving

$$\begin{aligned} &|\mathcal{M}|^2 \left(gg \to t\bar{t} \to W^+ b W^- \bar{b} \to (w_1 w_2) b (w_1' w_2') \bar{b}\right) \\ &= g_s^4 F_{w_1 w_2 b} F_{w_1' w_2' \bar{b}} \left(\frac{1}{6\tau_1 \tau_2} - \frac{3}{8}\right)\left(\tau_1^2 + \tau_2^2 + \rho - \frac{\rho^2}{4\tau_1 \tau_2}\right), \end{aligned} \quad (8.50)$$

with

$$\tau_i = \frac{m^2_{g_i w_1 w_2 b} - m^2_{w_1 w_2 b}}{m^2_{w_1 w_2 b w_1' w_2' \bar{b}}} \quad \text{and} \quad \rho = \frac{4m_t^2}{m^2_{w_1 w_2 b w_1' w_2' \bar{b}}}, \quad (8.51)$$

where g_i $(i = 1, 2)$ denotes the two incoming gluons. At the Tevatron, a $p\bar{p}$ collider with $\sqrt{s} = 1.96$ TeV, the fractions of top quark pair production from the $q\bar{q}$ and gg processes are, respectively,

$$\frac{\sigma^{\text{TeV}}_{q\bar{q} \to t\bar{t}}}{\sigma^{\text{TeV}}_{\text{total}}(t\bar{t})} \approx 85\% \quad \text{and} \quad \frac{\sigma^{\text{TeV}}_{gg \to t\bar{t}}}{\sigma^{\text{TeV}}_{\text{total}}(t\bar{t})} \approx 15\%, \quad (8.52)$$

with the total cross section at the Tevatron measured to be [6]

$$\sigma^{\text{TeV}}_{\text{total}}(t\bar{t}) = 7.2 \pm 0.7 \text{ pb} \tag{8.53}$$

while at the LHC, a pp-collider with $\sqrt{s} = 14$ TeV, the expected fractions and total cross section are given by

$$\frac{\sigma^{\text{LHC}}_{q\bar{q}\to t\bar{t}}}{\sigma^{\text{LHC}}_{\text{total}}(t\bar{t})} \approx 10\% \quad \text{and} \quad \frac{\sigma^{\text{LHC}}_{gg\to t\bar{t}}}{\sigma^{\text{LHC}}_{\text{total}}(t\bar{t})} \approx 90\%, \tag{8.54}$$

with

$$\sigma^{\text{LHC}}_{\text{total}}(t\bar{t}) \approx 900 \text{ pb}. \tag{8.55}$$

8.4.2 Electroweak Single Top Production

The heavy mass of the top quark introduces a substantial threshold for pair production via the strong interaction. The phase space trade-off of producing a single massive top quark is enough for electroweak production to be nonnegligible. Another factor reducing electroweak production is the need to go through a color-neutral propagator. At the Tevatron, the weak charged-current process for single top production, shown in figure 8.10,

$$u\bar{d} \to t\bar{b} \tag{8.56}$$

is approximately an order of magnitude smaller in cross section than strong pair production. The squared matrix element for subprocess (8.56) is given by

$$\overline{|\mathcal{M}|^2}\,(u\bar{d} \to t\bar{b}) = \frac{g^4 N_c^2}{4} |V^*_{ud} V_{tb}|^2 \frac{d \cdot (t \pm m_t \lambda)\,(2u \cdot b)}{(2u \cdot d - M_W^2)^2 + (M_W \Gamma_W)^2} \tag{8.57}$$

where λ is the spin vector of the top quark and the four-momenta of the quarks are represented by their letter: t, b, u, d. The electroweak couplings will produce a top quark that is 100% polarized along the incoming three-momentum of the d-antiquark. In the rest frame of the top quark, the spin of the top quark is in the same direction as the spatial part

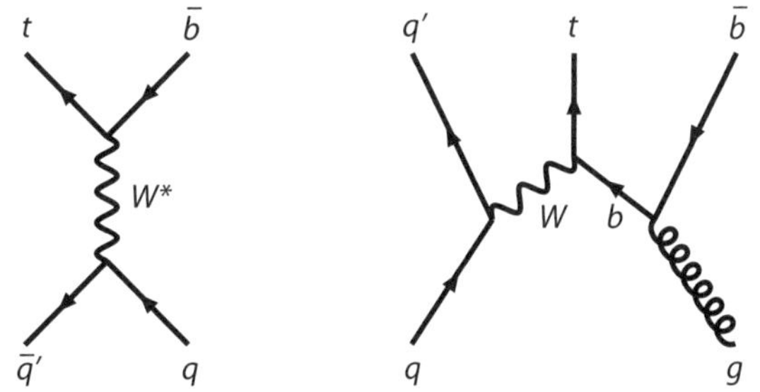

FIGURE 8.10. Single top quark production processes at a hadron collider showing s-channel on the left and W-gluon fusion on the right. At high-energy proton-proton colliders, the $W\,t$ process in figure 8.11 surpasses the s-channel contribution.

of $\lambda^{\mu} = (0, \boldsymbol{\lambda})$. Therefore, if the direction of the spatial part of λ is chosen to be along the d-antiquark three-momentum as seen in the rest frame of the top quark,

$$\boldsymbol{\lambda} = \boldsymbol{p}_d^* / |\boldsymbol{p}_d^*|, \tag{8.58}$$

then equation (8.57) will be zero for right-handed top quarks, corresponding to the plus (+) sign in the expression $t + m_t \lambda$. Note that only about 2% of events come from antiquarks originating in the proton. This indicates that in electroweak single top production, the top quarks are nearly 100% polarized and the direction of the spins are in the antiproton direction.

The W-gluon t-channel process, shown in figure 8.10, gives approximately twice the cross-section contribution of the s-channel electroweak production process at the Tevatron. The measured cross section for single top production for the sum of $s + t$ channels has been recently measured at the Tevatron in $p\bar{p}$ collisions at $\sqrt{s} = 1.96$ TeV [6]:

$$\sigma_{s+t}^{\text{TeV}}(\text{single top}) = 2.8 \pm 0.5 \text{ pb}. \tag{8.59}$$

At the LHC, the single top production $W\,t$ process becomes more important and surpasses the s-channel contribution, shown in figure 8.11.

8.4.3 Decay Properties

The top quark decays dominantly via the weak charged-current interaction to a b-quark and a W boson. The top quark decay matrix element is

$$\mathcal{M}(t \to bW^+) = \frac{gV_{tb}^*}{2\sqrt{2}} \epsilon_\mu^* \bar{u}_b \gamma^\mu (1 - \gamma_5) u_t. \tag{8.60}$$

Summing over all spins and dividing by 2 for an unpolarized top quark gives

$$\begin{aligned} \sum \overline{|\mathcal{M}|^2} = \frac{g^2 |V_{tb}|^2}{16} & \left[\sum_{\lambda=1,2,3} \epsilon_\mu(\lambda) \epsilon_\nu(\lambda)^* \right] \\ & \times \text{tr}\{(\not{p}_b + m_b)\gamma^\mu(1-\gamma_5)(\not{p}_t + m_t)\gamma^\nu(1-\gamma_5)\}. \end{aligned} \tag{8.61}$$

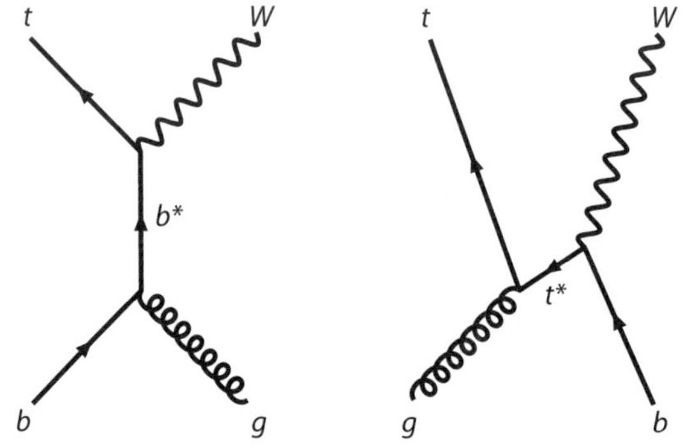

FIGURE 8.11. Flavor-excitation processes contribute significantly to single top quark production at the LHC.

If we ignore m_b, keep only terms with an even number of γ-matrices, and commute $1-\gamma_5$ through to the right, this gives

$$\sum \overline{|\mathcal{M}|^2} = \frac{g^2|V_{tb}|^2}{16}\left[\sum_{\lambda} \epsilon_\mu(\lambda)\epsilon_\nu(\lambda)^*\right] \mathrm{tr}\{\not{p}_b \gamma^\mu \not{p}_t \gamma^\nu (1-\gamma_5)\}. \tag{8.62}$$

The completeness relation for the W boson helicity states is

$$\sum_{\lambda=1,2,3} \epsilon_\mu(\lambda)\epsilon_\nu(\lambda)^* = -g_{\mu\nu} + \frac{p_\mu^W p_\nu^W}{M_W^2} \tag{8.63}$$

and given that this is a symmetric tensor, the γ_5 term in (8.62), which involves an asymmetric tensor, will give zero. Using the trace theorem for four γ-matrices reduces equation (8.61) to

$$\begin{aligned}\sum \overline{|\mathcal{M}|^2} &= \frac{g^2|V_{tb}|^2}{2}\left[(p_b \cdot p_t) + 2\frac{(p_W \cdot p_b)(p_W \cdot p_t)}{M_W^2}\right] \\ &\approx \frac{g^2|V_{tb}|^2}{2}(p_b \cdot p_t)\left[1 + \frac{2(p_W \cdot p_t)}{M_W^2}\right].\end{aligned} \tag{8.64}$$

In the top quark rest frame, we have, ignoring m_b,

$$\begin{aligned} p_b \cdot p_t &= m_t E_b = \frac{m_t^2 - M_W^2}{2}, \\ p_t \cdot p_W &= m_t E_W = \frac{m_t^2 + M_W^2}{2}, \end{aligned} \tag{8.65}$$

and, therefore,

$$\sum \overline{|\mathcal{M}|^2} = \frac{g^2}{4}(m_t^2 - M_W^2)\frac{(2M_W^2 + m_t^2)}{M_W^2}. \tag{8.66}$$

The total decay width of the top quark including the phase space factor predicts for $g^2/(4\pi) \approx 1/30$, $m_t = 172.0 \pm 1.6$ GeV, $|V_{tb}|^2 \approx 1$, and $M_W = 80.4$ GeV:

$$\Gamma_t = \frac{g^2|V_{tb}|^2}{4\pi}\frac{(m_t^2 - M_W^2)^2(m_t^2 + 2M_W^2)}{16M_W^2 m_t^3} \approx 1.4 \text{ GeV}, \tag{8.67}$$

consistent with current upper limits on the top quark width from the Tevatron. If we consider the separate contributions from the W helicities, we can see that two of three W helicities are allowed in top decay, as shown in figure 8.12. The left-handed and longitudinal contributions to the lepton angular distribution, from leptonic W decay, in the rest frame of the W boson produced in top quark decay are plotted in figure 8.13. The fraction of longitudinally polarized W bosons agrees with the estimate of $\approx$70% from equation (3.214).

8.4.4 Mass Measurement

Ultimately, everything we can predict about an elementary particle interaction comes from the Lagrangian of the theory and, subsequently, is part of the matrix element, phase

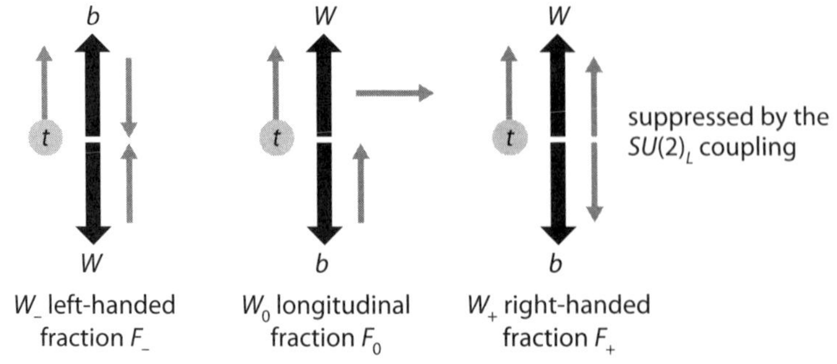

FIGURE 8.12. Helicity configurations of W bosons in top quark decay.

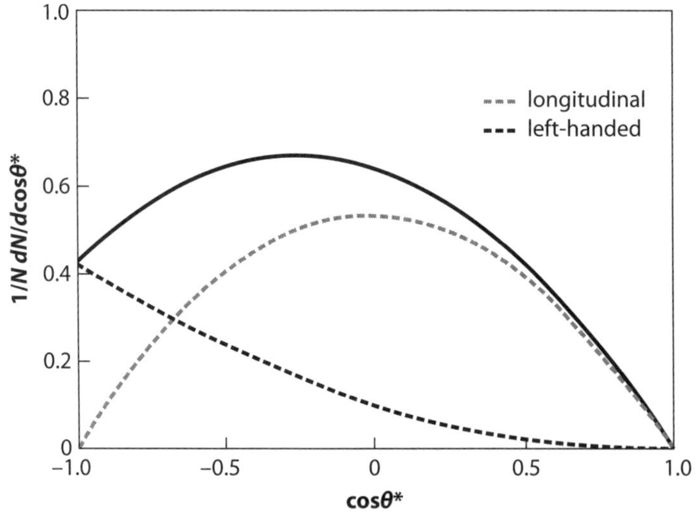

FIGURE 8.13. Lepton angular distribution from leptonically decaying W bosons produced in top quark decay. In the rest frame of the W boson, leptons from left-handed W bosons peak at $\cos\theta^* = -1$.

space, and parton luminosity integration that go into the calculation of the cross section of that process. The uncertainty in mapping measurements to theory comes primarily from those properties not directly observed, such as neutrino momenta and incoming parton types and momentum fractions, and those measurements that have large experimental resolutions or ambiguities, such as jet measurements, jet energy scale corrections, and jet-to-parton assignments. One of the most precise techniques for measuring the top mass makes direct use of the matrix element and differential cross section calculation to extract the top mass from the data. This technique is known as the *matrix element method.*

The method begins with computing the signal probability for the pair production of $t\bar{t}$ and the subsequent decay to a particular final state, which in this example is given by $W^+bW^-\bar{b} \to (\ell^+\nu_\ell)b(\bar{q}q')\bar{b}$ and other possible decays resulting in the topology with a lepton and missing transverse energy and four-jets. The probability density for a given partonic final state y, such as $\ell^+\nu_\ell b\bar{u}d\bar{b}$, to be produced in the hard scattering process is proportional to the differential cross section $d\sigma$ of the corresponding process, given by

$$d\sigma(q\bar{q} \to t\bar{t} \to y;\, m_t) = \frac{(2\pi)^4 |\mathcal{M}(q\bar{q} \to t\bar{t} \to y)|^2}{x_1 x_2 s}\, d\Phi_6. \tag{8.68}$$

Here, the symbol $\mathcal{M}$ denotes the matrix element for the process $q\bar{q} \to t\bar{t} \to (\ell^+\nu)b\,(\bar{q}q')\bar{b}$, s is the center-of-mass energy squared, x_1 and x_2 are the momentum fractions of the colliding partons (which are assumed to be massless) within the colliding proton and antiproton, and $d\Phi_6$ is an element of six-body phase space. To obtain the differential cross section $d\sigma(p\bar{p} \to t\bar{t} \to y;\, m_t)$ in $p\bar{p}$ collisions, the differential cross section from equation (8.68) is convoluted with the parton density functions for all possible flavor compositions of the colliding quark and antiquark,

$$d\sigma(p\bar{p} \to t\bar{t} \to y;\, m_t) = \int_{x_1,x_2} \sum_{\text{flavors}-i,j} dx_1 dx_2 f_i(x_1) f_j(x_2)\, d\sigma(q\bar{q} \to t\bar{t} \to y;\, m_t), \tag{8.69}$$

where $f_i(x)$ denotes the probability density to find a parton of given flavor i and momentum fraction x in the proton or antiproton.

The finite detector resolution is taken into account via a convolution with a *transfer function* $W(x, y; \text{JES})$ that describes the probability to reconstruct a partonic final state y as x in the detector where the relationship between y and x is parameterized by the jet energy scale (JES). The light-quark jet energy scale is determined intrinsically from the hadronic decay of the W boson from top quark decay. The b-jet energy scale is extrapolated. The unmeasured variables such as the neutrino longitudinal momentum are integrated out. Here, the transverse momentum of the $t\bar{t}$ system is approximated as zero. The differential cross section to observe a given reconstructed event x then becomes

$$d\sigma(p\bar{p} \to t\bar{t} \to x;\, m_t, \text{JES}) = \int_y d\sigma(p\bar{p} \to t\bar{t} \to y;\, m_t)\, W(x, y;\, \text{JES}). \tag{8.70}$$

In accounting for the acceptance in the integration, some obvious effects need to be taken into account. For instance, only events x that pass the *trigger conditions,* are inside the detector acceptance, and are reconstructed in the right topology are used in the measurement. The corresponding overall detector efficiency depends both on m_t and on the jet energy scale JES. This is taken into account in the cross section of $t\bar{t}$ events observed in the detector

$$\sigma_{\text{obs}}(p\bar{p} \to t\bar{t};\, m_t, \text{JES}) = \int_{x_1,x_2,x,y} d\sigma(p\bar{p} \to t\bar{t} \to y;\, m_t)\, W(x, y;\, \text{JES}) f_{\text{acc}}(x), \tag{8.71}$$

where $f_{\text{acc}} = 1$ for selected events and $f_{\text{acc}} = 0$, otherwise.

The differential probability to observe a $t\bar{t}$ event as x in the detector is then given by

$$P_{\text{sig}}(x;\, m_t, \text{JES}) = \frac{d\sigma(p\bar{p} \to t\bar{t} \to x;\, m_t, \text{JES})}{\sigma_{\text{obs}}(p\bar{p} \to t\bar{t};\, m_t, \text{JES})}, \tag{8.72}$$

where a summation is included over all possible jet-to-parton assignments.

In order to extract the top quark mass from a set of n measured events $x_1, \ldots, x_n$, a *likelihood function* is built from the individual event probabilities calculated according to equation (8.74) as

$$L(x_1, \ldots, x_n; m_t, \mathrm{JES}, f_{\mathrm{sig}}) = \prod_{i=1}^{n} P_{\mathrm{evt}}(x_i; m_t, \mathrm{JES}, f_{\mathrm{sig}}) \tag{8.73}$$

where the event probability P_{evt} can include a sum over signal and background probabilities that are summed according to a signal fraction f_{sig}. The probability P_{evt} for an event composed from probabilities for two processes, $t\bar{t}$ production and W+jets background events, is given by

$$P_{\mathrm{evt}}(x; m_t, \mathrm{JES}, f_{\mathrm{sig}}) = f_{\mathrm{sig}} \cdot P_{\mathrm{sig}}(x; m_t, \mathrm{JES}) + (1 - f_{\mathrm{sig}}) \cdot P_{\mathrm{bkg}}(x; \mathrm{JES}). \tag{8.74}$$

For every assumed pair of values (m_t, JES), the value f_{sig} that maximizes the likelihood is determined. The top quark mass and jet energy scale are then obtained by minimizing

$$-\ln L(x_1, \ldots, x_n; m_t, \mathrm{JES}, f_{\mathrm{sig}}(m_t, \mathrm{JES})) = -\sum_{i=1}^{n} \ln(P_{\mathrm{evt}}(x_i; m_t, \mathrm{JES}, f_{\mathrm{sig}}(m_t, \mathrm{JES}))) \tag{8.75}$$

with respect to m_t and JES, simultaneously. The Tevatron measurements achieve a precision of $\sim$1% on the total uncertainty on m_t with a current world average top quark mass of $m_t = 172.0 \pm 1.6$ GeV [6].

8.5 Trigger Rates and Thresholds

The Large Hadron Collider (LHC) will begin taking data with an initial target luminosity of $\mathcal{L} = 2 \times 10^{33}\,\mathrm{cm}^{-2}\mathrm{s}^{-1}$ for 14 TeV proton-proton collisions. As shown in figure 8.14, approximately 400 W and 100 Z bosons and 2 $t\bar{t}$ pairs will be produced per second. The rate for heavy-flavor production is in excess of 10^6 Hz. To set the scale of LHC operation, in the span of 15 minutes more b-quarks will have been produced than in all years of running at the e^+e^- collider B-factories. In approximately 6 hours, more top quarks will have been produced than from a decade of Tevatron operation. Within 1 day, more Z bosons will have been produced than from the entire LEP program. The rate of high-mass processes at the LHC is truly unprecedented.

Hadron collider experiments handle the large spectrum of rates with trigger systems. For instance, low-p_T dijet production is the dominant process at hadron colliders. At the same time, high-p_T dijets probe the smallest length scales. Measuring the QCD dijet cross section over a range of 10 orders of magnitude, as shown in figure 8.15, is an important test of pointlike structure and can be used to extract a running value of $\alpha_s(Q^2)$. The dijet spectrum as with similar types of steeply falling spectra is recorded with separate triggers, each of which has a different jet-p_T threshold and prescale factors (only one of every N events recorded on average).

The dijet cross section measurement is a piecewise assembly of several separate triggers. An example set of single-jet triggers is given in figure 8.16 with corresponding rate

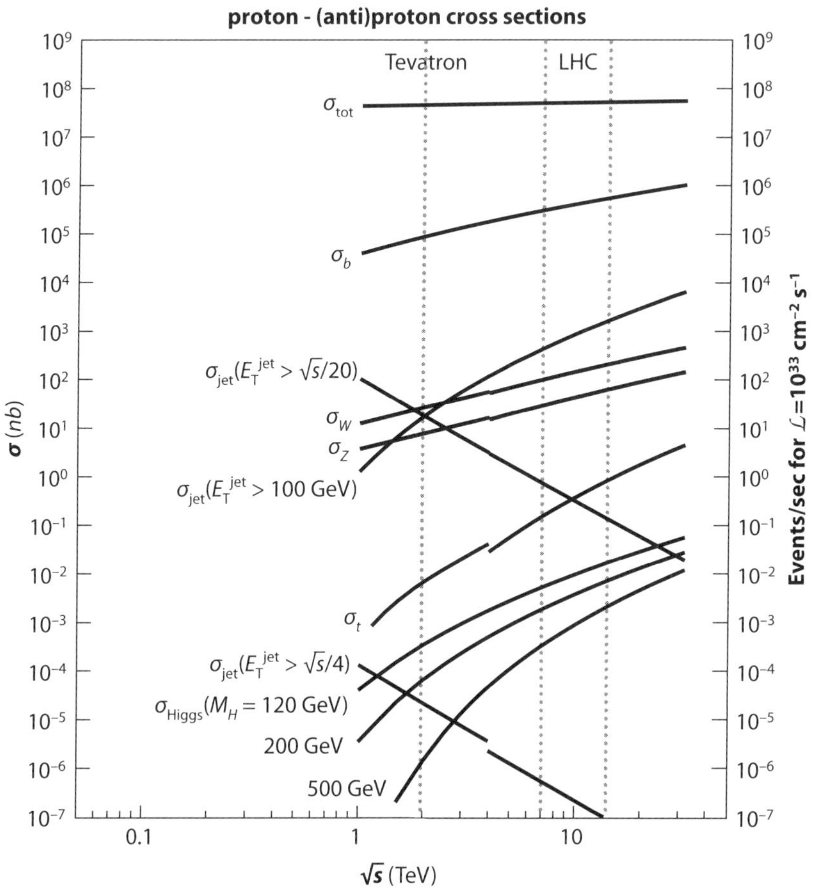

FIGURE 8.14. Cross sections and event rates at the Tevatron and LHC (Credit: W.J. Stirling).

distributions in figure 8.17. Only fully efficient regions of the trigger are kept and the lowest threshold trigger is dropped from the analysis. The trigger efficiency will transition from having zero efficiency for a low-p_T object to becoming fully efficient for a high-p_T object, given by the *trigger turn-on* curve as shown in figure 8.18. The turn-on curve for a trigger is a function of the jet resolution and detector uniformity. For this example, the dijet mass spectrum is covered down to 500 GeV with prescale factors, and only 2 TeV dijet resonances and higher are in unprescaled trigger regions. Figure 8.19 is the assembled cross-section measurement from the single-jet triggers.

Typically all datasets at hadron colliders are impacted by the trigger or triggers used to select them. The trigger path defines a narrow sieve through which a tremendous downpour of multijet QCD events can pass. Improbable conditions, such as a large missing transverse energy, are particularly sensitive to *instrumental backgrounds*. These appear as anomalous measurements and can be due to finite detector resolution, coverage, or false signals. Table 8.2 shows example estimates for the hardware and software online trigger thresholds, respectively, typical for an LHC general-purpose experiment.

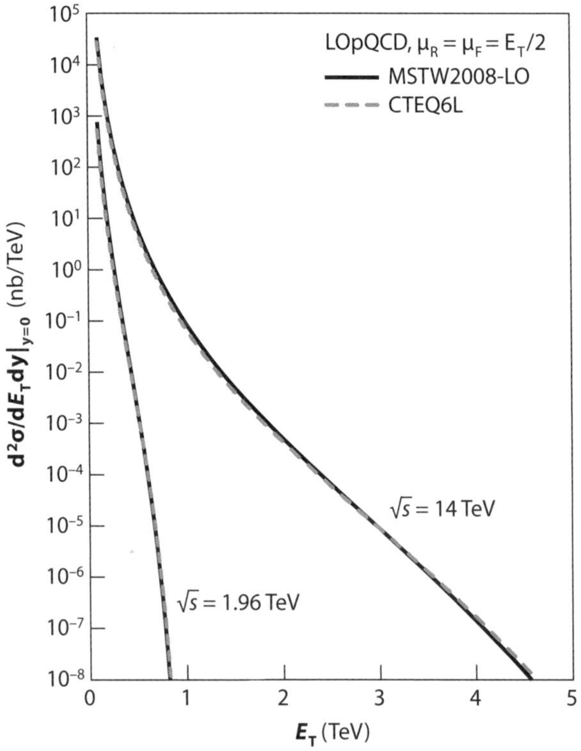

FIGURE 8.15. The expected dijet differential cross section for central rapidities at the LHC as compared with the Tevatron (Credit: W.J. Stirling).

Jet trigger path	**Hardware-level jet trigger** $P_T(E_T)$ (GeV)	prescale (1/*N*)	rate (KHz)	**Software-level jet trigger** $P_T(E_T)$ (GeV)	prescale (1/*N*)	rate (Hz)	**Jet total prescale**
High	177	1	~1	657	1	0.31	1
Med.	177	1	~1	350	30	0.33	30
Low	177	1	~1	180	600	0.45	600
Tiny	90	20	~1	90	600	0.48	12,000

FIGURE 8.16. Single-jet trigger prescale factors estimated for the hardware-level and software-level triggers (Credit: CMS) [7].

The total acceptance *rate to tape* for an LHC experiment is typically, but not constrainted to, 150–350 Hz (event sizes are ~1 Mbyte). At target luminosities, the rate of leptonic W decays alone would saturate the rate to tape. The rate of dileptons from Z-boson decay is ~10 Hz. If the lepton triggers only selected purely electroweak processes, then the trigger system could easily manage the rate. However, the rate of dijet $b\bar{b}$ production where at least one *B* hadron semileptonically decays is more than 10^5 Hz, exceeding the hardware trigger limitations. Dijet $b\bar{b}$ production at a hadron collider is a lepton factory. The semileptonic *B*-hadron decays swamp the trigger with low-p_T leptons. Therefore, the way to control the single and dilepton triggers is to require the leptons to be isolated from jets of hadrons, as

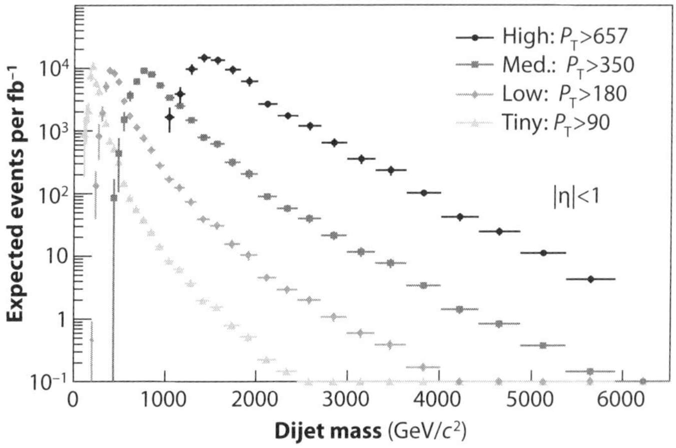

FIGURE 8.17. Dijet events from four single-jet triggers (Credit: CMS) [7].

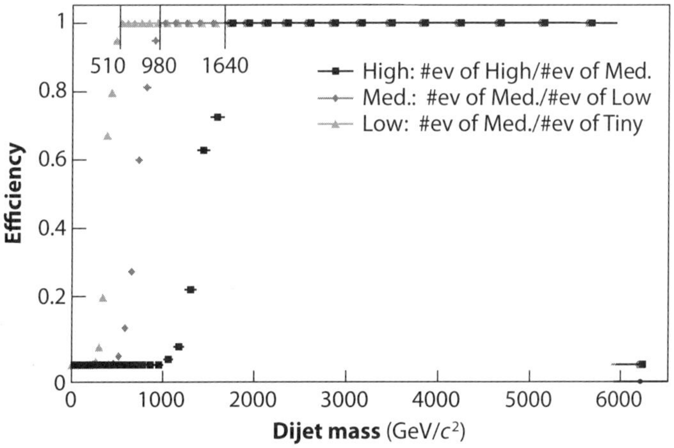

FIGURE 8.18. Single-jet trigger turn-on curves in dijet mass (Credit: CMS) [7].

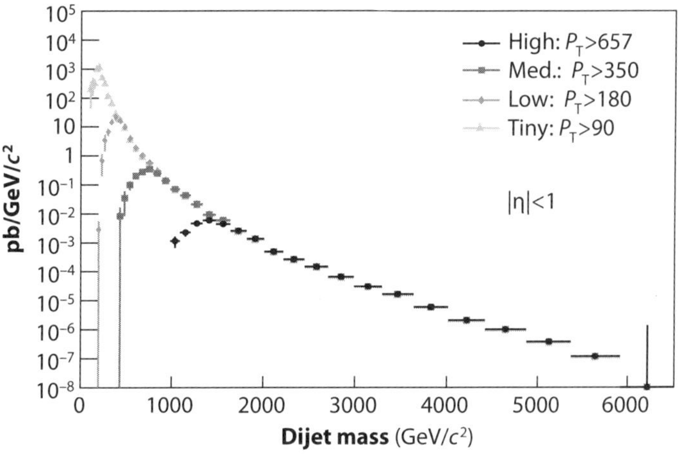

FIGURE 8.19. Dijet cross section assembled from multiple single-jet triggers (Credit: CMS) [7].

Table 8.2 Example estimates of primary LHC trigger paths. The actual trigger menu depends on the instantaneous luminosity and the balance of bandwidth allocated to different physics processes and searches. Some trigger paths use a common hardware-level trigger decision and then sharpen the requirements at the software-level trigger.

	Hardware-level trigger			Software-level trigger		
Trigger path	Threshold (GeV or GeV/c)	Rate (kHz)	Cumulative rate (kHz)	Threshold (GeV or GeV/c)	Rate (Hz)	Cumulative rate (Hz)
Inclusive electron	29	3.3	3.3	29	33	33
Inclusive photon	(29)	(3.3)	3.3	80	4	37
Dielectrons	17	1.3	4.3	17	1	38
Diphoton	(17)	(1.3)	4.3	40,25	5	43
Inclusive muon	14	2.7	7.0	19	25	68
Dimuon	3	0.9	7.9	7	4	72
Inclusive τ-jet	86	2.2	10.1	86	3	75
Di-τ-jets	59	1.0	10.9	59	1	76
1-jet OR 3-jets OR 4-jets	177,86,70	3.0	12.5	657,247,113	9	84
Inclusive b-jet	(177,86,70)	(3.0)	12.5	237	5	88
1-jet AND $E_{\mathrm{T}}^{\mathrm{miss}}$	86,46	2.3	14.3	180,123	5	93
Electron AND jet	21,45	0.8	15.1	19,45	2	95
Calibration (10%)	–	0.9	16.0	–	10	105
Total			16.0			105

most "soft-lepton"-tagged b-jets will contain the lepton within $\Delta R = 0.5$ separation of the jet axis. Jet directions are difficult to reconstruct at hardware level, so most online isolation criteria are based on hadronic activity in "hollow cones" centered on the lepton candidate direction, either in the calorimeter at hardware level or including the tracks at the software-level trigger. Lepton isolation criteria suffer instantaneous luminosity-dependent inefficiencies from underlying event, minbias pile-up, and magnetic field spreading of low-p_{T} pions.

The trigger rates at the LHC startup will be a major test of the preparatory work in progress at the CMS and ATLAS experiments. If the initial trigger rates exceed expectation, methods to tighten isolation criteria to control the rate will be invaluable. Directly raising the thresholds on jets and leptons will cut into vital Higgs boson signal efficiencies, as discussed in the chapter on the Higgs boson searches.

8.5.1 Multijet QCD Backgrounds

With such a high rate of multijet events, a single electron trigger, for example, will select some number of jets as electrons. This can be due to fragmentation or detector response fluctuations or other causes that are difficult to model with finite Monte Carlo simulation statistics. Data-driven estimates are required for singly important selection criteria, for example, a high-p_{T} isolated electron.

One technique to determine the normalization of the multijet background is to apply a *matrix method* from simultaneous equations sensitive to the differences in behavior when tightening electron identification criteria. One defines two samples, tight and loose. The efficiency for a true isolated electron to pass the tight cuts is given by ϵ_{sig} and for multijet background to pass a fake electron, the efficiency is ϵ_{QCD}. The set of linear equations is therefore given by

$$\begin{aligned} N_{loose} &= N^{sig} + N^{QCD}, \\ N_{tight} &= \epsilon_{sig} N^{sig} + \epsilon_{QCD} N^{QCD}. \end{aligned} \tag{8.76}$$

The efficiency ϵ_{sig} may be taken from Monte Carlo simulation after data-to-MC *scale factors* are applied to adjust for observed discrepancies. The procedure of using two-body correlated final states to extract data-to-MC corrections is known as the *tag and probe* method, as sketched in figure 8.20 for reconstruction and particle identification in $Z \to e^+e^-$ data. Corrections of this type are known to depend on the activity in the event such as the jet multiplicity.

The efficiency for fake electrons to pass the tight cuts is measured directly from data. For example, for signals with missing transverse energy, the low-E_T^{miss} region will be dominated by background, and ϵ_{QCD} is the ratio of tight-to-loose events in this background-enhanced sample, as shown in figure 8.21. The value of ϵ_{QCD} can depend strongly on isolation and threshold parameters in the trigger selection.

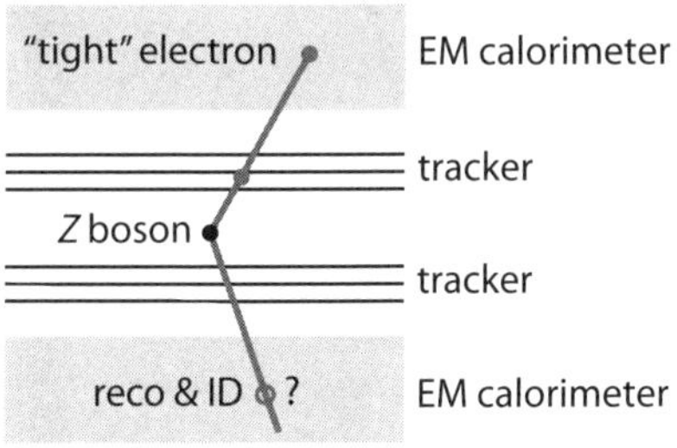

FIGURE 8.20. To verify the Monte Carlo (MC) simulation prediction for the electron signal efficiency ϵ_{sig}, the data-to-MC efficiencies are compared using a *tag and probe* method on $Z \to e^+e^-$ data. The tag electron is used to select or trigger the event and the probe electron is an unbiased estimator of quantities such as reconstruction and particle identification (ID) efficiencies.

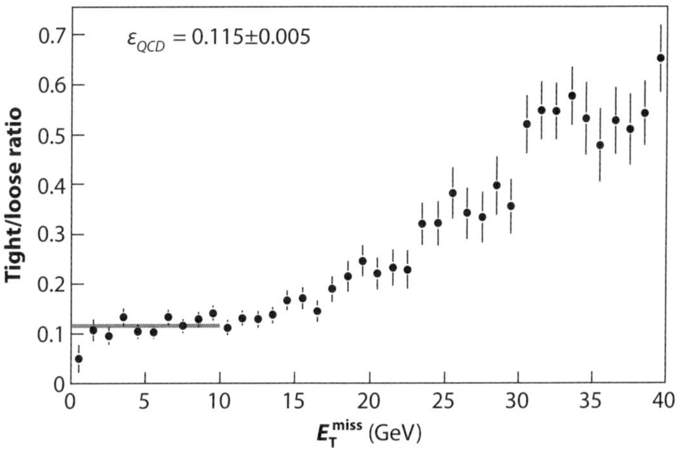

FIGURE 8.21. Ratio of tight-to-loose electron event selections as a function of E_T^{miss} where extrapolation to the low-E_T^{miss} region gives a stable value for the QCD fake rate for electrons ϵ_{QCD}.

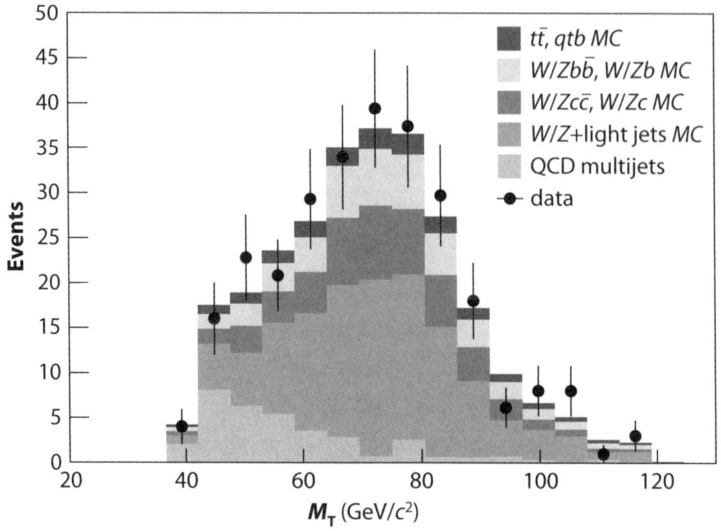

FIGURE 8.22. Example of the leptonic W transverse mass from the Tevatron in events with at least one b-tagged jet (Reprinted figure from [4] with permission by the APS). The background shape from QCD multijets is in gray with a normalization determined from the "matrix method."

The data-determined multijet background shape in a measurement is taken directly from the background-enhanced sample and added to the prediction with the matrix method normalization, as shown in figure 8.22 for the transverse mass spectrum of inclusive W boson plus heavy flavor jet production at a proton-antiproton collider. For proton-proton colliders, the rate of direct W^+ production exceeds that of direct W^- production due to the valence quark content of the proton, and hence the charge asymmetry in the production rates provides an additional handle for separating out the nearly charge-neutral multijet QCD backgrounds to the transverse mass spectrum.

8.6 Exercises

1. **Top quark production.** In 1995, the top quark was discovered in proton-antiproton ($p\bar{p}$) collisions at the Tevatron collider. The top quarks were discovered in pairs ($t\bar{t}$).

   (a) Write down the Feynman diagram of at least one of the dominant production mechanisms for s-channel top pair production in $p\bar{p}$ collisions. Label the initial, final, and intermediate articles in the diagram.

   (b) It is also possible to produce a single top quark through an s-channel process in $p\bar{p}$ collisions. Write down the Feynman diagram for this process. Why did this process need an order of magnitude more integrated luminosity to be observed experimentally compared to the initial top pair production discovery?

(c) For the *s*-channel single top production process, the squared matrix element is given by equation (8.57):

$$\overline{|\mathcal{M}|^2}(\pm) = \frac{g^4 N_c^2}{4} |V_{ud}^* V_{tb}|^2 \frac{d \cdot (t \pm m_t \lambda)(2u \cdot b)}{(2u \cdot d - M_W^2)^2 + (M_W \Gamma_W)^2}$$

where λ is the spin vector of the top quark and the four-momenta of the quarks are represented by their letter: t, b, u, d. Rewrite this expression in terms of the top quark velocity β and the cosine of the production angle of the top quark $\cos\theta^*$ measured with respect to the direction of the three-momentum of the d-quark in the rest frame of the top quark.

Write down the expressions for d, u, and b and expand out the expression for $\overline{|\mathcal{M}|^2}(\pm)$ in the top rest frame. Use $\cos\theta^*$, β, m_t, and s to express your answer.

(d) What is the value of $\overline{|\mathcal{M}|^2}(\pm)$ in the limit that β goes to 1? What does this imply about the interaction?

2. **γ+Jet production at hadron colliders.** Two leading-order $2 \rightarrow 2$ processes at hadron colliders produce a single high-p_T jet of hadrons recoiling off a photon in the final state.

(a) List the sets of partons (p_1, p_2, p_3) that give rise to the γ+jet parton-parton subprocess $p_1 p_2 \rightarrow \gamma p_3$. (*Hint:* There are two leading-order processes. One process is analogous to the QED pair annihilation process. The other process is analogous to the QED Compton scattering process.)

(b) A set of reference parton distribution functions for the proton are plotted in figure 8.23 for gluons, u-quarks, and $\bar{u}$-antiquarks. For a proton-antiproton collider with 1 TeV particle beams, photons are collected with a p_T of 40 GeV and a lab rapidity of $y_\gamma = 0$. What is the approximate value of y_{jet} at which the two γ+jet processes in part (a) are contributing equally? Use figure 8.23 to make the estimate. Ignore the d-quark and $\bar{d}$-antiquark contributions as well as other non-u-quark flavors.

(c) In a proton-proton collider with 7 TeV beam energies, what is the dominant final-state parton contributing to the γ+jet process?

3. Derive the *Jacobian peak* singularity in transverse mass m_T in the differential cross section for W boson production

$$\frac{d\sigma}{dm_T} \propto \left(M_W^2 - m_T^2\right)^{-1/2}.$$

4. **Massive vector resonance at the LHC.** A resonance is discovered at the LHC. However, the couplings to the elementary particles must be computed from the observed properties of the resonance and its decays. The known properties of the resonance are:

- The resonance decays some of the time into $\mu^+\mu^-$ with an invariant mass of $M = 1$ TeV and has a total width of Γ_{tot}. Assume that Γ_{tot} is an accurate estimate of the natural width of the resonance, where $\tau = 1/\Gamma_{\text{tot}}$ is the lifetime of the resonance.

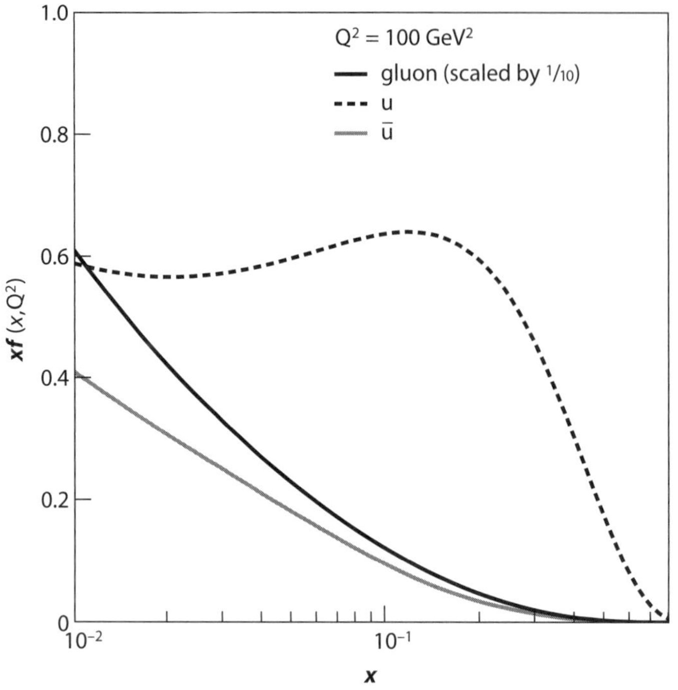

FIGURE 8.23. Proton parton distribution functions plotted as the product $xf(x)$ versus x for gluons, u-quarks, and $\bar{u}$-antiquarks. The gluon distribution is scaled down by a factor of 10 (Credit: HepData) [1].

- It is observed that the resonance decays to $\tau^+\tau^-$ at twice the rate compared to $\mu^+\mu^-$ decays. Namely, $\Gamma_{\tau\tau} = 2\Gamma_{\mu\mu}$.
- No decays to e^+e^- are observed, $\Gamma_{ee} = 0$.
- Based on the production cross section, it is unlikely that the resonance directly couples to the valence quarks in the proton.
- The decay rate to a pair of bottom quarks $b\bar{b}$, $\Gamma_{b\bar{b}}$, is equal to three times the decay rate to $\tau^+\tau^-$, as can be determined by tagging the b-jets using displaced vertices.
- The decay rate to a pair of top quarks $t\bar{t}$, $\Gamma_{t\bar{t}}$, is approximately 6% lower than the decay rate to $b\bar{b}$.
- Excluding the decays to top quarks, the partial decay width to hadrons (two jets) is 12 times the partial decay width to $\mu^+\mu^-$. Namely, $\Gamma_{\text{had}} = 12\Gamma_{\mu\mu}$, where Γ_{had} includes $b\bar{b}$ decays.
- The total observed width is equal to summing over $\Gamma_{\mu\mu}$, $\Gamma_{\tau\tau}$, Γ_{had}, and $\Gamma_{t\bar{t}}$. No other decays are directly observed. The difference between $\Gamma_{\text{tot}} - \Gamma_{\text{obs}} = 3\Gamma_{\mu\mu}$.

- The angular distribution of the fermions in all decay modes have a $1+\cos^2\theta$ distribution with no observed forward-backward asymmetry (no $\cos\theta$ term). Here θ is the angle between incoming W^+ boson and the outgoing positively charged fermion.

Assuming the resonance is due to a single massive vector particle that couples to fermions with a QED-like vertex, compute the absolute value of the coupling strength relative to the muon for each of the elementary fermions. The solution is not unique. Make a hypothesis that satisfies the constraints from the experimental data. Speculate based on the pattern observed in the fermion couplings that can be determined directly from the data as to what may be the couplings (relative to the muon) for the different neutrino flavors and different quark flavors.

8.7 References and Further Reading

A selection of reference texts with an emphasis on hadron collider physics can be found here: [8, 9, 10, 11].

A selection of Web sites of related experiments is found here: [3, 5, 7, 12, 13, 14].

A selection of Monte Carlo event generators can be found here: [15, 16, 17, 18].

Compilations of patron distribution functions can be found here: [1].

[1] Durham HepData Project. http://durpdg.dur.ac.uk/hepdata/pdf3.html.

[2] Martin-Stirling-Thorne-Watt patron distribution functions. http://projects.hepforge.org/mstwpdf.

[3] D0 Experiment at the Tevatron. http://www-d0.fnal.gov.

[4] D0 Collaboration, *Phys. Rev. Lett.* **94**, 161801 (2005). http://arxiv.org/abs/hep-ex/0410078.

[5] CDF Experiment in the Tevatron. http://www-cdf.fnal.gov.

[6] K. Nakamura et al. (Particle Data Group). *J. Phys. G* **37**, (2010). http://pdg.lbl.gov.

[7] CMS Experiment at the LHC. http://cms.web.cern.ch/cms.

[8] Dan Green. *High p_T Physics at Hadron Colliders.* Cambridge University Press, 2005. ISBN 0-521-83509-7.

[9] Donald Perkins. *Introduction to High Energy Physics.* Cambridge University Press, 2000. ISBN 0-521-62196-8.

[10] R. K. Ellis, W. J. Stirling, and B. R. Webber. *QCD and Collider Physics.* Cambridge University Press, 1996. ISBN 0-521-54589-7.

[11] Vernon D. Barger and Roger J. N. Phillips. *Collider Physics.* Westview Press, 1991. ISBN 0-201-14945-1.

[12] ATLAS Experiment at the LHC. http://atlas.web.cern.ch/Atlas.

[13] H1 Experiment at HERA (1992–2007). http://www-h1.desy.de.

[14] ZEUS Experiment at HERA (1992–2007). http://www-zeus.desy.de.

[15] ALPGEN Monte Carlo. http://mlm.home.cern.ch/mlm/alpgen.

[16] HERWIG++ Monte Carlo. http://hepwww.rl.ac.uk/theory/seymour/herwig.

[17] PYTHIA Monte Carlo. http://home.thep.lu.se/~torbjorn/Pythia.html.

[18] SHERPA Monte Carlo. http://projects.hepforge.org/sherpa.

9 Higgs Physics

An introduction to the history of Higgs boson searches can be found in reference [1]. Some of the main arguments and results are repeated here.

The Standard Model of electroweak and strong interactions describes the present observations remarkably well. Its electroweak sector relies on the gauge symmetry group $SU(2)_L \times U(1)_Y$. The gauge invariance of the theory requires that vector bosons be massless. The fact that this symmetry couples differently to left- and right-handed fermion fields also forbids fermion mass terms, which also contradicts observations.

From another standpoint, several obstacles prevent the massive intermediate vector boson theory, an extension of the Fermi model, from being a satisfactory theory of the weak interaction. On one hand the cross section for the process $W^+W^- \rightarrow W^+W^-$ fails unitarity in perturbation theory above a finite scattering energy and on the other hand the presence of massive vector bosons prevents loop corrections to masses and couplings from being finite.

The Higgs mechanism, which consists of a spontaneous breaking of the electroweak symmetry, is an elegant solution to these problems. This subtle mechanism allows gauge bosons to be massive while their interactions are still described by the $SU(2)_L \times U(1)_Y$ gauge group. The general concept is that the symmetry is hidden or, in other words spontaneously broken, leaving the $U(1)_{\text{EM}}$ symmetry apparent to describe the electromagnetic interaction with a massless gauge boson: the photon. In this mechanism, the electroweak symmetry is spontaneously broken by the introduction of an $SU(2)_L$ doublet of scalar complex fields ϕ with the potential $V(\phi) = -\mu^2\phi^2 + \lambda\phi^4$. The spontaneous breaking occurs when the vacuum state of the scalar theory falls in a nontrivial minimum of the Higgs potential $V(\phi)$. In the ground state, the complex scalar doublet ϕ then has a nonzero vacuum expectation value v and the W and Z bosons thus acquire masses by absorbing three of the four initial degrees of freedom of the complex scalar doublet. The remaining degree of freedom is an elementary physical scalar state whose mass is not predicted at tree level by the theory: the Higgs boson. Its presence is not only a signature of the Higgs mechanism but also ensures the unitarity of the $W^+W^- \rightarrow W^+W^-$ process, if its mass does not exceed $\sqrt{4\pi\sqrt{2}/3G_F}$ (approximately 700 GeV). The Higgs mechanism also allows fermion masses through their Yukawa coupling to the Higgs doublet.

The Higgs mechanism is amazingly predictive. It relates the masses of the gauge bosons to the electromagnetic (e) and $SU(2)_L$ (g) coupling constants and predicts

$$\sin^2\theta_W = 1 - \frac{m_W^2}{m_Z^2} = \frac{e^2}{g^2}. \tag{9.1}$$

This tree-level prediction is verified in the present data, since the measurement of $\rho \equiv m_W^2/(m_Z^2\cos^2\theta_W)$ is found to be in good agreement with its expected value of unity. The couplings of the Higgs boson to fermions are governed by the same Yukawa couplings that generate fermion mass terms and are thus proportional to the masses of the fermions. Therefore, the signature of the Higgs mechanism will be clear once the decay modes of the scalar boson are measured.

The mass of the Higgs boson can be expressed as a function of its quartic coupling λ, the mass of the W boson and g:

$$m_H^2 = \frac{4\lambda m_W^2}{g^2}. \tag{9.2}$$

Although the mass of the Higgs boson is a free parameter of the theory, the running of the quartic coupling λ can infer both lower and upper bounds to the mass of the Higgs boson according to two precepts. The first is the stability of the vacuum, which requires $\lambda > 0$ within the domain in which the theory is valid (below a given energy scale Λ above which new physics appears superseding the standard theory), or else the Higgs potential is unstable. The second is directly inferred by the running of λ. To illustrate this in a simple case, a pure ϕ^4 theory, the running of λ is given by

$$\frac{1}{\lambda(\nu)} = \frac{1}{\lambda(\Lambda)} + \frac{3}{16\pi^2}\ln\frac{\Lambda}{\nu}, \tag{9.3}$$

which implies that

$$\lambda(\nu) < \frac{16\pi^2}{3\ln(\Lambda/\nu)} \qquad \text{or} \qquad m_H < \frac{4\pi\nu}{3\sqrt{\ln(\Lambda/\nu)}}. \tag{9.4}$$

This argument implies that $\lambda \to 0$ as $\Lambda \to \infty$, and thus is called *triviality* and yields an upper bound on the mass of the Higgs boson. The bounds inferred by these two arguments are shown in figure 9.1 where, in particular, the contribution of the top quark to the running of λ, which is essential to lead λ toward negative values, is taken into account.

The arguments of unitarity and triviality both tend to indicate that if the Higgs boson exists, it should not have a mass exceedingly high.

Although the Higgs boson resolves most of the dilemmas of the electroweak theory, it also creates a serious *naturalness* problem. Corrections to the Higgs boson mass, such as those illustrated in figure 9.2, are quadratically divergent. The contribution of these diagrams is of the order

$$\Delta m^2 \propto \int^{\Lambda} \frac{d^4k}{(2\pi)^4}\frac{1}{k^2} \sim \frac{\Lambda^2}{16\pi^2}. \tag{9.5}$$

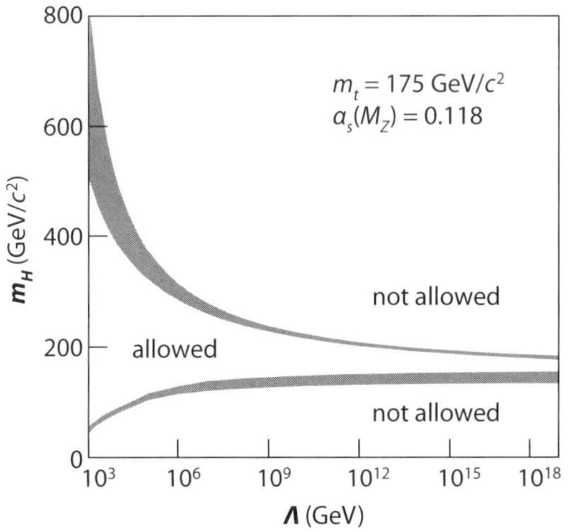

FIGURE 9.1. Lower and upper limits on the Higgs boson mass as a function of the cut-off energy scale Λ, relying on the vacuum stability (lower band) and triviality (upper band) arguments [1].

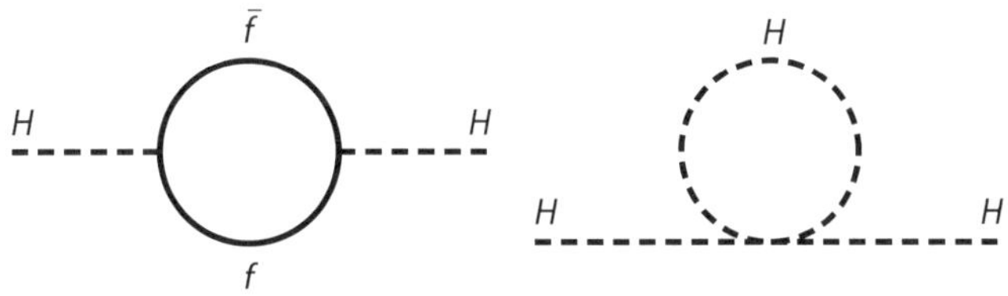

FIGURE 9.2. One-loop quadratically divergent corrections to the Higgs boson mass.

Assuming that the cutoff (Λ) is large, typically the *Planck scale* (10^{19} GeV), the mass of the Higgs boson should naturally be of the same scale, as otherwise the bare mass would need to be *fine-tuned* to compensate its corrections. Knowing from the unitarity and triviality arguments that the mass of the Higgs boson should be smaller than 1 TeV/c^2, the bare mass must be fine-tuned to a precision of over 16 orders of magnitude in order to yield such a low value. The same problem appears at all orders of perturbation and renders the Standard Model conceptually unnatural.

The Standard Model of electroweak interactions would work beautifully with a low-mass Higgs boson. However, if it does exist, a theory beyond the Standard Model is needed to solve the naturalness problem.

9.1 Pre-LEP Searches

Before the LEP start in 1989, Higgs boson masses below 5 GeV/c^2 were thought to be unlikely. The lack of certainty in this result resides in the fact that it relied on the

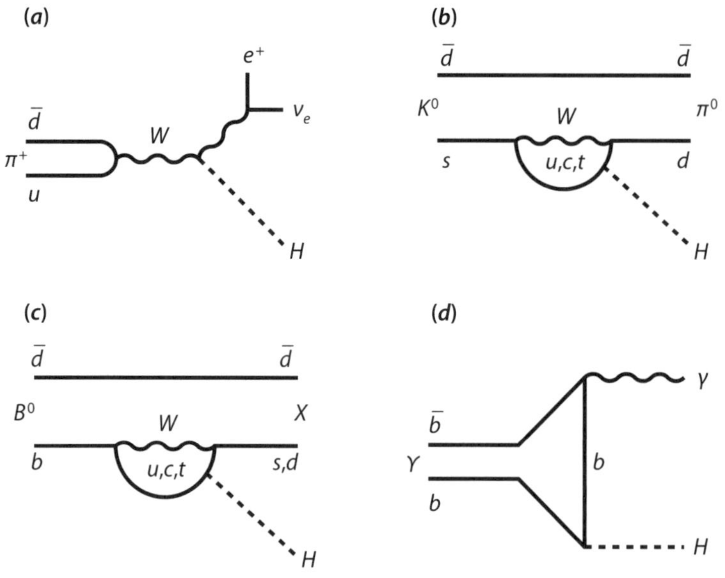

FIGURE 9.3. Light Higgs boson production via W boson or flavor-changing neutral currents.

combination of numerous experiments that were subject to large theoretical uncertainties on the production cross sections and branching fractions. To illustrate some of these searches, a few examples corresponding to the diagrams shown in figure 9.3 are given. For very low Higgs boson masses, the SINDRUM spectrometer experiment at the PSI 590 MeV proton cyclotron investigated the decay of the pion to an electron, an electron neutrino, and a Higgs boson decaying in turn to a pair of electrons (figure 9.3a). The negative result of this search led to an exclusion of the mass domain 10 MeV/$c^2 < m_H <$ 110 MeV/c^2. The CERN-Edinburgh-Mainz-Orsay-Pisa-Siegen collaboration at the CERN SPS also searched for the decay of a Higgs boson into a pair of electrons in the decay of $K_L^\circ \to \pi^\circ H$ (figure 9.3b). These searches severely constrained the Higgs boson mass in the domain below 50 MeV/c^2 by inferring an upper limit on the product of the branching ratios BR $(K_L^\circ \to \pi^\circ H) \times \mathrm{BR}(H \to e^+e^-)$ of approximately 2×10^{-8}. Before 1989, CLEO investigated decays of the Higgs boson into a pair of muons, pions, and kaons produced through the flavor-changing neutral current decay $\mathrm{B} \to K^\circ H$ (figure 9.3c). They found no evidence for a Higgs boson and succeeded in excluding the mass range 0.2–3.6 GeV/c^2. This exclusion relied on the *B*-to-Higgs-boson branching evaluation, which is subject to a large theoretical uncertainty. Finally, the CUSB collaboration investigated the radiative decay of the Υ into a Higgs boson (figure 9.3d). Their search for a monochromatic photon sample from the decay $\Upsilon \to \gamma + X$ allowed them to exclude the range from $2m_\mu$ up to 5 GeV/c^2. All these searches were sensitive to potentially large QCD corrections, thus justifying the importance of unambiguous searches in the low-mass region.

9.2 LEP1 Era

Because of the large production cross section in Z boson decays at low m_H, LEP provides a good environment to further exclude light Higgs boson masses.

9.2.1 *Production Mechanisms at the Z Resonance*

At LEP1 the Bjorken process $e^+e^- \to HZ^* \to Hf\bar{f}$, illustrated in figure 9.4, is the dominant production mechanism. Although to a lesser extent, the Wilczek process, proceeding through a top quark loop, $e^+e^- \to H\gamma$ shown in figure 9.4, could also contribute to the search of the Higgs boson at LEP. Not only is the production rate through the Wilczek process much smaller than that expected through the Bjorken process, but backgrounds such as $e^+e^- \to q\bar{q}\gamma$ or $e^+e^- \to q\bar{q}g$, where one jet hadronizes to an energetic π^0, greatly weaken the search potential of this channel. Only the Bjorken process will be considered in the following.

9.2.2 *Low-Mass Searches*

For Higgs boson masses such that $m_H < 2\ m_e$, the Higgs boson can decay to a pair of photons only via a loop of W bosons or top quarks and is thus long-lived. For Higgs boson masses below $2m_\mu$ the Higgs boson will essentially decay to a pair of electrons; above $2m_\mu$ but less than $2m_\pi$ it will predominantly decay to a pair of muons. As Higgs boson masses above the $2m_\pi$ threshold are tested, the situation becomes slightly more intricate. For masses below 2–3 GeV/c^2 and above $2m_\pi$ the Higgs boson will decay to a pair of hadrons via its interaction with two gluons through a top quark loop or its interaction with quarks. The hadronization of these gluons will be increasingly complex with higher Higgs boson masses. The branching fractions of the Higgs boson in this "nonperturbative QCD" mass range are depicted in figure 9.5.

The transition to perturbative QCD is suggested by the smooth variation of the branching ratios above ~2 GeV/c^2 in figure 9.5. The Higgs boson mass at which the transition to the perturbative domain occurs was estimated by requiring that the branching ratios into strangeness and the final-state multiplicities be identical. This transition occurs for

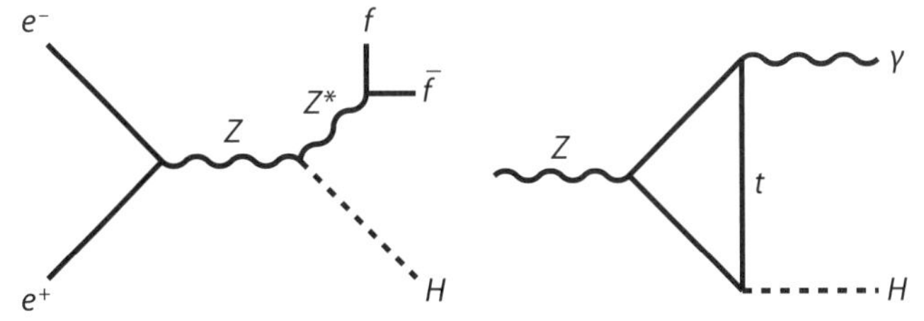

FIGURE 9.4. Higgs boson production processes in Z decays at LEP1. The Bjorken (left) and Wilczek (right) processes.

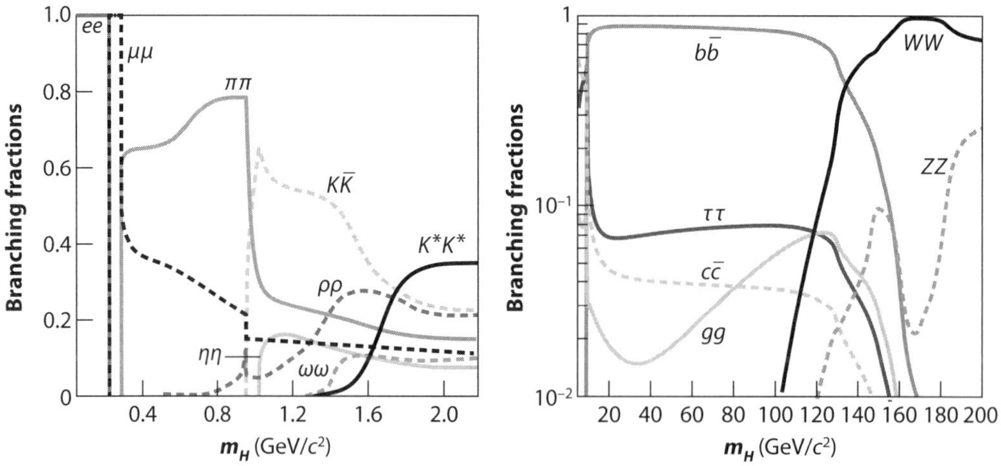

FIGURE 9.5. Higgs boson branching fractions in (left) the "non-perturbative QCD" low-mass range and (right) for heavier mass hypotheses [1].

a Higgs boson mass of ~2 GeV/c^2. Within the "perturbative QCD" domain and below the $b\bar{b}$ threshold, the decays into the heaviest available fermion pair ($c\bar{c}$ or $\tau^+\tau^-$) prevail via its direct coupling to the fermion pair or through its coupling to a gluon pair. Above the $b\bar{b}$ threshold and for Higgs boson masses reachable at LEP (below ~115 GeV/c^2), the branching fractions are dominated by the decay $H \to b\bar{b}$ (~85%), as shown in figure 9.5.

For a Higgs boson mass below 20 GeV/c^2, more than 10,000 events were expected to be produced at LEP1 from the Higgsstrahlung process (the production cross section rapidly decreases at higher masses). Such large rates simplified the search. Three topologies were included in the search:

1. The acoplanar lepton-pair topology with missing energy and momentum where the leptons are assumed to originate from the decay of the Z^* boson. The search in this channel is essentially background free and covers the mass domain below and up to $2m_e$ where the Higgs boson is long-lived and thus escapes detection.

2. The acoplanar pair topology where a pair of charged particles resulting from the decay of the Higgs boson recoil to missing energy and momentum corresponding to the decay of the Z^* boson into a pair of neutrinos.

3. The monojet topology covering the hadronic decays of the Higgs boson with an intermediate mass and with a more intricate fragmentation process, where a single jet recoils to a Z^* boson decaying to a pair of neutrinos.

The last two topologies, contrary to the first, are slightly affected by the $e^+e^- \to \gamma^* Z$ background process where the Z boson decays to a pair of neutrinos. The number of observed events in these channels were in good agreement with the Standard Model background expectation. The mass domain $m_H < 20$ GeV/c^2 was, therefore, excluded at much more than 95% confidencel level (CL).

In the higher Higgs boson mass domain, the Higgs boson decays dominantly to a pair of b-quarks. The topologies investigated at LEP were thus determined by the decays of the Z boson. The overwhelming background from the hadronic decays of the Z boson on the one hand and the rather small production rate (for $m_H = 65$ GeV/c^2, approximately 40 events are expected) on the other did not allow the investigation of topologies involving Z^* decays to hadrons or taus. The two sole channels used at LEP1 in this mass range were those where the Z^* boson decays to a pair of neutrinos or charged leptons (electrons or muons). These two topologies represent ~25% of all final states. The small number of events expected (altogether ~10 events for $m_H = 65$ GeV/c^2) to be found among the 13 million hadronic Z decays collected at LEP by all four experiments required more sophisticated analyses. In total 13 events were observed in slight deficit, though still compatible with the 20.6 events expected from the Standard Model background. The negative results of these searches in all four experiments were combined to yield a lower limit on the mass of the Higgs boson of 65.6 GeV/c^2 [2].

9.3 Higgs Boson Production and Decay

The trilinear couplings of the Higgs boson are given in table 9.1. These predict the following tree-level invariant decay amplitudes:

$$\begin{aligned}
\mathcal{M}(H \to \mathrm{f\bar{f}}) &= -i[(\sqrt{2}\,G_F)^{1/2}]m_f \bar{u} v, \\
\mathcal{M}(H \to W^+W^-) &= i[(\sqrt{2}\,G_F)^{1/2}]2M_W^2 \epsilon_\mu^*(\lambda_1)\epsilon^\mu(\lambda_2), \\
\mathcal{M}(H \to ZZ) &= i[(\sqrt{2}\,G_F)^{1/2}]2M_Z^2 \epsilon_\mu^*(\lambda_1)\epsilon^\mu(\lambda_2).
\end{aligned} \tag{9.6}$$

For the Higgs-to-fermion decays, the partial width is computed by starting with the matrix element (9.6)

$$\begin{aligned}
\overline{|\mathcal{M}|^2} &= \sqrt{2}\,G_F m_f^2 \sum \mathrm{tr}\{\bar{v} v \bar{u} u\} \\
&= \sqrt{2}\,G_F m_f^2 4 p_1 \cdot p_2 \\
&= \sqrt{2}\,G_F m_f^2 2 m_H^2
\end{aligned} \tag{9.7}$$

Table 9.1 Higgs trilinear couplings.

Vertex	Vertex factor
$H(W^+)^\mu(W^-)^\nu$	$igM_W g_{\mu\nu} = i[(\sqrt{2}\,G_F)^{1/2}]2M_W^2\, g_{\mu\nu}$
$HZ^\mu Z^\nu$	$ig\,\frac{M_Z}{\cos\theta_W}\, g_{\mu\nu} = i[(\sqrt{2}\,G_F)^{1/2}]2M_Z^2\, g_{\mu\nu}$
$H\mathrm{f\bar{f}}$	$-i\frac{\lambda_f}{\sqrt{2}} = -i[(\sqrt{2}\,G_F)^{1/2}]m_f$
HHH	$-i6\lambda_H v = -i[(\sqrt{2}\,G_F)^{1/2}]3m_H^2$

for m_f small compared to m_H. The two-body decay phase space integrates out to give $\Gamma = \overline{|\mathcal{M}|^2} p_1^*/(8\pi m_H^2)$. Therefore, the partial width is (for $p_1^* = m_H/2$)

$$\Gamma(H \to f\bar{f}) = \frac{N_c \sqrt{2}\, G_F m_f^2 m_H}{8\pi} \tag{9.8}$$

where N_c is the number of colors. For a Higgs mass of $m_H \sim m_Z$, the heaviest fermion is the b-quark with a mass of $m_b \sim 4.2$ GeV/c^2 and the partial width is enhanced by the $N_c = 3$ factor. The predicted total Higgs width for $m_H \sim m_Z$ is roughly $\Gamma \sim 4$ MeV in this case, and, therefore, it is much narrower than experimental resolutions on the measurement of b-jets. For the vector boson decays, the boson masses are nonnegligible and the three polarizations are summed over. The three polarizations are given by

$$\begin{aligned} \epsilon^\mu(L,R) &= \frac{1}{\sqrt{2}}(0, 1, \mp i, 0), \\ \epsilon^\mu(S) &= \frac{1}{M}(|\boldsymbol{p}|, 0, 0, E). \end{aligned} \tag{9.9}$$

Given that $p_{W^+} + p_{W^-} = p_H$, we get $E_W = \sqrt{|\boldsymbol{p}_W|^2 + M_W^2} = m_H/2$, in the narrow-width approximation for the W boson masses. Therefore, the W boson momentum in the Higgs boson rest frame is

$$|\boldsymbol{p}_W| = \frac{m_H}{2}\sqrt{1 - \frac{4M_W^2}{m_H^2}}. \tag{9.10}$$

The summed squared matrix element is therefore

$$\begin{aligned} \overline{|\mathcal{M}|^2} &= \sqrt{2}\, G_F 4M_W^4 \sum_{\lambda_1,\lambda_2 = (L,L),(R,R),(S,S)} |\epsilon_\mu^*(\lambda_1)\,\epsilon^\mu(\lambda_2)|^2 \\ &= \sqrt{2}\, G_F 4M_W^4 \left[1 + 1 + \frac{1}{M_W^4}(|\boldsymbol{p}_W|^2 + E_W^2)^2\right] \\ &= \sqrt{2}\, G_F 4M_W^4 \left[2 + \frac{1}{M_W^4}\left(M_W^2 + \frac{m_H^2}{2}\left(1 - \frac{4M_W^2}{m_H^2}\right)\right)^2\right] \\ &= \sqrt{2}\, G_F 4M_W^4 \left[2 + \frac{1}{M_W^4}\left(\frac{m_H^2}{2} - M_W^2\right)^2\right] \\ &= \sqrt{2}\, G_F 4M_W^4 \left[3 - 4\frac{m_H^2}{4M_W^2} + 4\frac{m_H^4}{16M_W^4}\right] \\ &= \sqrt{2}\, G_F 4M_W^4 \frac{1}{x_W^2}(3x_W^2 - 4x_W + 4) \end{aligned} \tag{9.11}$$

with $x_W = 4M_W^2/m_H^2$. Using $\Gamma = \overline{|\mathcal{M}|^2} p_W^*/(8\pi m_H^2)$, we get

$$\begin{aligned} \Gamma(H \to W^+W^-) &= \frac{\sqrt{2}\, G_F M_W^4}{4\pi m_H} \frac{\sqrt{1 - x_W}}{x_W^2}(3x_W^2 - 4x_W + 4) \\ &= \frac{\sqrt{2}\, G_F M_W^2 m_H}{16\pi} \frac{\sqrt{1 - x_W}}{x_W}(3x_W^2 - 4x_W + 4). \end{aligned} \tag{9.12}$$

For $H \to ZZ$, we replace x_W with $x_Z = 4M_Z^2/m_H^2$ and divide by 2 to account for identical particles in the final state:

$$\Gamma(H \to ZZ) = \frac{\sqrt{2}\,G_F M_Z^2 m_H}{32\pi} \frac{\sqrt{1-x_Z}}{x_Z}(3x_Z^2 - 4x_Z + 4). \tag{9.13}$$

Notice that for large m_H, the width of the Higgs decaying to bosons is given by

$$\Gamma(H \to W^+W^-) + \Gamma(H \to ZZ) = \frac{3\sqrt{2}\,G_F m_H^3}{32\pi} \quad \text{for } m_H \gg M_W, M_Z. \tag{9.14}$$

9.3.1 Higgsstrahlung at an e^+e^- Collider

For Higgs production via Higgsstrahlung, the process $e^+e^- \to ZH \to \mathrm{f\bar{f}}H$ can be split into a two-stage process. The first stage is $e^+e^- \to ZH$, which is a two-body final state with matrix element

$$\mathcal{M}_{ZH} = \frac{-ie^2 M_Z \bar{v}_2 \gamma_\mu [C_R^e(1+\gamma^5) + C_L^e(1-\gamma^5)] u_1 \epsilon_\lambda^{*\mu}}{2\sin^2\theta_W \cos^2\theta_W [(s - M_Z^2) + iM_Z\Gamma_Z]} \tag{9.15}$$

where $\epsilon_\lambda^{*\mu}$ is the polarization vector of the outgoing Z boson. The second stage is $Z \to \mathrm{f\bar{f}}$ where the outgoing fermion is taken to be massless. The complete matrix element for Higgsstrahlung can be written

$$\mathcal{M}(e^+e^- \to ZH \to \mathrm{f\bar{f}}\,H) = \mathcal{M}_{ZH}\, iG_V(p_Z^2)\, \mathcal{M}(Z \to \mathrm{f\bar{f}}) \tag{9.16}$$

where $iG_V(p_Z^2)$ is the Z propagator including the finite width term. The differential cross section in terms of the invariant mass m of the Z boson against which the Higgs boson recoils is given by

$$\frac{d^2\sigma}{d\cos\theta\, dm^2} = \frac{|\mathcal{M}|^2 p_H}{(2\pi)^3 32 s^{3/2}}. \tag{9.17}$$

Putting the terms (9.15), (9.16), and (9.17) together, gives

$$\begin{aligned} \frac{d^2\sigma}{d\cos\theta\, dm^2} &= \left(\frac{d\sigma_{ZH}}{d\cos\theta}\right) \frac{|G_V(p_Z^2)|^2 |\mathcal{M}(Z \to \mathrm{f\bar{f}})|^2}{16\pi^2} \\ &= \frac{\overline{|\mathcal{M}_{ZH}|^2}\, p_H m \Gamma_{\mathrm{f\bar{f}}}}{16\pi^2 s^{3/2}[(m^2 - M_Z^2)^2 + M_Z^2\Gamma_Z^2]} \end{aligned} \tag{9.18}$$

where $\Gamma_{\mathrm{f\bar{f}}}$ is the $Z \to \mathrm{f\bar{f}}$ decay width and the spin-averaged (unpolarized beams) matrix element for $\mathcal{M}_{ZH}$ summing over Z boson polarizations is given by

$$\overline{|\mathcal{M}_{ZH}|^2} = \frac{e^4 s M_Z^2 (C_R^{e2} + C_L^{e2})}{2\sin^4\theta_W \cos^4\theta_W [(s - M_Z^2)^2 + M_Z^2\Gamma_Z^2]} \left(1 + \frac{p_H^2 \sin^2\theta}{2m^2}\right). \tag{9.19}$$

Therefore, the final cross section depends strongly on m for Higgs production on the Z peak and strongly on $\sqrt{s} - M_Z$ through the Higgs momentum for production above the Z peak.

9.3.2 Vector Boson Fusion at a Hadron Collider

High-energy proton-proton colliders are sometimes referred to as WW colliders when considering purely electroweak scattering. This notion comes from the application of the *effective W approximation*. In the proton-proton scattering process, a pair of colliding quarks can each initiate the scattering by branching into a quark and a W boson. The probability for this branching and the momentum fraction distribution of W bosons can be treated as an effective luminosity for the WW subprocess scattering using techniques similar to that of the parton model. The effective vector boson approximation applies for subprocess center-of-mass energies where $\hat{s} \gg M_V^2$. Therefore, if we are considering the Higgs vector boson fusion (VBF) production process

$$q_1 q_2 \to q_1' q_2' V_1 V_2 \to q_1' q_2' H, \tag{9.20}$$

then the effective W/Z approximation only applies for Higgs boson masses significantly heavier than the W and Z masses, $m_H \gg m_W, m_Z$. For lighter Higgs boson masses, the scattered valence quarks q_1', q_2' will have a transverse momentum comparable to half the W boson mass, $p_T \sim 40$ GeV/c. At the LHC, these weakly scattered partons will form narrow energetic forward jets that are measured in forward calorimeters covering the pseudorapidity range $\eta = 2$–5. The identification of the forward jets is a method of tagging the Higgs production process and constraining the production kinematics.

The matrix element for VBF vector-vector scattering $q_1 q_2 \to q_1' q_2' V_1 V_2 \to q_1' q_2' H$ valid for the full range of Higgs masses is given by

$$\mathcal{M}_{WW} = \frac{g^3 M_W}{2} \frac{\bar{u}(q_1')\gamma^\mu P_L u(q_1)\,\bar{u}(q_2')\gamma_\mu P_L u(q_2)}{((q_1' - q_1)^2 - M_W^2 + iM_W\Gamma_W)((q_2' - q_2)^2 - M_W^2 + iM_W\Gamma_W)} \tag{9.21}$$

for WW scattering with $P_L = (1 - \gamma^5)/2$; for ZZ scattering, we insert the Z couplings to quarks

$$\mathcal{M}_{ZZ} = \frac{g^3 M_W}{\cos^4\theta_W} \frac{\bar{u}(q_1')\gamma^\mu(g_V^{q_1} + g_A^{q_1}\gamma^5)u(q_1)\bar{u}(q_2')\gamma_\mu(g_V^{q_2} + g_A^{q_2}\gamma^5)u(q_2)}{((q_1' - q_1)^2 - M_Z^2 + iM_Z\Gamma_Z)((q_2' - q_2)^2 - M_Z^2 + iM_Z\Gamma_Z)}. \tag{9.22}$$

Squaring up the matrix element, summing over final spins, and averaging over initial spins gives for WW scattering

$$\overline{|\mathcal{M}_{WW}|^2} = g^6 M_W^2 \frac{(q_1 \cdot q_2)(q_1' \cdot q_2')}{(((q_1' - q_1)^2 - M_W^2)^2 + M_W^2\Gamma_W^2)(((q_2' - q_2)^2 - M_W^2)^2 + M_W^2\Gamma_W^2)}. \tag{9.23}$$

For ZZ scattering, we get

$$\overline{|\mathcal{M}_{ZZ}|^2} = \frac{4g^6 M_W^2}{\cos^8\theta_W} \frac{(C_{LL}^2 + C_{RR}^2)(q_1 \cdot q_2)(q_1' \cdot q_2') + (C_{LR}^2 + C_{RL}^2)(q_1 \cdot q_2')(q_1' \cdot q_2)}{(((q_1' - q_1)^2 - M_Z^2)^2 + M_Z^2\Gamma_Z^2)(((q_2' - q_2)^2 - M_Z^2)^2 + M_Z^2\Gamma_Z^2)} \tag{9.24}$$

with the coefficients given by

$$\begin{aligned} C_{LL,RR} &= (g_V^{q_1} \mp g_A^{q_1})(g_V^{q_2} \mp g_A^{q_2}), \\ C_{LR,RL} &= (g_V^{q_1} \mp g_A^{q_1})(g_V^{q_2} \pm g_A^{q_2}). \end{aligned} \tag{9.25}$$

The short-distance differential cross section $d\hat{\sigma}$ is given by the standard form

$$d\hat{\sigma} = \frac{1}{2\hat{s}} \frac{d^3 q_1'}{(2\pi)^3 2E_{q_1'}} \frac{d^3 q_2'}{(2\pi)^3 2E_{q_2'}} \frac{d^3 p_H}{(2\pi)^3 2E_H} (2\pi)^4 \delta^{(4)}(q_1 + q_2 - q_1' - q_2' - p_H) \sum \overline{|\mathcal{M}_{VV}|^2}, \qquad (9.26)$$

which is then integrated over the valence quark parton luminosities to determine the $pp \to jjH + X$ production cross section.

9.4 From LEP2 to the Tevatron and LHC

An interesting comparison can be made between the experimental approach to the Higgs boson searches at LEP2 and those at the Tevatron and LHC. After early experimental measurements at the Z peak closed the door to the possibility of a Standard Model Higgs boson with a mass in the range 0–65.6 GeV/c^2, a new era of direct searches for on-shell diboson (ZH) production began in 1996 when LEP increased the center-of-mass energy above ~160 GeV and continued through 2000 to the ultimate LEP energy reach of $\sqrt{s} = 209$ GeV. During this time, the event signatures were categorized according to the decay modes of the predicted Higgs boson and the well-studied decay modes of the Z boson, as shown in figure 9.6. The Standard Model Higgs was searched for in its dominant decays into third generation fermions, i.e., τ-leptons and b-quark pairs.

Nearly all possible Higgs boson production modes and decay channels were covered by the LEP searches (~90%) achieving a maximum detection efficiency with varying degrees of sensitivity in each final state. The closest "Higgs bosonlike" backgrounds came from ZZ production with at least one Z boson decaying to $b\bar{b}$, given the finite and non-Gaussian detector resolution on the measurement of the dijet $b\bar{b}$-mass. An example four-jet event is shown in figure 9.7.

In the missing energy channel of the Higgs search, some contribution to the sensitivity came from WW fusion, a process whereby both the incoming positron and electron convert to neutrinos radiating W bosons which fuse to form the Higgs boson, as shown in

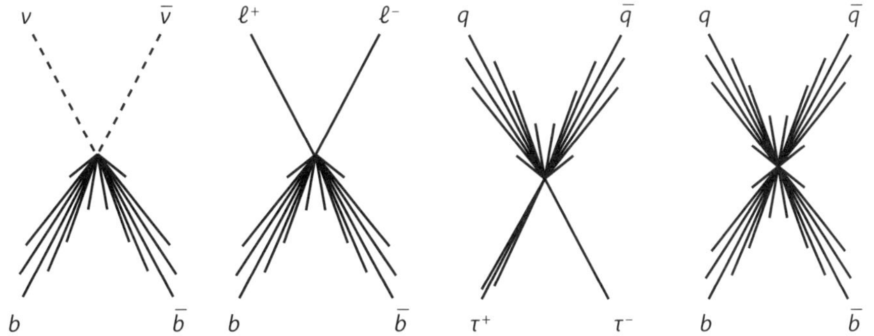

FIGURE 9.6. Topologies involved in the search for the Standard Model Higgs boson at LEP2, missing energy, lepton pairs, $H \to \tau^+\tau^-$ and hadronic Z boson decays, four-jets.

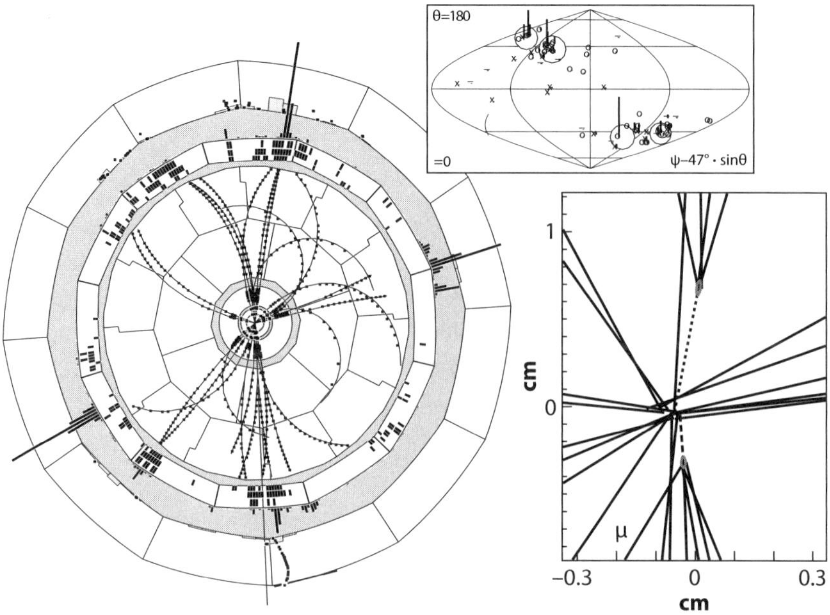

FIGURE 9.7. Four-jet event recorded by the ALEPH experiment. The Higgs boson candidate in this event has a mass of 114.3 GeV/c^2. The two displaced vertices indicate a $b\bar{b}$-dijet system (Credit: ALEPH).

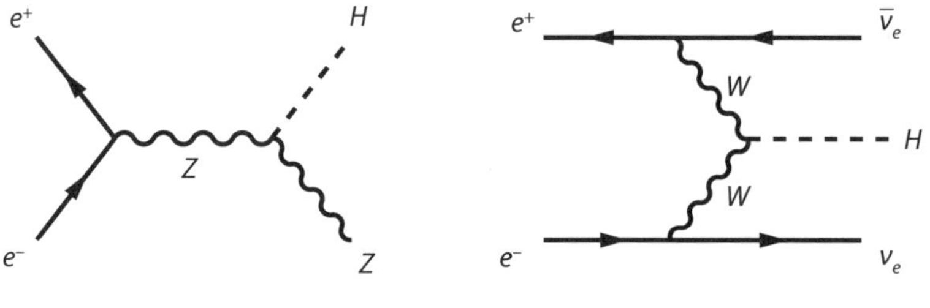

FIGURE 9.8. Diagrams of the Higgsstrahlung (left) and weak boson fusion (right) processes of Higgs boson production at LEP2.

figure 9.8. This exceptional process becomes a larger fraction of the total Higgs production rate with increasing beam energy. An example missing energy event is shown in figure 9.9.

Many analysis techniques were used to quantify the Higgs boson search background predictions from Standard Model processes. In all cases background measurements and predictions were found to agree exceptionally well with relatively small theoretical and experimental uncertainties, for example, as shown in figure 7.1. Therefore, background systematic uncertainties had little effect on the overall search sensitivity. An important aspect of the LEP searches was the use of event-by-event discrimination that had greater sensitivity to event properties than simple event counting alone. The final LEP measure of the Higgs search data quantified both mass and rate information, as shown in figure 9.10, where a Higgs boson signal, if present, will form a minimum in the observed likelihood

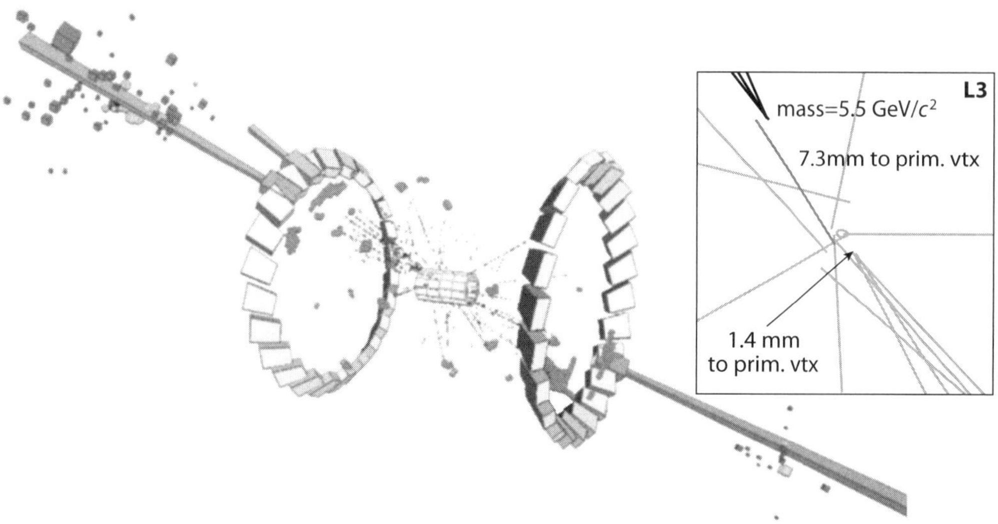

FIGURE 9.9. Missing energy event recorded by the L3 experiment. The Higgs boson candidate in this event has a mass of 115.0 GeV/c^2. The two displaced vertices indicate a $b\bar{b}$-dijet system (Credit: L3).

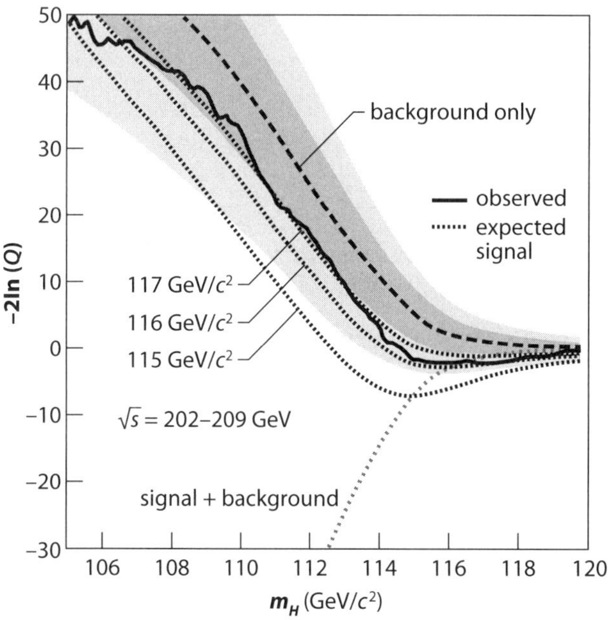

FIGURE 9.10. The combined LEP Standard Model Higgs boson search data for $\sqrt{s}$ = 202–209 GeV with the solid line indicating the observed data in comparison to background-only and signal+background hypotheses (Credit: LEP HWG) [2]. The bands indicate one and two standard deviation intervals on the background estimate with respect to the mean expectation.

ratio at the true Higgs boson mass with a dip amplitude consistent with the expected Standard Model production rate.

The ZH cross section turns on rapidly above the threshold for Higgs boson production for a given mass m_H while at the same time the magnitude of the cross section is comparatively small, as shown in figure 7.1. This made the LEP Higgs search a combined optimization of the highest center-of-mass energy for a given expected amount of integrated luminosity at that energy delivered to each of the four LEP experiments, ALEPH, DELPHI, L3, and OPAL. With four LEP experiments, sensitivity to a ~60 fb cross section at the kinematic limit of $m_H = \sqrt{s} - m_Z$ was achieved with ~200 pb^{-1} delivered per experiment. A summary of the LEP Higgs search sensitivity is given by the expected significance of a signal versus mass, as shown in figure 9.11 for the entire LEP2 dataset. The LEP Higgs Working Group set a 95% confidence-level lower limit of 114.4 GeV on the Standard Model Higgs boson mass [2].

At the LHC, the LEP approach of searching in all available final-state topologies is still a valuable starting point for the construction of the analyses. However, the high-rate hadron collider backgrounds overwhelm Higgs decays into b-quark pairs and the low-mass inclusive τ-lepton pair decays. Judicious choice of final-state search topologies must also accommodate the triggering capabilities of the experiment. The trigger limitations impact final states with low-p_T single leptons or jets. At the LHC, there are few cases of low-background search channels, and more of the leverage in terms of search sensitivity comes from high-resolution mass reconstruction or large multilepton rates. In particular, search channels

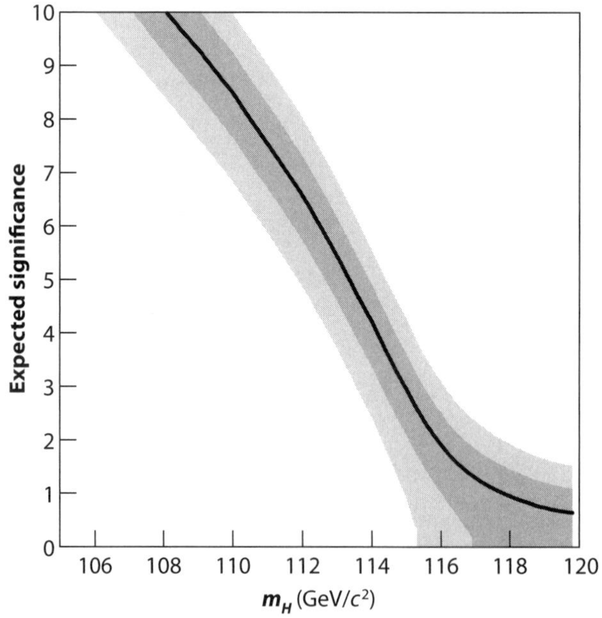

FIGURE 9.11. Expected significance in number of standard deviations from background for a Standard Model Higgs boson with the entire LEP2 dataset (Credit: LEP HWG) [2]. The bands indicate one and two standard deviation intervals on the significance estimate with respect to the mean expectation.

that require counting a fixed number of jets to isolate a process are very inefficient due to a combination of low jet reconstruction efficiency at low p_T and the abundance of low-p_T jets from initial-state gluon radiation, minbias pile-up, and the underlying event. Search channels with significant multijet QCD backgrounds tend to be limited in search sensitivity. Even in the lepton channels, most multijet-induced backgrounds are necessarily estimated using data-based samples as opposed to Monte Carlo simulation. We will find that these and other aspects of hadron collider physics will greatly influence the focus of the Standard Model Higgs search at the LHC. The many possible manifestations of the Higgs boson discovery and crucial first measurements are discussed in the following section.

9.5 Standard Model Higgs Search

The Standard Model Higgs production cross sections at the LHC and branching fractions are plotted in figure 9.12.

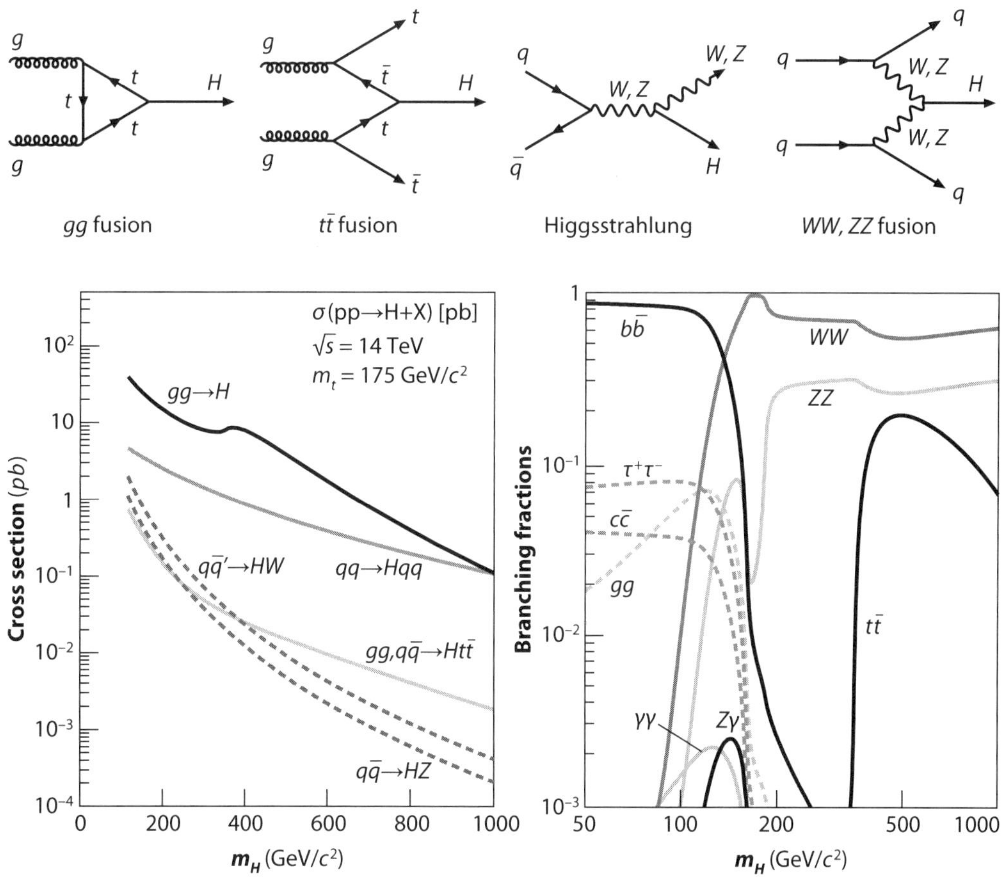

FIGURE 9.12. Standard Model Higgs boson production diagrams at the LHC (top) with corresponding cross sections (left) and branching fractions (right) as a function of Higgs boson mass.

At low Higgs mass, $m_H \approx 115$ GeV/c^2, the dominant production process is gluon-gluon fusion with a cross section 1000 times larger than the corresponding LEP Higgs search. The *WW* fusion process is much stronger at the LHC due to the relatively low mass scale of the Higgs boson relative to the proton beam energy. The next largest relevant production rates are due to Higgsstrahlung processes from the heaviest known elementary particles: WH, ZH, and $t\bar{t}$H. In terms of decay modes, a dramatic transition occurs in branching fractions from third-generation fermion-dominated decays to diboson-dominated decays at $m_H \sim 130$ GeV/c^2.

By folding the Higgs production cross sections and branching fractions, shown in figure 9.12, against the trigger and selection efficiencies, a preliminary list of relevant search channels for a low-mass Higgs search can be formed for Higgs boson masses below the *WW* threshold. This is given in table 9.2. Of the channels in the leftmost column, only the diboson decays of the Higgs are sufficiently clean to be detected inclusively within corresponding specific trigger paths. The columns to the right are a set of exclusive decay channels where identification of associated production particles gives at least an order of magnitude improvement in signal-to-background separation, relative to the inclusive searches. The exclusive channels have unique sensitivities to third-generation Higgs couplings, Higgs couplings to electroweak bosons, and precise mass and partial decay width measurements. The possibility of continuing the low-mass Tevatron Higgs search in the *WH*/*ZH* processes at the LHC with the Higgs boson decaying to $b\bar{b}$ is not excluded. However, initial studies show that mainly highly boosted Higgs bosons, $p_T > 200$ GeV, decaying to $b\bar{b}$ produced in association with a leptonically decaying W or Z boson are sufficiently well discriminated from backgrounds, which as at the Tevatron are from $t\bar{t}$, QCD multijet, *V*+heavy flavor jets, and *VV* production. Progress is being made in using low Higgs boson p_T production by exploiting decay angular distributions and momentum constraints from $Z \to \ell^+\ell^-$ decays in the *ZH* process.

Table 9.2 LHC production modes and decay channels for the Standard Model Higgs boson search for m_H below the WW threshold. The checks highlight the highest sensitivity search channels.

Production vs. decay	Inclusive (including gg fusion)	Weak boson fusion	WH/ZH	$t\bar{t}$H
$H \to \gamma\gamma$	√	√	√	√
$H \to b\bar{b}$			(√)	√
$H \to \tau^+\tau^-$		√		
$H \to WW^*, W \to \ell\nu_\ell$	√	√	√	
$H \to ZZ^*, Z \to \ell^+\ell^-$	√			
$H \to Z\gamma, Z \to \ell^+\ell^-$	very low			

In the 20 GeV/c^2 mass range between the WW and ZZ thresholds, the inclusive WW channel is the dominant decay mode with substantial statistics to form a transverse mass measurement of the Higgs. This range of masses, however, has been nearly excluded by the Tevatron searches as will be summarized at the end of this chapter. Above the ZZ threshold, the four-lepton decay is the golden channel for Higgs discovery with low backgrounds and high-resolution mass reconstruction in a mixture of pairs of dielectron and dimuon decays.

At the highest masses, the dropping production cross sections are compensated by the addition of hadronic W and Z decay modes. The high-p_T boson signature has lower backgrounds and the dijets begin to merge, providing a massive monojet signature. Similarly, the neutrino decays of high-p_T Z bosons provide a substantial missing transverse energy. These highly boosted diboson decays provide Higgs boson search coverage up through 1 TeV/c^2 where the width of the Higgs becomes comparable to its mass. In the absence of a Higgs boson, the electroweak scattering of massive weak bosons will begin to form resonances in a semistrong coupling regime. Thus, 1 TeV/c^2 marks the upper limit to the production of a meaningful particle excitation of the Standard Model Higgs field.

9.5.1 *High-Resolution Search Channels*

The subthreshold decay of the Higgs boson to ZZ^* is kinematically similar to a semileptonic b-quark decay in that dominantly one Z boson is nearly on-shell and the second Z boson has a mass corresponding to the remaining Q^2 of the decay. Therefore, a 130 GeV/c^2 Higgs boson will decay into an ~90 GeV/c^2 and a <40 GeV/c^2 pair of Z bosons. This effect is demonstated in figure 9.13. The soft Z boson decay into leptons is problematic in terms of lepton backgrounds and reconstruction efficiency. Ultimately, low-p_T lepton detection and diminishing ZZ^* branching fraction limit this channel to above $m_H = 130$ GeV/c^2.

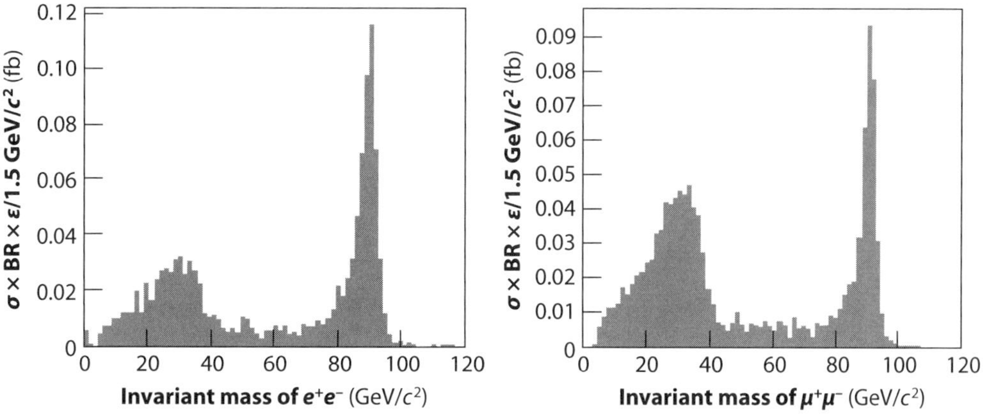

FIGURE 9.13. Dilepton mass distributions for $H \to ZZ^* \to 2e2\mu$, $m_H = 130$ GeV/c^2, on the left for dielectrons and on the right for dimuons. The low-mass dimuons are enhanced because of the high muon reconstruction efficiency at low p_T. The vertical axis is the product of the cross section (σ) times the branching fraction (BR) to leptons times the selection efficiency (ϵ) (Credit: CMS) [3].

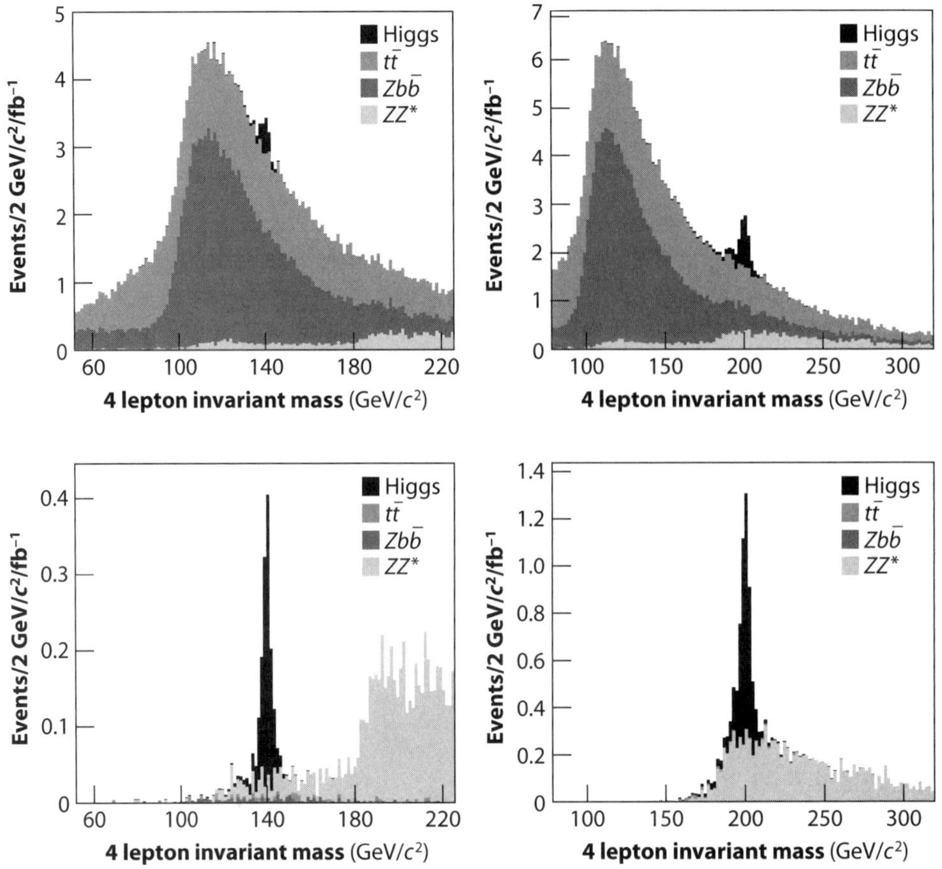

FIGURE 9.14. The $H \to ZZ^* \to 2e2\mu$ mass distributions before (top) and after (bottom) final selections for m_H = 140 GeV/c^2 (left) and m_H = 200 GeV/c^2 (right) (Credit: CMS) [3].

Background to $H \to 4\ell$ comes from $t\bar{t}$ dilepton decays with both b-jets producing leptons from semileptonic decays. Similarly, the process $Zb\bar{b}$ is also a background with a Z decaying to leptons and each b-jet producing a lepton (for background studies replace b everywhere with lepton). Note, leptons from B-hadron semileptonic decay can be partially removed using impact parameter techniques (4ℓ vertex probability) as well as lepton isolation. The above-threshold $H \to ZZ$ decay has irreducible background from on-shell ZZ production, as shown in figure 9.14.

With a mass resolution of 1.0–1.2%, the direct measurement of the Higgs boson width Γ_H becomes possible above ~200 GeV/c^2 as shown in figure 9.15. Note, the growth of the Higgs boson width beyond 10 GeV/c^2 is a feature of the Standard Model Higgs boson where for large m_H the total width is dominated by the longitudinally polarized diboson decays. An anomalously narrow high-mass Higgs would indicate a preference for fermionic decay modes, as predicted in several beyond the Standard Model Higgs sectors.

In addition to being a dominant discovery mode, the $H \to ZZ \to 4\ell$ decay mode is a powerful probe of the spin, parity, and CP. The most general HZZ vertex factor can be written

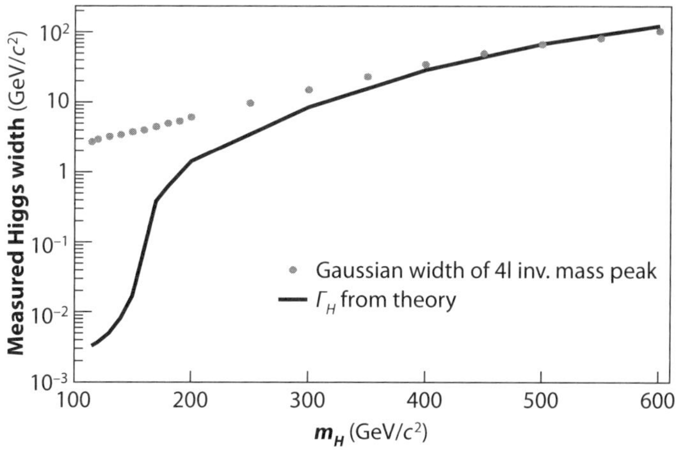

FIGURE 9.15. Direct measurement of the Higgs boson width becomes possible above $m_H \sim 200$ GeV/c^2 in the $H \to 2e2\mu$ channel (Credit: CMS) [3].

in terms of the four-momenta of the Z bosons, k_1 and k_2, and the four-momentum of the Higgs boson, $p \equiv k_1 + k_2$, and is given by

$$\frac{igM_Z}{\cos\theta_W}\left(ag_{\mu\nu} + b\frac{p_\mu p_\nu}{M_Z^2} + c\epsilon_{\mu\nu\rho\sigma}\frac{k_1^\rho k_2^\sigma}{M_Z^2}\right) \tag{9.27}$$

where the first term corresponds to the SM scalar, the second to a non-SM scalar, and the third to a non-SM pseudoscalar. *CP* violation would be present for admixtures of nonzero *a*, *b*, and *c* in equation (9.27). The experimental observables are the azimuthal and polar angular distributions. The azimuthal angle ϕ is measured between the two planes defined by the leptonic decays of the two Z bosons in the Higgs rest frame, as shown in figure 9.16. This yields a distribution

$$F(\phi) = 1 + \alpha\cos\phi + \beta\cos 2\phi. \tag{9.28}$$

The polar angle is defined in the rest frame of each of the Z bosons and is the angle between the negatively charged lepton and the direction of motion of the Z boson in the Higgs boson rest frame, as shown in figure 9.16. The polar angle distribution has the form

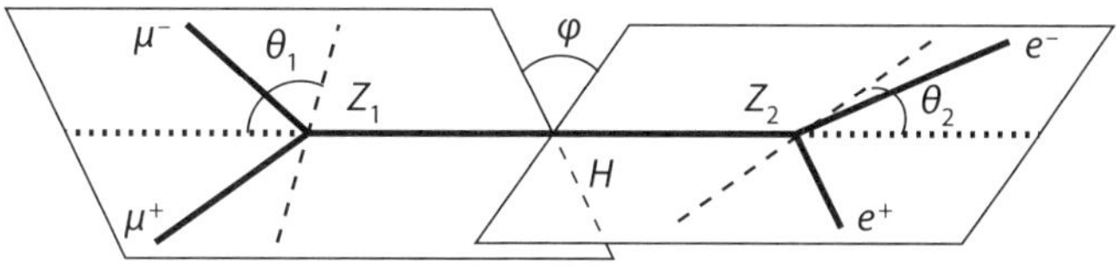

FIGURE 9.16. Definition of the $H \to ZZ \to \mu^+\mu^-e^+e^-$ decay angles in the Higgs boson rest frame.

$$G(\theta_{1,2}) = T(1+\cos^2\theta_{1,2}) + L\sin^2\theta_{1,2}. \tag{9.29}$$

One defines the asymmetry $R = (L - T)/(L + T)$ to better distinguish the models. The helicity amplitude of the $|ZZ\rangle$ state to be in the $|LL\rangle$ state is predicted to be

$$T_{LL} = \frac{m_H^2 - 2M_Z^2}{2M_Z^2} \quad \text{for } m_H > 2M_Z. \tag{9.30}$$

The value of R for a $S^{CP} = 0^+$ state is $R = (T_{LL}^2 - 1)/(T_{LL}^2 + 1)$ and therefore has a value of $R \approx 0.2$ at the ZZ threshold, increasing asymptotically to unity for large m_H. For a $S^{CP} = 0^-$ state, the decays are predominantly transverse, giving a value of $R = -1$.

The channel with the highest resolution of the high-resolution Higgs boson search channels is the $H \to \gamma\gamma$ decay. If the high-energy photons from the Higgs boson decay propagate through the beam pipe and tracker volume without conversion and are stopped by a high-resolution electromagnetic calorimeter, then an event-by-event mass measurement resolution below 1 GeV/c^2 can be achieved. The $H \to \gamma\gamma$ search is relevant for masses between 110 and 140 GeV/c^2 and is typically the single most powerful low-mass search channel. An event display from a simulation in the CMS detector is shown in figure 9.17.

On careful inspection of figure 9.17 in the vicinity of the photons, it can be seen that all channels of the electromagnetic calorimeter are readout. This is due to the overwhelming importance of photon identification criteria in the Higgs analysis. Currently, the most sensitive technique for separating high-energy photons from energetic π^0's or, in

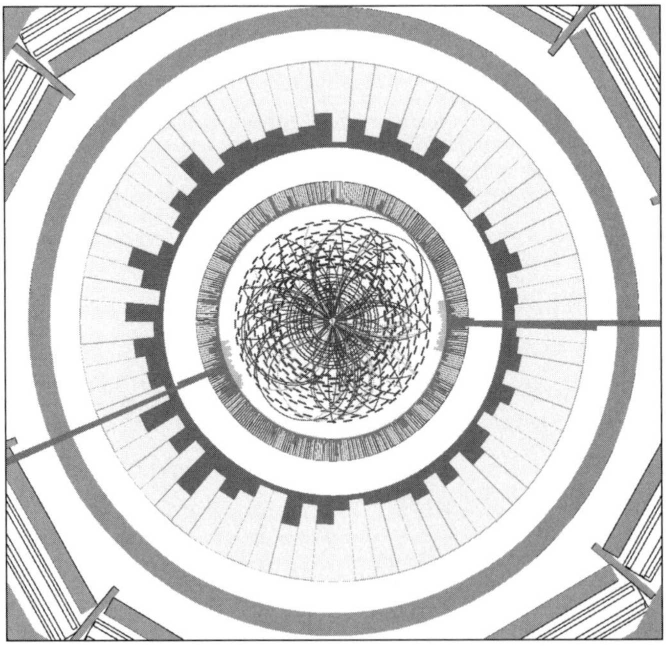

FIGURE 9.17. Simulation of a $H \to \gamma\gamma$ event in the CMS detector. The two large spikes indicate the experimental signature for high-energy photons from Higgs boson decay interacting with the electromagnetic calorimeter (Credit: CMS) [3].

general, from narrow neutral QCD jets is to require a narrow single-particle electromagnetic shower profile, isolated from other activity in the electromagnetic calorimeter, hadronic calorimeter, and tracker. For the two general-purpose LHC detectors, ATLAS and CMS, the calorimeters use different technologies to detect and measure photons. In the case of the CMS experiment, all of the energy of the photon is stopped and fully absorbed within a single longitudinal segmentation, while the lateral shower is measured precisely (the rear hadronic calorimeter serves mainly as a veto). For the ATLAS experiment, the accordion liquid argon calorimeter samples a fraction of the total photon energy, but in both the lateral and longitudinal directions. Equally important for photon measurements is the amount of material, measured in radiation lengths, located in front of the calorimeter. An estimate of the *material budget*, as it is called, for the CMS experiment is shown in figure 9.18 corresponding to a photon survival probability of ~60% in the central $|\eta| < 1.4$ and ~40% in the forward regions. Nearly all converted photons are recovered and included in the Higgs search, but with some loss in resolution which in some cases is offset by improved π^0 rejection and vertex pointing from the pair of charged tracks from the converted photon.

The $H \rightarrow \gamma\gamma$ branching fraction is naturally expected to rise and then drop off on approaching the WW threshold. This can be seen from the loop diagrams of figure 9.19, which have $Q^2 \approx 4M_W^2$. The negative interference between the top (fermion) and W boson loops decreases the partial decay width by ~10%.

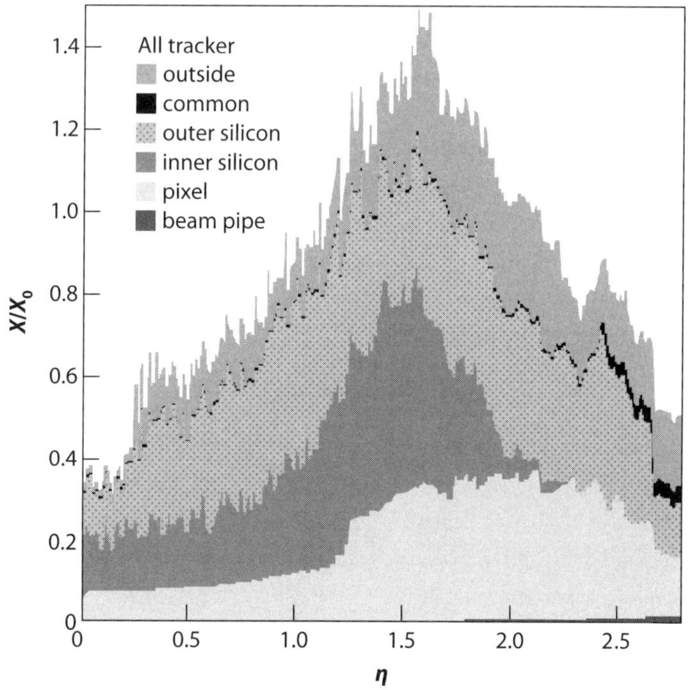

FIGURE 9.18. Estimated material budget in radiation lengths from the tracking detectors in front of the CMS electromagnetic calorimeter. The locations of the inner tracking systems relative to the calorimeter are shown in figure 5.8 (Credit: CMS) [3].

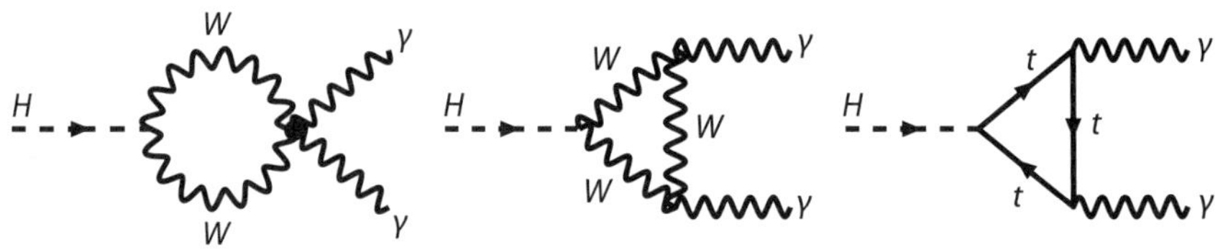

FIGURE 9.19. Loop diagrams for $H \to \gamma\gamma$ decay.

Table 9.3 Expected cross section and branching fraction for $H \to \gamma\gamma$ from a typical analysis.

Higgs mass m_H	115 GeV/c^2	120 GeV/c^2	130 GeV/c^2	140 GeV/c^2	150 GeV/c^2
σ (gg fusion)	39.2 pb	36.4 pb	31.6 pb	27.7 pb	24.5~pb
σ (weak boson fusion)	4.7 pb	4.5 pb	4.1 pb	3.8 pb	3.6 pb
σ ($WH, ZH, t\bar{t}H$)	3.8 pb	3.3 pb	2.6 pb	2.1 pb	1.7 pb
Total σ	47.6 pb	44.2 pb	38.3 pb	33.6 pb	29.7 pb
$\mathrm{Br}(H \to \gamma\gamma)$	2.08×10^{-3}	2.20×10^{-3}	2.24×10^{-3}	1.95×10^{-3}	1.40×10^{-3}
Inclusive $\sigma \times \mathrm{Br}(H \to \gamma\gamma)$	99.3 fb	97.5 fb	86.0 fb	65.5 fb	41.5 fb

The expected contributions to the total Higgs production cross section at $m_H = 120$ GeV/c^2 are given in table 9.3 for $\sqrt{s}$ = 14 TeV. The irreducible backgrounds, i.e., those with two real photons in the final state, come from:

- $gg \to 2\gamma$ (loop diagram),
- $q\bar{q} \to 2\gamma$ (direct production diagrams),
- $pp \to 2\gamma + jets$ with two prompt photons,

while the reducible backgrounds come from neutral jets (or hard π^0's) that fake one or two photons in the processes $pp \to \gamma +$ jet and $pp \to$ jets, respectively. The process of two neutral jets faking a pair of high-energy photons cannot be reliably simulated with Monte Carlo techniques. Data-driven techniques will be needed to estimate these backgrounds through the process of tightening photon isolation variables and using sideband regions of the reconstructed Higgs boson mass.

The overall selection efficiency for Higgs boson decays to photons at $m_H = 120$ GeV/c^2 is estimated to be ~32% with contributions listed in table 9.4. In an example analysis, for 1 fb^{-1} of integrated luminosity and a 2.5 GeV mass window, the expected number of $H \to \gamma\gamma$ events is 27 for a background of 445 events, giving an $S/B \sim 6\%$. For comparison, at LEP, cutting out search regions with $S/B < 5\%$ had little effect on the search sensitivity, indicating a stark contrast with the LHC Higgs boson search.

The predicted $H \to \gamma\gamma$ search sensitivity depends strongly on the electromagnetic calorimeter energy resolution, as shown in figure 9.20. Part of the resolution contribution is from primary vertex assignment, which correctly assigns the signal vertex in ~81% of the

Table 9.4 Estimate of the Higgs boson selection efficiency and accepted cross section for $H \rightarrow \gamma\gamma$, $m_H = 120$ GeV/c^2.

After photon selection	+ Tracker isolation	+ Calorimeter isolation	Accepted cross section	±1 GeV mass window
50%	35%	32%	31 fb	17.6 fb

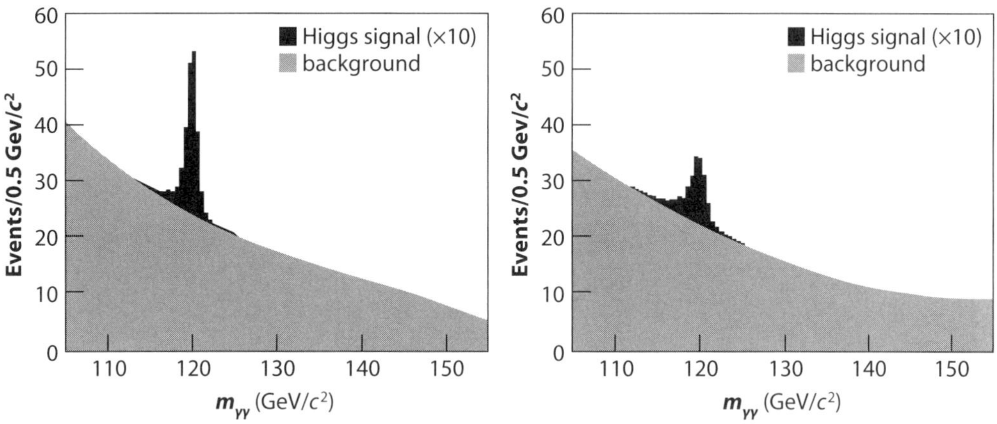

FIGURE 9.20. Comparison of signal and background distributions in $H \rightarrow \gamma\gamma$ search for a calorimeter mass resolution of $\sigma_{m_H} = 700$ MeV/c^2 (left) and $\sigma_{m_H} = 900\$$ MeV/c^2 (right) at $m_H = 120$ GeV/c^2 (Credit: CMS) [3].

events at low luminosity. The technique consists of assigning the vertex with the highest p_T track in the event as the $H \rightarrow \gamma\gamma$ signal vertex. Other techniques that have been investigated include the p_T-balance method, i.e., projecting the momenta of all tracks from a vertex along the candidate Higgs boson momentum direction and comparing with the expected recoil momentum, and the highest track multiplicity vertex selection. For comparison, one of the Tevatron techniques is to compare the p_T spectra of all tracks in the triggered event with those from minbias triggers to assign the least probable minbias vertex. Without an event-by-event vertex assignment, the invariant mass of the two-photon state is smeared by an additional 1.5 GeV/c^2, substantially reducing the sensitivity of the search channel.

The gluon-gluon fusion loop diagram for Higgs production is necessarily a high-Q^2 process due to the top quark mass. The high-Q^2 of the loop and the fact that gluons have a relatively large color charge increase the probability that a high-p_T jet will be produced recoiling off the Higgs boson. It was proposed in reference [4] that the search for $pp \rightarrow H \rightarrow \gamma\gamma$+jet should have a higher signal-to-background ratio than the inclusive search by enhancing the contributions at high effective center-of-mass energies $\sqrt{\hat{s}}$, as shown in figure 9.21. The backgrounds are suppressed with a cut on $\sqrt{\hat{s}} > 300$ GeV, as in figure 9.22. This channel also benefits from a more efficient primary-vertex assignment, using the tracks of the high-p_T jet to assign the signal vertex. The reduced background level in $H \rightarrow \gamma\gamma$+jet and in the weak boson fusion Higgs production will provide a flatter

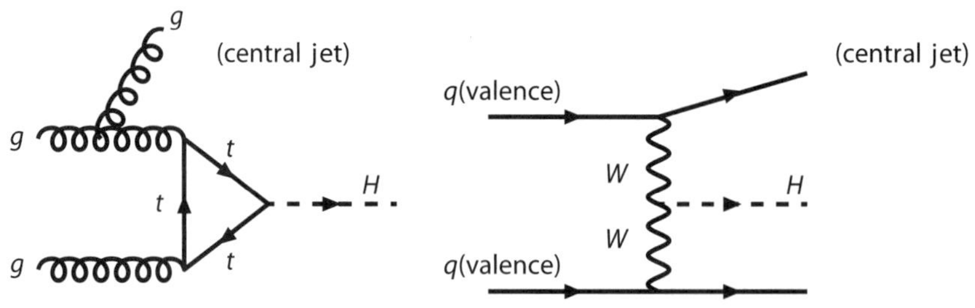

FIGURE 9.21. Two leading diagrams for $H \to \gamma\gamma$ + jet production. The diagram on the left is gluon-gluon fusion Higgs boson production with initial-state radiation, and the diagram on the right is qqH weak boson fusion production of the Higgs boson.

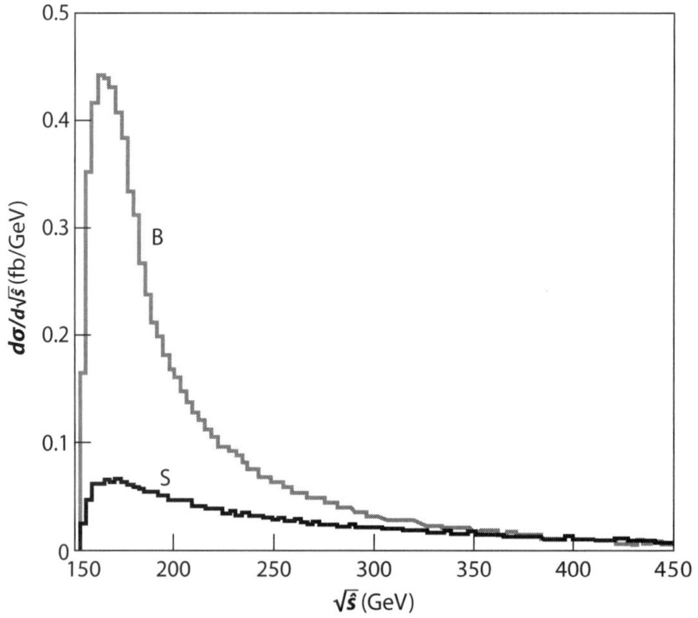

FIGURE 9.22. Distribution of the differential cross section in effective center-of-mass energy $\sqrt{\hat{s}}$ for the $H \to \gamma\gamma$ + jet signal (S) and background (B) processes at the LHC (Credit: M. Dubinin) [4].

background shape on which to measure the mass of a light Higgs boson, as shown in figure 9.23.

9.5.2 *JetMET-Oriented Low-Mass Channels*

Three exclusive channels in the low-mass Higgs boson search compete in sensitivity with the $H \to \gamma\gamma$ search channel. These are $t\bar{t}H$ ($H \to b\bar{b}$), qqH ($H \to \tau\tau$), and qqH ($H \to WW^* \to 2\ell 2\nu$), as shown in figure 9.24. All of these channels involve jets and missing transverse energy in the final state and are, therefore, "JetMET"-related analyses. The resolution of jet-p_T measurements is roughly an order of magnitude worse than

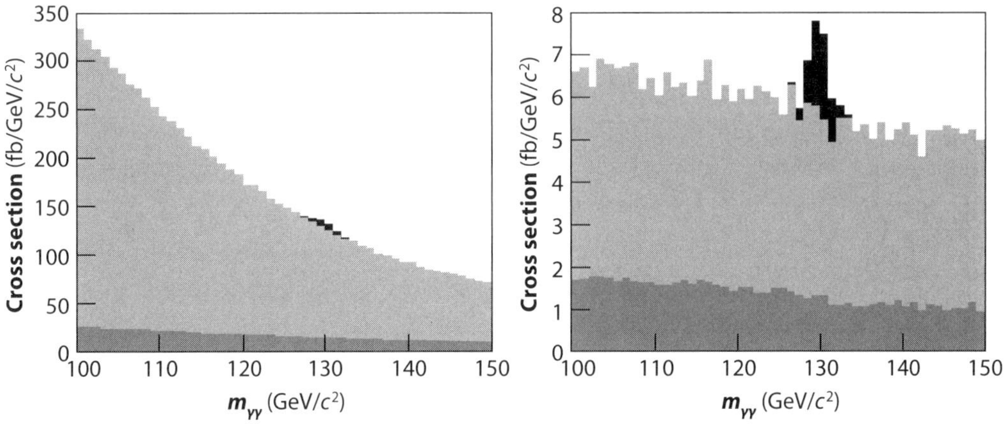

FIGURE 9.23. Comparison of signal and background distributions in the inclusive $H \to 2\gamma$ search (left) and $H \to 2\gamma\gamma$ + jet search (right), $m_H = 130$ GeV/c^2 (Credit: ATLAS) [5].

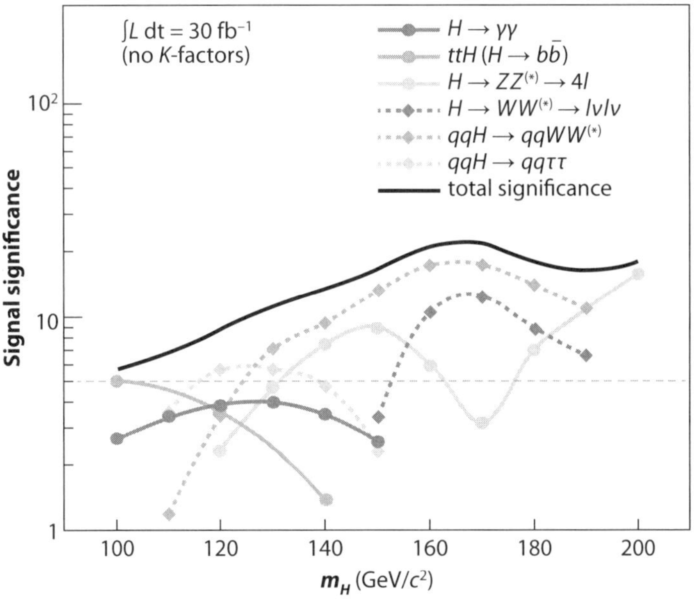

FIGURE 9.24. Estimated Standard Model Higgs boson search sensitivities in the low-mass region for $\sqrt{s} = 14$ TeV at the LHC (Credit: ATLAS) [5].

that of corresponding electron, muon, or photon measurements. The relatively poor measurement resolution reduces the reconstruction efficiency of low-p_T jets and introduces fake jets formed from low-energy background sources. Thus, defining a search for final states involving a fixed number of low-p_T jets has relatively large inefficiencies and uncertainties.

The search for the Higgs boson in the process of Higgsstrahlung from a top quark has always been considered an important test of Standard Model Higgs-fermion theory.

The cross section for $t\bar{t}$ production at the LHC is approximately 900 pb, i.e., more than a factor of 100 times larger than at the Tevatron, and therefore it is an abundant source of events. The cross section for associated Higgs production $t\bar{t}H$ is ~0.7 pb, three orders of magnitude smaller. While at first sight the eight-fermion final state of $t\bar{t}H$ with $H \to b\bar{b}$ and $t\bar{t} \to \ell$+jets appears to be exceptionally clean to identify and rare compared to background processes, especially from the heavy-flavor content, this is not the case in practice. The $t\bar{t}H$ search channel suffers from several sources of efficiency loss from branching fractions for $t\bar{t} \to \ell$+jets and $H \to b\bar{b}$ (total branching fraction ~20%), single-lepton trigger inefficiency, event selection when requiring exactly six jets, jet reconstruction efficiency, combinatorics of b-jet assignment, dijet $b\bar{b}$ mass resolution for defining a mass window, and b-tagging efficiency to the fourth power, $\epsilon_b^4 \approx 11\%$. Figure 9.25 shows a simulated reconstruction turn-on efficiency for jet reconstruction at an instantaneous luminosity of $\mathcal{L} = 2 \times 10^{33}$ cm^{-2}s^{-1} indicating low reconstruction efficiency for 10–15 GeV jets. At parton level, the efficiency for the $t\bar{t}H$ channel decreases from 89%, 60%, 30%, to 11% for corresponding p_T cuts on all partons of 10, 20, 30, and 40 GeV. Therefore, the use of tracking to enhance the calorimeter jet reconstruction and other techniques to sharpen the jet reconstruction efficiencies are essential to maintain low thresholds.

An example distribution of the reconstructed Higgs boson mass in the $H \to b\bar{b}$ decay mode in $t\bar{t}H$ events is shown in figure 9.26. The background shoulder is partially due to the $t\bar{t}Z$ background with $Z \to b\bar{b}$. The $b\bar{b}$-dijet mass resolution is expect to be ~12–15%; therefore, separating m_Z and m_H will be limited by detector energy resolutions and

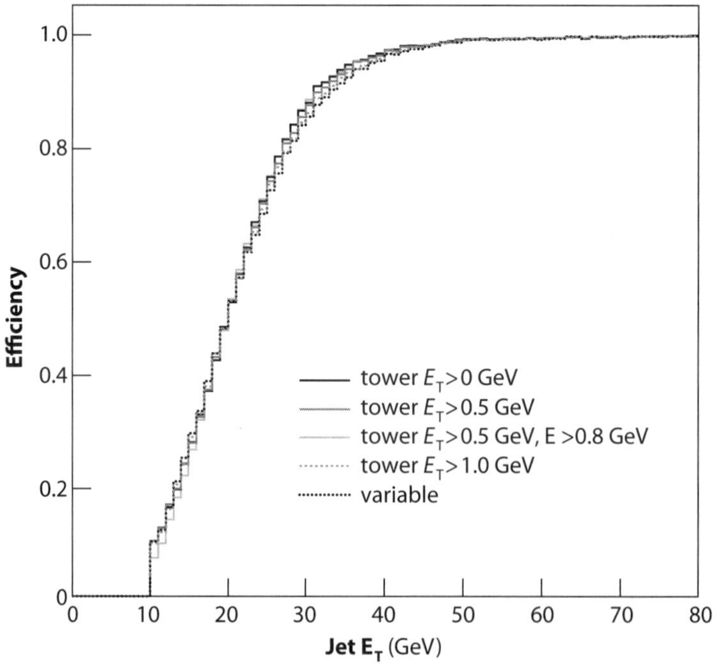

FIGURE 9.25. Estimated turn-on curve of jet reconstruction efficiency for calorimeter jets (Credit: CMS) [3].

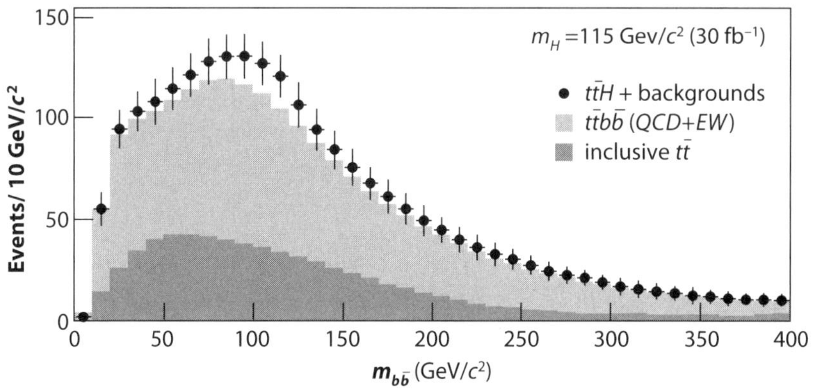

FIGURE 9.26. Higgs boson mass reconstruction in an example $t\bar{t}H$ search using simulation. The Higgs boson signal is located between 80–150 GeV/c^2 appearing in the same invariant mass range where the background distribution peaks.

calibrations. There is a broad contribution to the mass distribution from $t\bar{t}b\bar{b}$ production from QCD interactions, having more theoretical uncertainties in the shape and cross section than the $t\bar{t}Z$ process. Thus, the search in the $t\bar{t}H(\rightarrow b\bar{b})$ channel has relatively low sensitivity.

Another JetMET-related channel is the weak boson fusion production qqH of the Higgs boson with the $H\rightarrow\tau\tau$ decay. In the decay of the Higgs boson to τ leptons, a mass reconstruction technique involving the missing transverse energy aides in the separation of the large background from $Z\rightarrow\tau\tau$+jets. Only ~80% of the τ-lepton energy in hadronic-τ decays is visible in the calorimeter, the remainder is lost to neutrinos. Figure 9.27 shows how the missing transverse energy can be projected onto the τ-lepton momentum directions to recover part of the undetected τ-lepton energy.

The effectiveness of using the measured $E_{\mathrm{T}}^{\mathrm{miss}}$ depends on the $E_{\mathrm{T}}^{\mathrm{miss}}$ resolution and resolution tails. Generally, the technique in figure 9.27 requires a large Higgs boson p_{T} in order to avoid false solutions, typically only 40% of $H\rightarrow\tau\tau$ events have physical solutions using the $E_{\mathrm{T}}^{\mathrm{miss}}$ projection technique.

The effect of resolution tails is to throw intrinsically balanced background events into the event selection. The instrinsic Gaussian resolution of the missing transverse energy,

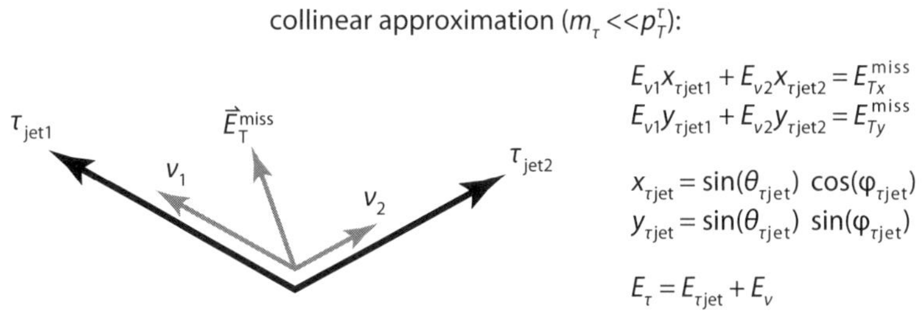

FIGURE 9.27. Higgs boson mass reconstruction technique for boosted $H\rightarrow\tau^+\tau^-$ decays.

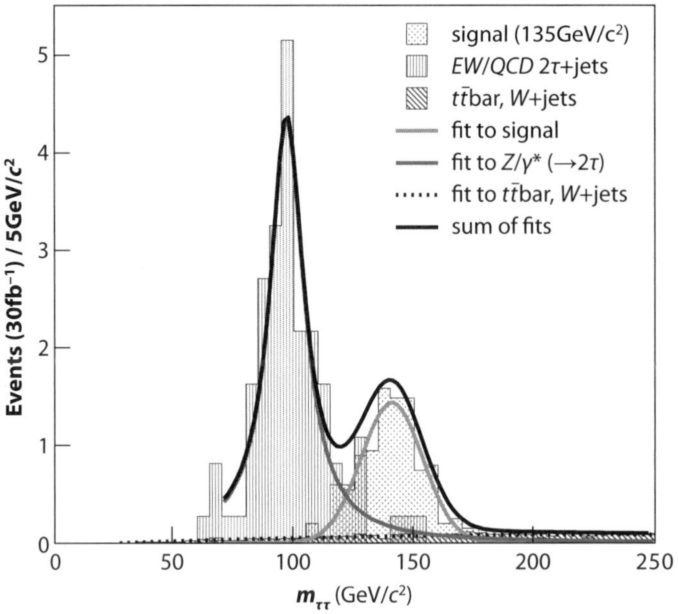

FIGURE 9.28. Reconstructed $\tau\tau$-mass using the E_T^{miss} projection technique in the $H \to \tau^+\tau^-$ channel, $m_H = 135$ GeV/c^2, from an example analysis. The Z boson decay $Z \to \tau^+\tau^-$ is to the left of the Higgs boson signal (Credit: CMS) [3].

on the other hand, is the primary parameter separating low-mass Higgs boson events from the large $Z \to \tau\tau$ background, as seen in figure 9.28. For events with no intrinsic E_T^{miss}, the direction of the measured E_T^{miss} is correlated to the jet and lepton directions, as the mismeasurement of energetic objects is a leading source of p_T-imbalance. The construction of an E_T^{miss} significance, or a requirement on the minimum angular separation of the missing energy direction and the nearest or back-to-back jet or lepton, can substantially reduce backgrounds from high-rate intrinsically low-E_T^{miss} processes.

The selection efficiency for a pair of τ leptons, such as from $H \to \tau\tau$ decay, has a substantial contribution from hadronic τ-lepton decays, as shown in table 9.5. The identification of hadronic τ-lepton decays relies on the narrowness and isolation of the τ-jet. Figures 5.15 and 5.16 show the definition and performance, respectively, of a "shrinking cone" method for separating τ-jets from QCD jets. Due to the finite mass of the τ lepton, a higher p_T boost will result in a narrower jet of particles in the lab frame from the τ-lepton decay.

Weak boson fusion production of the Higgs boson is predicted to improve the identification of Higgs boson events through the detection of the forward-tagging jets. The triggering of qqH with $H \to \tau\tau$ relies on high-efficiency, high-purity hardware-level techniques for τ-lepton isolation. Nominally, the signal signature is a four-jet final state with low E_T^{miss} and, therefore, it has a potentially overwhelming multijet QCD background. However, the kinematics of the forward jets and the narrowness of the τ-jets suppresses the multijet background. The pseudorapidity distributions of the jets and τ leptons are plotted in figure 9.29.

Table 9.5 Final states of τ-lepton pairs consisting of mixtures of electron (e), muon (μ), and hadronic (had) τ-lepton decay modes.

τ-Pair final states	Branching fractions
$\tau_e\tau_{had}$	22%
$\tau_\mu\tau_{had}$	22%
$\tau_{had}\tau_{had}$	41%
$\tau_e\tau_\mu$	6%
$\tau_e\tau_e$	3%
$\tau_\mu\tau_\mu$	3%

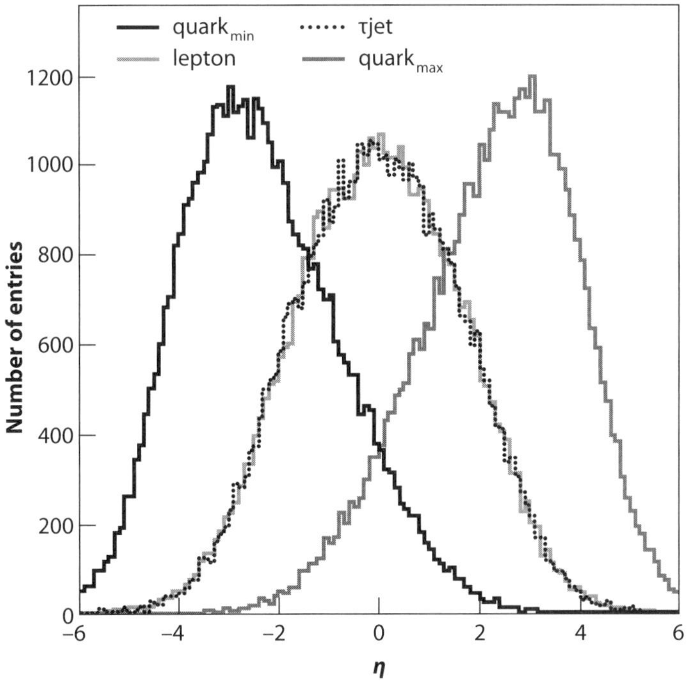

FIGURE 9.29. Expected forward-jet and τ-lepton pseudorapidity distributions at the LHC in the qqH process with $H \to \tau^+\tau^-$ (Credit: CMS) [3].

An important background for the qqH search in the decay $H \to \tau\tau$ are $t\bar{t}$ events in the dilepton decay mode. The jets coming from this process are more central than forward-tagging jets from weak boson fusion and can be b-tagged if they are within the tracker acceptance, i.e., $|\eta| < 2.4$ for a typical LHC detector. Forward-tagging jets are light-quark jets that originate from valence quark scattering, as shown in figure 9.21. Therefore, the use of a central jet veto can be a powerful technique for reducing $t\bar{t}$ backgrounds. However, with ~5–20 or more multiple interactions per LHC bunch crossing at high instantaneous luminosities, there are a number of jets that originate from minbias interactions in every event with a p_T spectrum extending above 30 GeV/c. To maintain an effective central jet veto, the jets must be assigned to a vertex using the tracking information. The tracks of the central τ-jets define the primary vertex; therefore, the jets from the nonsignal vertices are ignored in the central jet veto. Jet-vertex assignment will be important for all LHC measurements to reduce the effects of minbias pile-up.

9.5.3 *Inclusive Dilepton Analysis*

The $H \to WW$ decay mode has a branching fraction of ~8% at $m_H = 115$ GeV/c^2 that grows steadily to a maximum of 96% right below the ZZ threshold. In this range of m_H, only dilepton final states are used for WW event selection. The light-quark jets from W decay are overwhelmed by low-p_T jet background. However, dilepton production from Standard Model cross sections is sufficiently suppressed at a hadron collider that the trigger system is able to maintain a high efficiency for a wide range of processes with two energetic leptons in the final state, the main exception being the high rates at low-p_T (<5 GeV/c) leptons coming from inclusive J/Ψ production and B-hadron decay. Higgs boson search backgrounds involving leptons from hadron decay can be controlled to some degree by isolation requirements in the tracker and calorimeter. The additional requirement of prompt leptons, i.e., no significant displacement from the primary vertex, can be used to significantly suppress leptons from B-hadron decay. Table 9.6 gives a list of Standard Model processes contributing to the dilepton channel at the LHC.

The leptons from the $H \to WW$ decay tend to point in the same direction, as shown in figure 9.30. This provides a natural search variable, $\Delta\phi_{\ell\ell}$ the azimuthal angle between the leptons. The high opening-angle backgrounds come from $Z \to \tau\tau$ where both τ leptons

Table 9.6 Standard Model processes contributing to inclusive dilepton production and the corresponding cross section times branching ratio at the LHC. The $b\bar{b}$ contribution is estimated using asymmetric p_T thresholds of (20, 10) GeV for the lepton pair, but is largely removed from the final analysis with isolation criteria.

Inclusive dilepton processes	$\sigma \times Br(pb)$
$b\bar{b} \to 2\ell$	~3000 (high-p_T inclusive)
$Z/\gamma^* \to 2\ell$	~165
$t\bar{t} \to 2\ell$	39.2
$qq(gg) \to WW \to 2\ell$	11.7(0.5)
$tW \to 2\ell$	3.4
$ZW \to 3\ell$	1.6
$ZZ \to 2\ell$	1.5

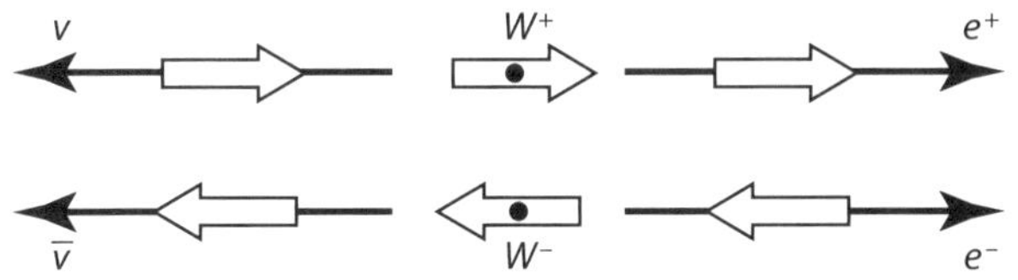

FIGURE 9.30. Leptons from $H \to W^+W^-$ decay tend to have three-momenta that point in the same direction.

decay leptonically. The backgrounds in the reconstructed dilepton mass $m_{\ell\ell}$ depend on the flavor combination of the leptons, where the $e\mu$ channel has substantially lower backgrounds from Drell-Yan and $Z \to \ell\ell$.

The $t\bar{t}$, WW, and $Z \to \tau\tau$ backgrounds to the $H \to WW$ search have differing dilepton kinematics, jet multiplicities, and missing transverse energy distributions. As it will be particularly difficult to fully separate samples with differing jet multiplicities and $E_{\mathrm{T}}^{\mathrm{miss}}$ distributions, an inclusive approach is to simultaneously fit for the contributions of multiple dileptons processes. In addition, the LHC will have new backgrounds such as flavor-excitation production of $gb \to tW$, as shown in figure 8.11. Without direct mass reconstruction, as here only the transverse mass is available, a successful search in the dilepton channel requires an accurate normalization of Standard Model contributions to the dilepton event selection and well-described signal kinematics. The transverse mass distribution for $H \to WW$ at $m_H = 160$ GeV is shown in figure 9.31. The effect from higher-order corrections on signal kinematics is evident in the Higgs boson p_{T}-distribution in gluon-gluon fusion production, shown in figure 9.32 and is a source of uncertainty on the dilepton distributions.

In the subthreshold region of the $H \to WW^*$ decay, the requirement of forward-jet tagging, according to the kinematics of qqH production, results in a factor of 10 reduction in background, thereby opening up the low-mass search in this channel. Similarly, if the second lepton p_{T} is too low, then the forward jets must be included in the trigger decision in order to reduce events from W+jets and multijet QCD background as these would dominate in a single-lepton trigger. Forward jets have a transverse momentum of roughly half the W boson mass, as expected from massive W boson propagators in the fusion process. The tagging jets occur in the pseudorapidities $1.5 < |\eta| < 4.5$ as shown in figure 9.29 with corresponding energies of $\sim$300 GeV or higher.

The qqH channel with $H \to WW^*$ loses sensitivity when the effective W^* mass goes below $\sim M_W/2$ resulting in softer second-leading lepton-p_{T} spectra. To continue the $H \to WW^*$ search down to lower masses, the WH production channel can be used. In this channel, there are two nearly on-shell W bosons in the $WH \to WWW^*$ process. The hard dilepton-p_{T} spectrum in this channel increases trigger efficiencies and opens up the possibility of a *same-sign* dilepton search for half the sample. This signature has relatively rare Standard Model backgrounds especially for dissimilar lepton flavors in the $e^{\pm}\mu^{\pm}$ channel. The soft third lepton may be above reconstruction thresholds, especially

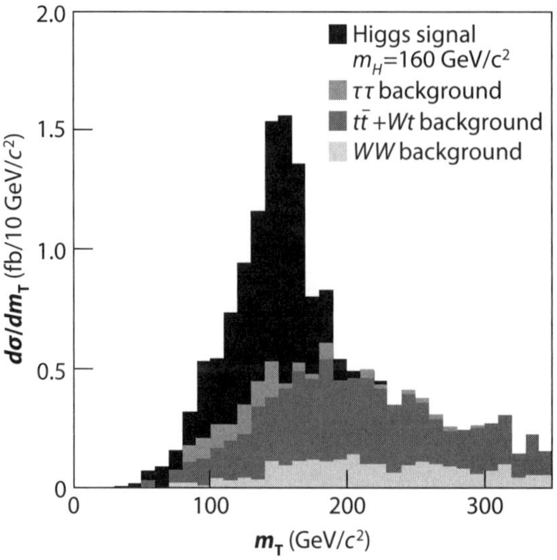

FIGURE 9.31. Transverse mass distribution in the $H \rightarrow W^+W^- \rightarrow \epsilon\mu\nu\nu$ channel with forward-tagging jet requirements from LHC simulations (Credit: CMS) [3].

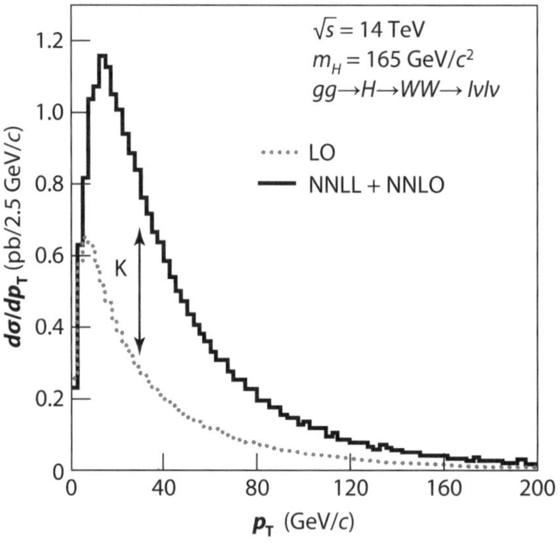

FIGURE 9.32. Leading-order event simulations of gluon-gluon fusion production of the Higgs boson at the LHC are reweighted for higher-order corrections to the event kinematics (Credit: CMS) [3].

for muons, thus introducing the *trilepton* signature. The trilepton and same-sign dilepton searches are high-purity analyses and have mainly diboson and instrumental backgrounds. The rejection of the Z peak in the WWW^* signature is effective in removing diboson backgrounds. The ZH production mechanism is another source of enhanced triple-boson production.

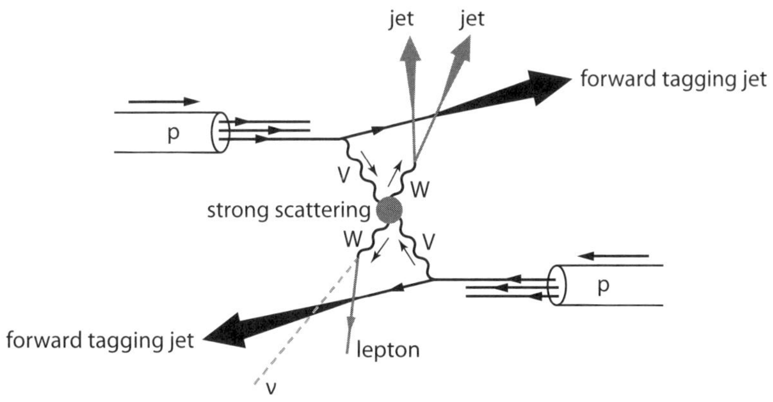

FIGURE 9.33. Heavy Higgs boson production and strong *WW* scattering event signature.

9.5.4 *Boosted Dibosons from Heavy Higgs Decay*

As shown in figure 9.15, the natural width of the Higgs boson exceeds 100 GeV for a mass of $m_H = 600$ GeV and above, and the total production cross section drops to below 1 pb with a significant fraction of the total rate coming from qqH production. To cover this heavy mass region, additional decay channels of $H \to WW$ and $H \to ZZ$ need to be utilized. The highly boosted W and Z bosons coming from an 800 GeV Higgs boson decay will have collimated decay products, as drawn in figure 9.33 for the strong WW scattering process. The hadronic decays appear as single jets in the detector. The mass of the single jet can be used to identify the jet as originating from a W or Z boson. The Higgs search in the qqH production mode with $H \to WW \to \ell^{\pm}\nu jj$ and $H \to ZZ \to \ell^{+}\ell^{-}jj$ decay modes benefits from the higher branching ratios of the hadronic boson decay, and the identification of a massive single jet is an effective part of the event signature. A similar high-branching-ratio decay channel is the $H \to ZZ \to \ell^{+}\ell^{-}\nu\bar{\nu}$ in qqH production. The boosted $Z \to \nu\bar{\nu}$ decay provides a large missing transverse energy signature.

9.5.5 *From 115 GeV/c² to 1 TeV/c²*

An overview of the Standard Model Higgs search sensitivity with the CMS experiment is shown in figures 9.34 and 9.35. The LHC Higgs search has a broader variety than LEP of possible analyses due to the increased mass range and higher production cross sections. Furthermore, the experimental challenges are starkly different in the different search topologies and mass ranges. The high mass resolution channels depend strongly on the calibration and alignment of the tracking, calorimeters and muon spectrometers and are subject to intense commissioning efforts. The JetMET-related search channels must contend with the low-p_T jet reconstruction efficiencies and fake rates. The use of the missing transverse energy in these channels is also subject to commissioning and the effects of detector noise, jet energy mismeasurement tails, and minbias pile-up. For weak boson fusion search

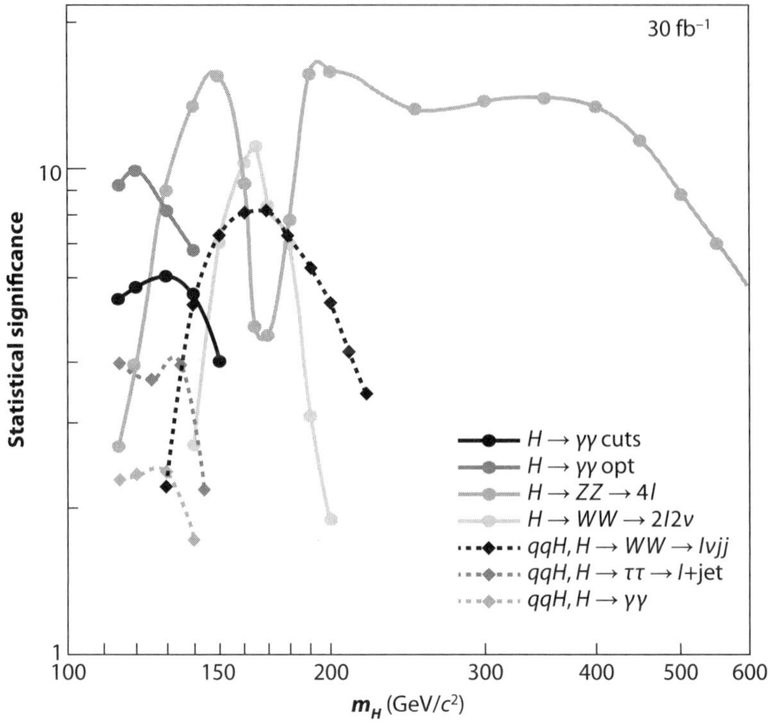

FIGURE 9.34. Estimates of the statistical significance of the Standard Model Higgs boson searches with 30 fb^{-1} of integrated luminosity for $m_H < 200$ GeV/c^2 for $\sqrt{s} = 14$ TeV at the LHC (Credit: CMS) [3].

topologies, the forward-tagging jets will be nearly the highest pseudorapidity jets used for analysis in hadronic collisions, posing an uncertainty in the reliability of simulations of the jet backgrounds in the forward regions. In cases where the search channel depends on efficient τ-lepton triggering or on single-lepton triggers plus jets, techniques for data-driven background calculations must be applied to correctly account for multijet QCD backgrounds and instrumental backgrounds enhanced by the trigger selection. Inclusive analyses, such as the dilepton analysis, will have to address a wide variety of Standard Model processes, some never measured before, such as flavor-excitation single top production. The same-sign dilepton and trilepton searches will address rare background processes with unknown contributions from fake and misreconstructed leptons, requiring data-derived estimates. Searches for heavy Higgs bosons will require dedicated treatment of highly boosted W and Z boson decays. In general, the understanding of the searches is expected to start at high p_T and move toward lower masses. It would not be surprising in the LHC Standard Model Higgs search to see a rapid exclusion for Higgs boson masses above $m_H \approx 150$ GeV and extending up to 500 GeV for datasets as small as 10 fb^{-1}, and then followed by more detailed longer-term experimental efforts, requiring an order of magnitude more data, as discussed here, to cover the low-mass region down to $m_H = 115$ GeV and the high-mass region up to 1

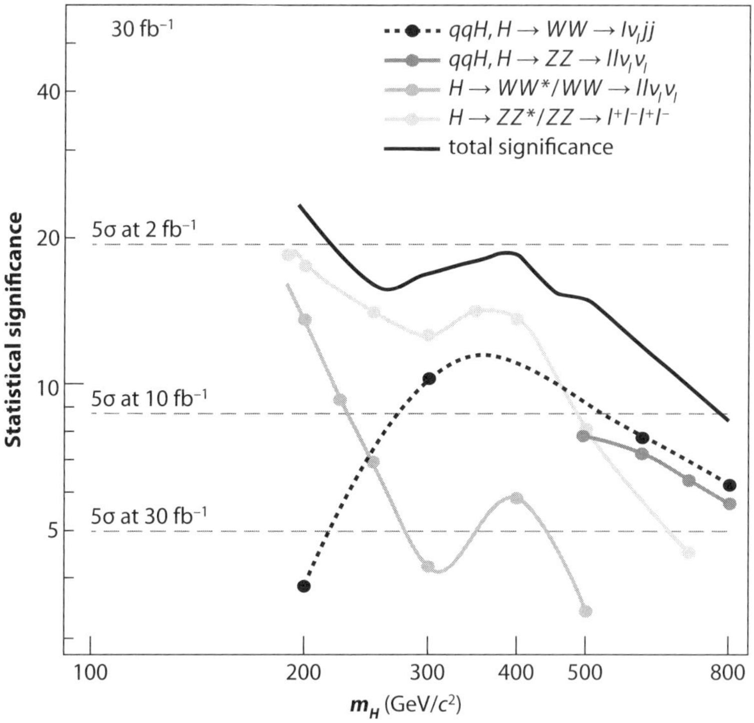

FIGURE 9.35. Estimates of the statistical significance of the Standard Model Higgs boson searches with 30 fb^{-1} of integrated luminosity for $m_H > 200$ GeV/c^2 for $\sqrt{s} = 14$ TeV at the LHC (Credit: CMS) [3].

TeV. The first major step of Higgs boson mass exclusion above $m_H = 115$ GeV was recently achieved at the Tevatron, as shown in figure 9.36, excluding the mass range $m_H = 158$–175 GeV [6, 7]. In the coming years, the Tevatron has the opportunity to extend the search to cover the entire Higgs boson mass region up to the upper board of $m_H = 185$ GeV allowed by the Standard Model theory according to global fits to electroweak data [8].

The current Tevatron search yields a mass exclusion in the $H \to WW$ channel. At the same time, the search sensitivity continues to grow in the $m_H = 115$ GeV mass region with WH and ZH production searches in the $H \to b\bar{b}$ decay mode. The Tevatron search channels cover what are perhaps the two most important mass regions in the Standard Model Higgs boson search. If the entire Higgs boson mass region is excluded below 185 GeV, then the quantum corrections to observables at the electroweak scale are hiding new physics contributions that require higher energy or higher precision experiments to explain.

The Standard Model Higgs boson search has begun at the LHC with $\sqrt{s} = 7$ TeV data. Looking to the longer-term future, the LHC Higgs boson search has a rich landscape and is the culmination of nearly all we know about high-p_T Standard Model physics. MultiTeV-scale physics may have many surprises in store, and the Higgs search may need to dramatically adapt in the LHC energy frontier.

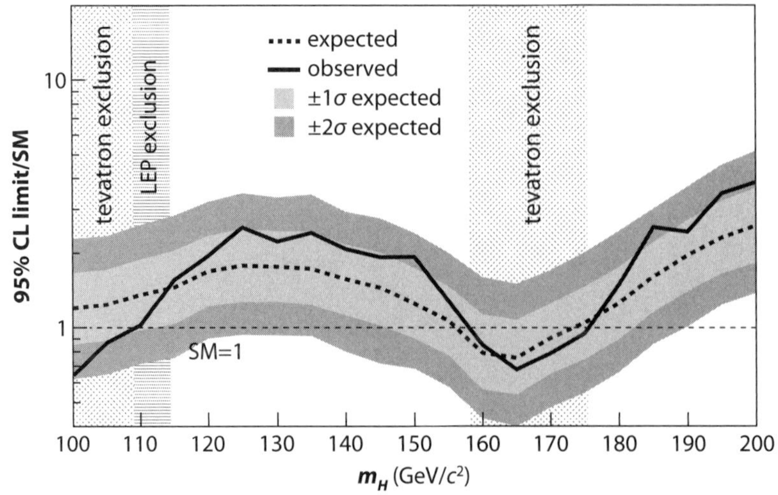

FIGURE 9.36. The combined Tevatron Standard Model Higgs boson search [6] using an average integrated luminosity of 5.9 fb^{-1} excludes the mass range $m_H = 158$–175 GeV/c^2 and mass regions below the existing LEP2 limits (Reprinted figure from [7] with permission by the APS). The vertical axis is the ratio of the 95% confidence-level exclusion limit on the Higgs boson production cross section to the predicted Standard Model cross section.

9.6 Exercises

1. **Fermiophobic Higgs.** In this problem we investigate the physical processes predicted by the Yukawa couplings of the Higgs field to the elementary particles. Here we consider the Standard Model Higgs field in which there is only one physical scalar Higgs particle in the theory. The interaction Lagrangians are given by

$$\begin{aligned} \mathcal{L}_{\text{gauge}} &= (\sqrt{2}\,G_F)^{\frac{1}{2}}(2M_W^2 HW_\mu^+ W^{-\mu} + M_Z^2 HZ_\mu Z^\mu), \\ \mathcal{L}_{\text{fermions}} &= -(\sqrt{2}\,G_F)^{\frac{1}{2}} m_f Hf\bar{f}. \end{aligned} \tag{9.31}$$

   (a) For an e^+e^- collider with a center-of-mass energy of 200 GeV, draw the three most significant tree-level Feynman diagrams that contribute to Higgs production. (*Hint:* Consider only the direct coupling of the Higgs field to the W and Z bosons. These diagrams are called the Higgsstrahlung and fusion diagrams.) Make an order-of-magnitude estimation of how much smaller is the cross section of the third most significant diagram relative to the second most significant. These numbers may be helpful in the estimate: $\Gamma_W = 2.1$ GeV, $\Gamma_Z = 2.5$ GeV, $\mathrm{Br}(Z \to e^+e^-) \approx 3\%$, $M_W = 80.4$ GeV, and $M_Z = 91.2$ GeV. Do these three diagrams interfere? In which final states?

   (b) Consider the decay mode $H \to \gamma\gamma$. What are the two dominant 1-loop triangle diagrams contributing to this decay? Do these diagrams interfere constructively or destructively?

(c) Take the Standard Model Higgs field and set the Yukawa couplings to the fermions to zero. Consider only the three most significant decay modes of the Higgs and give a rough sketch of the branching ratio plot of Higgs decay versus the Higgs mass from $m_H = 0$ to $m_H = 250$ GeV. Indicate the location of kinematical thresholds. The branching ratio plot should go from 0 to 1 on the vertical access, and the sum of the curves for the three decay modes should sum to unity.

2. **Two-photon production of the Higgs boson.** High-energy photons can be produced through the inverse Compton scattering of laser light off a high-energy electron beam.

   (a) Two high-energy photon beams are directed toward a single interaction region to form a photon-photon collider. Suppose the two photon beams collide with a center-of-mass energy $\sqrt{s} = m_H c^2$, the rest energy of the massive neutral scalar particle the Higgs boson. What Feynman diagrams are responsible for Higgs boson production? Plot the shape of the Higgs production cross section versus center-of-mass energy in the vicinity of $\sqrt{s} = m_H c^2$.

   (b) Suppose the photon beams collide at a center-of-mass energy, $\sqrt{s} = m_Z c^2$, the rest energy of the massive vector particle the Z boson. Do we expect Z bosons to be produced? Why?

3. From figures 9.34 and 9.35, list the Higgs couplings involved for Higgs production and decay at $m_H = 115, 130, 150, 170, 200, 300, 500, 800\ \text{GeV}/c^2$.

4. **Higgs boson pair production.** In the process of weak boson fusion, an incoming proton beam traveling in the $+z$ direction radiates a W boson that collides with a W boson radiated by a proton beam traveling in the opposite direction ($-z$). The two W bosons collide, annihilating into a neutral Higgs boson (spin-0). If the center-of-mass energy of the W^+W^- collision is at or above twice the mass of the Higgs boson, then the Higgs boson propagator can decay into a pair of Higgs bosons through a triple Higgs vertex. In this problem, assume that two on-shell W bosons in the initial state collide at a center-of-mass energy, denoted $\sqrt{s}$, and that the center-of-mass is at rest in the lab frame.

   (a) Write down the s-channel Feynman diagram for this process and write down each of the vertex terms.

   (b) Compute the squared invariant amplitude $\overline{|\mathcal{M}|^2}$ for this process for unpolarized W^+W^- beams. Indicate the configurations of initial spin states of the W bosons that contribute to the cross section.

   (c) Compute the differential cross section and give the explicit form of the angular distribution in the lab frame, where θ is the angle between one of the outgoing Higgs bosons relative to the incoming W^+ boson.

5. **Higgs decays to W_LW_L, W_RW_R, and top quarks.** Suppose that the Higgs boson is forbidden to decay into a pair of longitudinally polarized W bosons by some unknown symmetry.

(a) Compute the partial decay width of a Higgs boson decaying into pairs of transversely polarized W bosons, (L, L) and (R, R).

(b) Compute the partial decay width of a Higgs boson decaying into a pair of massive top quarks, and include the top quark mass dependence in the phase space.

(c) For a Higgs boson mass of 500 GeV/c^2, compute an approximate value for the ratio of $\Gamma_{t\bar{t}}$ to the sum of $\Gamma_{W_L W_L} + \Gamma_{W_R W_R}$.

9.7 References and Further Reading

A selection of reference texts and review articles on Higgs physics can be found here: [1, 9]

Public results from the ATLAS and CMS Experiments at the LHC can be found here: [3, 5]

Compilations of electroweak fit and Higgs boson search results can be found here: [2, 6, 8]

A selection of useful computational tools for Feynman diagrams and event generation can be found here: [10, 11, 12, 13, 14].

[1] Marumi Kado and Christopher G. Tully. The Searches for Higgs Bosons at LEP. *Ann. Rev. Nucl. Part. Sci.* **52** (2002) 65–113, doi:10.1146/annurev.nucl.52.050102.090656.

[2] LEP Higgs Working Group. http://lephiggs.web.cern.ch/LEPHIGGS.

[3] CMS Experiment Public Results. https://twiki.cern.ch/twiki/bin/view/CMSPublic/Physics Results.

[4] S. Abdullin et al. *Phys. Lett.* B **431** (1998) 410–419. doi:10.1016/S0370-2693(98)00547-4.

[5] ATLAS Experiment Public Results. https://twiki.cern.ch/twiki/bin/view/Atlas/AtlasResults.

[6] Tevatron New Phenomena and Higgs Working Group. http://tevnphwg.fnal.gov.

[7] CDF and D0 Collaborations, *Phys. Rev. Lett.* **104**, 061802 (2010).

[8] LEP Electroweak Working Group. http://lepewwg.web.cern.ch/LEPEWWG.

[9] John F. Gunion, Howard E. Haber, Gordon Kane, and Sally Dawson. *The Higgs Hunter's Guide.* Westview Press, 1990. ISBN 0-7382-0305-X.

[10] MADGRAPH/MADEVENT. http://madgraph.hep.uiuc.edu.

[11] CompHEP. http://comphep.sinp.msu.ru.

[12] CalcHEP. http://theory.sinp.msu.ru/~pukhov/calchep.html.

[13] FeynArts. http://www.feynarts.de.

[14] FeyneRules. http://europa.fyma.ucl.ac.be/feynrules.

Appendix Standard Model Interactions and Their Vertex Forms

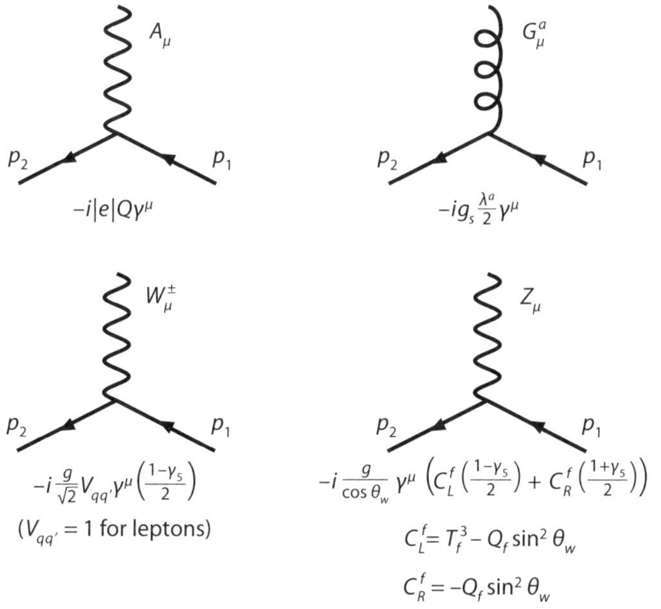

FIGURE A.1. Minimal gauge couplings.

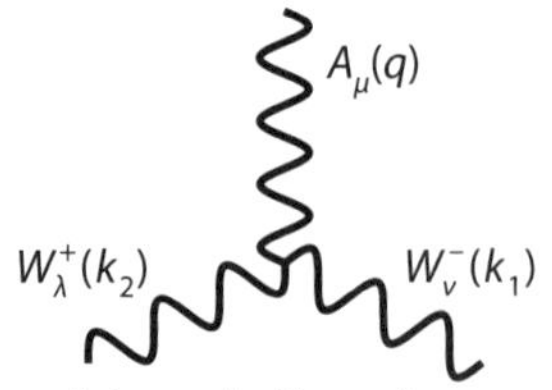

$$ie\,[g^{\nu\lambda}(k_1 - k_2)^\mu + g^{\lambda\mu}(k_2 - q)^\nu + g^{\mu\nu}(q - k_1)^\lambda]$$

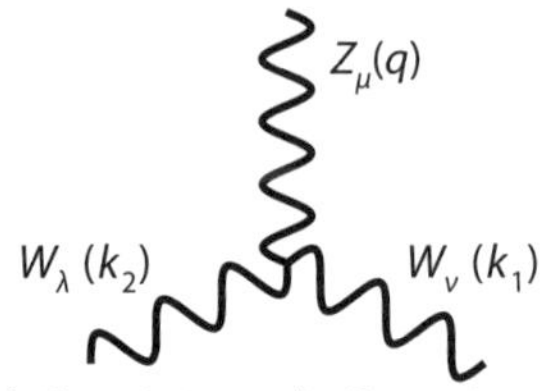

$$ig\cos\theta_W\,[g^{\nu\lambda}(k_1 - k_2)^\mu + g^{\lambda\mu}(k_2 - q)^\nu + g^{\mu\nu}(q - k_1)^\lambda]$$

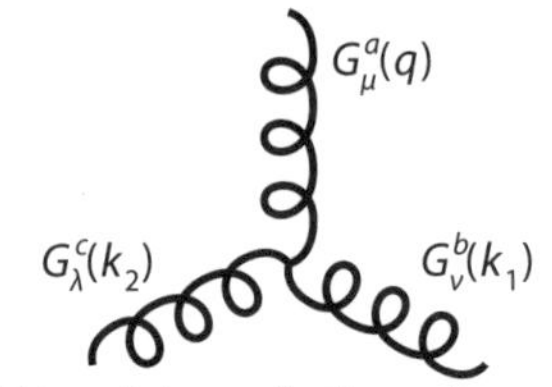

$$-g_s f_{abc}\,[g^{\nu\lambda}(k_1 - k_2)^\mu + g^{\lambda\mu}(k_2 - q)^\nu + g^{\mu\nu}(q - k_1)^\lambda]$$

FIGURE A.2. Triple-gauge boson couplings.

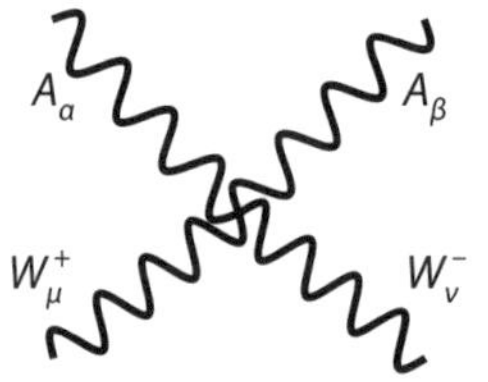

$$-ie^2(2g^{\alpha\beta}g^{\mu\nu} - g^{\alpha\mu}g^{\beta\nu} - g^{\alpha\nu}g^{\beta\mu})$$

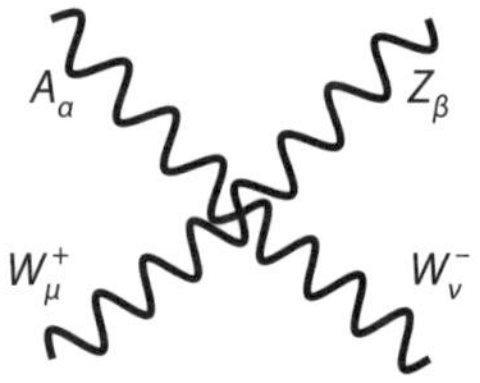

$$-ieg\cos\theta_W(2g^{\alpha\beta}g^{\mu\nu} - g^{\alpha\mu}g^{\beta\nu} - g^{\alpha\nu}g^{\beta\mu})$$

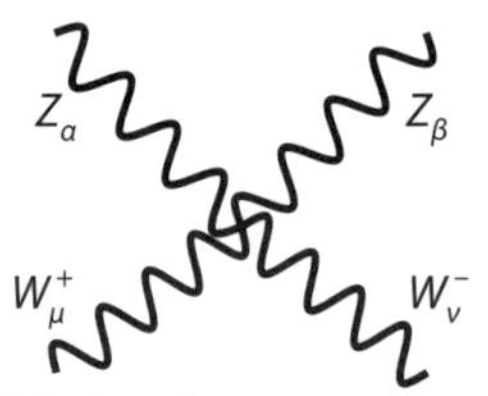

$$-ig^2\cos^2\theta_W(2g^{\alpha\beta}g^{\mu\nu} - g^{\alpha\mu}g^{\beta\nu} - g^{\alpha\nu}g^{\beta\mu})$$

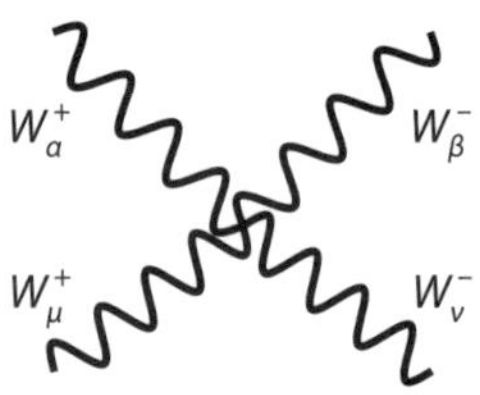

$$-ig^2(g^{\alpha\beta}g^{\mu\nu} - 2g^{\alpha\mu}g^{\beta\nu} - g^{\alpha\nu}g^{\beta\mu})$$

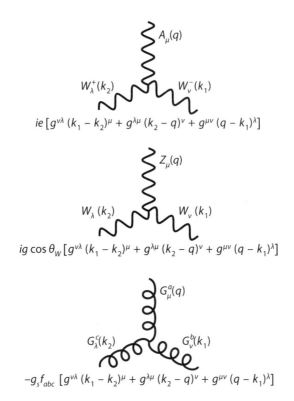

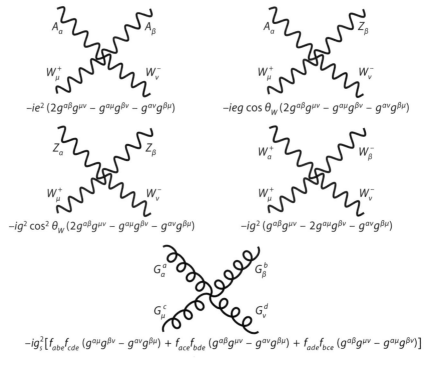

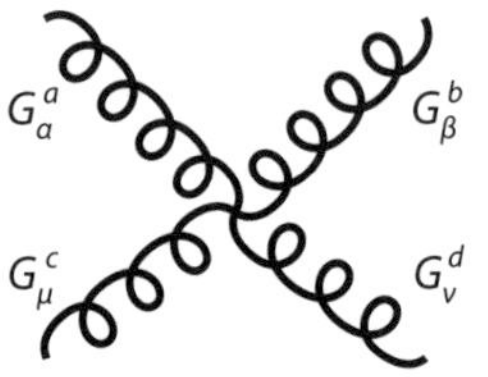

$$-ig_s^2[f_{abe}f_{cde}(g^{\alpha\mu}g^{\beta\nu} - g^{\alpha\nu}g^{\beta\mu}) + f_{ace}f_{bde}(g^{\alpha\beta}g^{\mu\nu} - g^{\alpha\nu}g^{\beta\mu}) + f_{ade}f_{bce}(g^{\alpha\beta}g^{\mu\nu} - g^{\alpha\mu}g^{\beta\nu})]$$

FIGURE A.3. Quartic-gauge boson couplings.

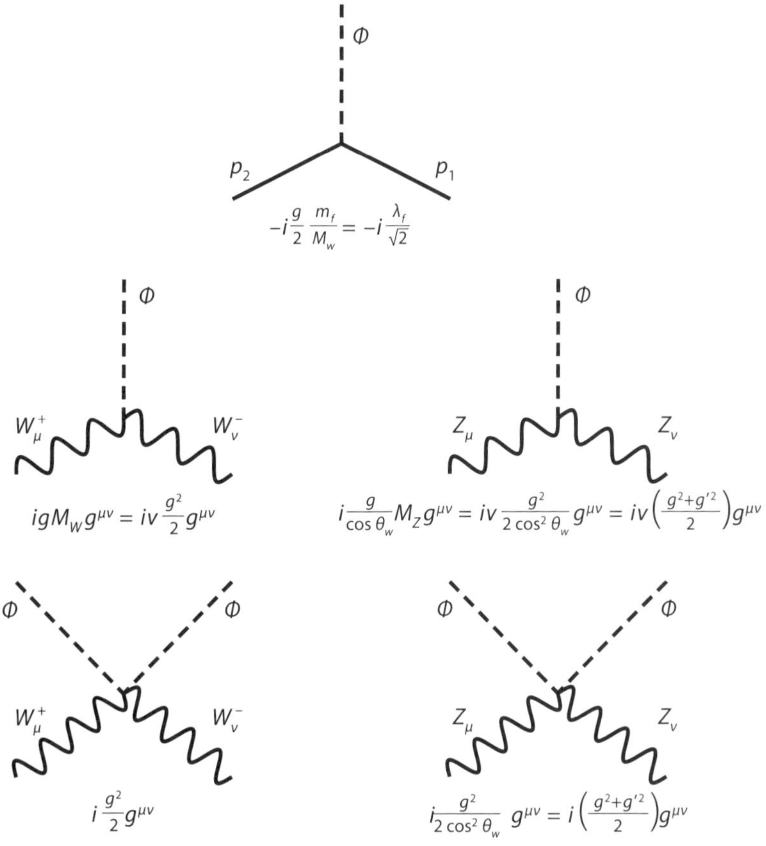

FIGURE A.4. Higgs-fermion Yukawa couplings and Higgs scalar couplings to the gauge bosons.

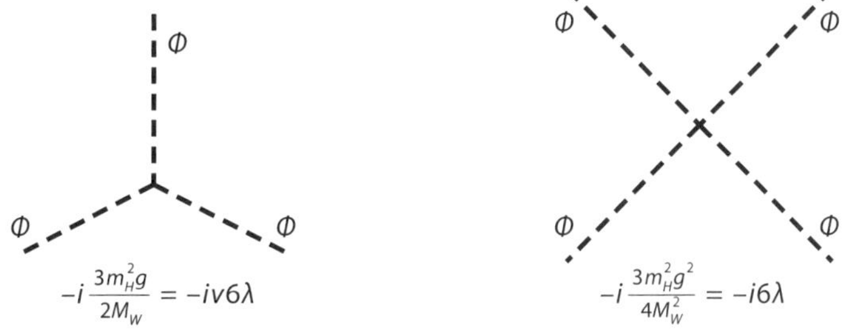

FIGURE A.5. Higgs scalar self-couplings.

Index